CISM COURSES AND LECTURES

Series Editors:

The Rectors of CISM
Sandor Kaliszky - Budapest
Mahir Sayir - Zurich
Wilhelm Schneider - Wien

The Secretary General of CISM
Giovanni Bianchi - Milan

Executive Editor
Carlo Tasso - Udine

The series presents lecture notes, monographs, edited works and
proceedings in the field of Mechanics, Engineering, Computer Science
and Applied Mathematics.
Purpose of the series is to make known in the international scientific
and technical community results obtained in some of the activities
organized by CISM, the International Centre for Mechanical Sciences.

INTERNATIONAL CENTRE FOR MECHANICAL SCIENCES

COURSES AND LECTURES - No. 388

WHYS AND HOWS
IN UNCERTAINTY MODELLING

PROBABILITY, FUZZINESS AND ANTI-OPTIMIZATION

EDITED BY

ISAAC ELISHAKOFF
FLORIDA ATLANTIC UNIVERSITY

Springer Wien New York

This volume contains 153 illustrations

In order to make this volume available as economically and as
rapidly as possible the authors' typescripts have been
reproduced in their original forms. This method unfortunately
has its typographical limitations but it is hoped that they in no
way distract the reader.

ISBN 3-211-83155-X Springer-Verlag Wien New York

PREFACE

Recognition of the need to introduce the ideas of uncertainty reflects in part some of the profound changes in engineering over the last two decades. The natural question arises: how to deal with uncertainty? Since one of the meanings of uncertainty is randomness, a natural answer to this question was and often appears to be to apply the theory of probability and random processes. This idea was spectacularly successful, and many scientists even refer to a "probabilistic revolution".

However, as was established recently the probabilistic methods are accompanied with serious difficulties during their implementation process in engineering applications. This gave rise to other, alternative treatments of uncertainty. This volume deals with both probabilistic and non-probabilistic aspects of uncertainty modelling. In particular, the fuzzy-set-based analysis and the anti-optimization methods are pursued, in addition to recent developments in the stochastics theory and applications. The emphasis is placed on applications in civil, mechanical and aerospace engineering.

The aim of this volume is to present to researchers, engineers and graduate students working on and interested in the problems concerned with the mechanics of solids and structures, a unified view on uncertainty. The current state of the development and applications of uncertainty modelling and analysis will be reviewed from alternative and seemingly opposing points of view. Indeed, the present state of affairs can be characterised as that of Tower of Babel: There is no discussion between various researchers, actively embracing different points of view.

The book aims to disseminate recent stochastic and, especially, non-stochastic approaches to model the uncertainty. It poses questions which many engineers were interested to ask, but were afraid to do. Answers to these questions and in-depth discussions should appeal in the first place to practicing engineers who must constantly widen their perspectives, researchers will find the lecturers most illuminating for the book accommodated alternative and often opposing points of view. Pragmatic approach allows one to choose modern techniques which most suit to the practical situations. This is not just one more book about uncertainty. This is a collection of the latest thoughts of engineers and scientists from all over the world about many, often overlooked and yet to be recognized facets of ever-present uncertainty.

It was a pleasure to edit this volume. The cooperation of all authors is gratefully appreciated. Most unfortunately, Dr. Rudolf Lohner was unable to submit his contribution on the interval analysis. We were most fortunate, that Professor Ulrich Kulisch graciously agreed to contribute to this volume, on this topic. We record our heartfelt thanks to CISM International Centre for

Mechanical Sciences for both approving the course and providing the most pleasant atmosphere during its conducting in the beatiful palace in Udine. Rare courtesies extended by the Secretary General, Professor Giovanni Bianchi and Rector, Professor Sandor Kaliszky must be recorded. Reliable and encouraging assistance as well as the extremely kind patience of Professor Carlo Tasso of the University of Udine are most kindly acknowledged. The staff of Mrs. Elsa Venir Burti provided most reliable and timely service throughout the course, and created an atmosphere of exceptional dedication and work including nearly instant copying service, efficient communication between the lecturers and the attendees, library facilities, and solution of numerous problems of various magnitudes; all of these seemingly small gestures and good will and courtesy truly make CISM a unique and great institution that must be cherished, preserved and possibly expanded for the benefit of engineering sciences. Last but not least, our thanks are to all the attendees of the course: While teaching them, all lecturers learned a lot, not excluding the ways how the material should be presented more efficiently. The responsibility on any misconceptions or typographical errors lies upon ourselves, the authors themselves. We will be most indebted if you could communicate them to us by electronic mail at ielishak@me.fau.edu. or by the FAX of 561-297-2825.

Isaac Elishakoff

CONTENTS

CHAPTER 1

HOW TO UTILIZE THE ANTI-OPTIMIZATION ANALYSIS TO TREAT UNCERTAINTY IN SEISMIC EXCITATION?

A. Baratta and G. Zuccaro
University of Naples "Federico II", Naples, Italy

ABSTRACT

In the paper an approach to treat uncertainty in the response of structures under the action of earthquakes is presented. The main problem is focused in the unpredictability of the seismic accelerograms, and the research effort points at identifying the set of admisible quakes by a few basic parameters (duration, peak acceleration, gross information on the power spectrum, ...). The influence of the details of ground shaking, that really have a very significant influence on the structure's performance, is approached by the institution of a worst-case scenario. The basic idea is to build up a consistent model for the seismic hazard at the site, and to set up some rules which ground shaking must fit. Hence, the worst combination of details of the quake is sought by a search procedure, aiming at identifying the earthquake producing the extremum of some response parameters, in the set of admissible quakes. In the paper a number of r5eference models are set up, and results referred to a particular seismic area, the Campania region in Italy, are drawn, proving that the procedure is efficient and, to some extent, practical.

1) INTRODUCTION

The safety and economics of structures in seismic areas are highly dependent on the calibration of the appropriate seismic action. This is certainly one of the most interesting subjects in aseismic engineering. The extremely unpredictable variability of the proceeding of an earthquake and its effects at a given site, does not allow to define the solution of the problem in a definite, unique, way in a deterministic context. Uncertainty involves basically, from a structural point of view, the time-history of ground oscillation, since it, apart from some overall features, seems quite unpredictable while, on the other side, the structures' response is very sensitive even to small and apparently not significant details of the base vibration (see i.e. Ref.[1], where it is proved that a 2÷10% random perturbation of the ordinates of an accelerogram yields 50% and more variation in the response of th P-Δ elastic-plastic oscillator). The intrinsically hazardous nature of the problem suggests that a *probabilistic* or *possibilistic* point of view must be taken, aiming at keeping the hazard as small as possible, as far as this is economically compatible. On the other hand, the lack of instrumental records of significant seismic events at any site has addressed the researchers to make reference to some abstract model as a reference point for the synthesis of earthquake-type accelerograms. This approach requires that fundamental characters of seismic shaking at the site are known or, at least, can be reasonably predicted. In this regard, it has been recognised all over the world that some properties of recorded accelerograms, like peak acceleration, duration, build-up and decay, power distribution on the frequency range and so on, are strictly related to the localization of the epicenter and to the geological constitution of the site. This means that some *central characters* of the forcing function can be identified at every site, while a lot of many *other details* remain free to vary in a less controllable way, practically *random or indeterminate* to within some given, possibly *fuzzy*, boundary. Therefore, methods for the synthesis of earthquake-type accelerograms mostly rely on the simulation of a large number of random parameters correlated to the accelerogram ordinates, plus a finite, smaller number of *local* constants which represent the parameters conditioning the basic properties of the accelerograms to be generated. According to this philosophy, design accelerograms are taken more or less randomly from the whole population that can be potentially generated on the basis of the rationale one refers to; safety relies mainly on the statistical managing of the results and can be guaranteed only up to a given extent.

The problem that is illustrated in the present chapter is concerned with the case when *high reliability* is required; in this case one may need to select the most severe shape of the forcing function in the set of site-compatible accelerograms, reserving randomness, and consequent statistical treatment of uncertainty, to basic properties like peak acceleration and energy. Baratta and Zuccaro (see Baratta [2,3,4], Zuccaro [5], Baratta and Zuccaro [6,7,8]) developped a technique to produce the maximum theoretical values of the structural response under seismic load at given site. They pursued the goal combining available techniques for synthesis of random accelerograms (Ruiz-Penzien [9,10]) with optimization procedures capable to maximize some parameters, significant for aseismic design, in the respect of the constraints represented by the basic values characterizing the shaking properties at the site, as will be explained in Sec. 4. The introduction of an optimal criterion for the selection of time-histories that are the most dangerous with respect to a given response parameter, leads to the idea that the compatibility criterion for accelerograms to be synthesized can be formulated in some simpler way, that lends itself to optimization more naturally than random generators do. The first proposals on this line were advanced by Drenick [11,12,13,14] and Shinozuka [15] in the early 70's, and later on by Elishakoff and Pletner [16] and Baratta et al. [17,18]. All these approaches utilized an alternative, non-probabilistic, avenue. Drenick [11,12] used a constraint on the total energy which the earthquake is likely to develop at a certain site, as a description of uncertainty. He used the Cauchy-Schwarz inequality to determine the maximum response of the system to such an excitation. In the opinion of several investigators such a bound was too conservative.

Shinozuka [15] has suggested to characterize the earthquake uncertainty by specifying an envelope of the Fourier amplitude spectrum. Numerical calculations have demonstrated that the maximum response of the structure predicted by this method is less than that predicted in Refs. [11,12]. Elishakoff and Pletner [16] investigated the modification of the response prediction when the global information on the excitation is increased. In particular, the maximum possible response, which the structure may develop, was evaluated under the assumption that only the bound on base acceleration is known; then the maximum response was modified under the assumption that in addition to the base acceleration bound, the bounds on base velocity and/or displacement were specified. Finally, an application to earthquake engineering of *ellipsoidal modelling*, proposed as a general model for uncertainty .in mechanical problems by Elishakoff and Ben-Haim in 1990 [19], and first introduced in the theory of control by Schweppe [20] and previously applied to deal with geometrical imperfections in structural analysis [21], is developed by Baratta et al. in Refs [17, 18], and is fully illustrated in Sec.5. The term *antioptimization* was coined by Elishakoff (1991) for more general, than convex, uncertainties.

2) BASIC EQUATIONS AND BEHAVIOUR OF THE SDOF SHEAR-TYPE FRAME

2.1) ELASTIC BEHAVIOUR

Consider the simple linear SDOF shear frame in Fig. 2.1, with natural frequency ω_o, damping ratio ζ, and inertial mass m, and let u be the horizontal displacement of the beam with respect to the base of the column, (positive as shown). under the action of a ground acceleration a(t).

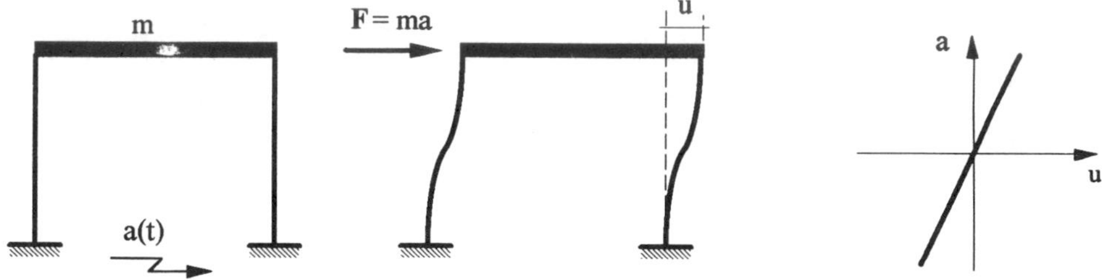

Fig. 2.1: The sdof shear-frame in the elastic range

The equation of motion is

$$\ddot{u} + 2\zeta\omega_o\dot{u} + \omega_o^2 u = a(t) \tag{2.1}$$

with initial conditions

$$u(0) = 0 \quad ; \quad \dot{u}(0) = 0 \tag{2.2}$$

The solution, as well known, is given by

$$u(t) = \int_0^t a(\tau) h(t - \tau) d\tau \tag{2.3}$$

where $h(\cdot)$ is the impulse response function

$$h(x) = \begin{cases} \dfrac{1}{\omega_d} \exp(-\zeta \omega_o x) \sin(\omega_d x) & \text{if} \quad x \geq 0 \\[2mm] 0 & \text{if} \quad x < 0 \end{cases} \tag{2.4}$$

with $\omega_d = \omega_o \sqrt{1 - \zeta^2}$.

2.2) *PERFECTLY PLASTIC BEHAVIOUR WITH DUCTILITY CONTROL*

2.2.1) *Basic statements*

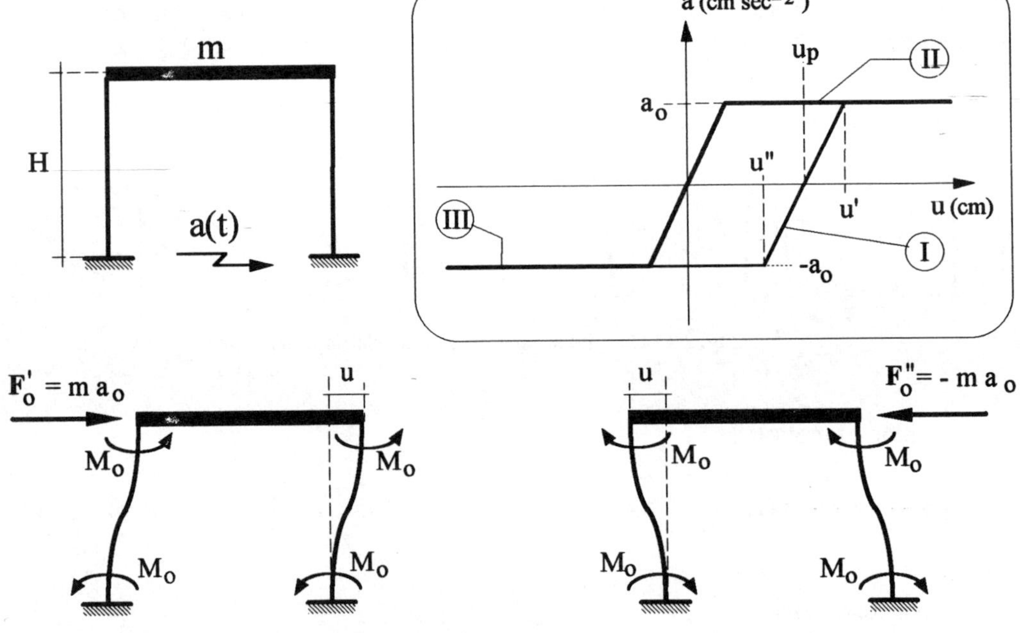

Fig. 2.2: The shear type sdof system under consideration

One considers the shear type frame (STF) in Fig. 2.2. under the action of a ground acceleration $a(t)$, and denotes by M_o the limit plastic moment in the columns.

The condition of the rotational equilibrium leads to

$$M = \frac{FH}{4} \qquad (2.5)$$

where M denotes the bending moment at the top and at the bottom of the piles, and F denotes the total horizontal force on the transverse beam.

Putting $M = M_o$ in eq. (2.5), one gets the total limit shear force

$$F_o = \frac{4M_o}{H} = ma_o \qquad (2.6)$$

where the strength of the system is expressed in term of acceleration by

$$a_o = \frac{F_o}{m} \qquad (2.7)$$

The characteristic diagram of the structure is shown in Fig.2.2; where u_p is the instantaneous value of the plastic displacement after the elastic thresholds $a'_o = a_o > 0$ and/or $a''_o = -a_o < 0$ have been violated.

The equations of the lines I, II and III shown in Fig. 2.2 are

$$
\begin{aligned}
\text{line I:} \quad & a_o(u) = \omega_0^2(u - u_p) \\
\text{line II:} \quad & a_o(u) = a'_o \\
\text{line III:} \quad & a_o(u) = a''_o
\end{aligned}
\qquad (2.8)
$$

2.2.2 The elastic-plastic equation of motion

One studies the dynamic behaviour of the system considered by the following elastic-plastic equation of the motion

$$\ddot{u}(t) + f_r(u, \dot{u}, u_p, \dot{u}_p) = a(t) \qquad (2.9)$$

where $f_r(u, \dot{u}, u_p, \dot{u}_p)$ denotes the restoring force that the numerical procedure updates during the elastic-plastic hysteresis; it can be expressed in the following form:

$$f_r(u,\dot{u},u_p,\dot{u}_p) = \begin{cases} 2\zeta\omega_0(\dot{u}-\dot{u}_p)+\omega_0^2(u-u_p) & \text{if} \begin{cases} a_o'' < \omega_o^2\left[u(t)-u_p(t)\right] < a_o' & (2..10) \\ \omega_o^2\left[u(t)-u_p(t)\right] = a_o' \text{ and } \dot{u}(t)\leq 0 \\ \omega_o^2\left[u(t)-u_p(t)\right] = a_o'' \text{ and } \dot{u}(t)\geq 0 \end{cases} \\[2em] 2\zeta\omega_0(\dot{u}-\dot{u}_p)+a_o' & \text{if } \omega_o^2\left[u(t)-u_p(t)\right] = a_o' \text{ and } \dot{u}(t)>0 \qquad (2..11) \\[1em] 2\zeta\omega_0(\dot{u}-\dot{u}_p)+a_o'' & \text{if } \omega_o^2\left[u(t)-u_p(t)\right] = a_o'' \text{ and } \dot{u}(t)<0 \qquad (2..12) \end{cases}$$

where ζ is the damping coefficient and u_p and $\dot{u}_p$ are given by

$$\dot{u}_p(t) = \begin{cases} \dot{u} \text{ if } \left[\omega_o^2(u-u_p)=a_o' \text{ and } \dot{u}(t)>0\right] \text{ or } \left[\omega_o^2(u-u_p)=a_o'' \text{ and } \dot{u}(t)<0\right] \\ 0 \hspace{8em} \text{otherwise} \end{cases} \qquad (2.13)$$

$$u_p(t) = \int_0^t \dot{u}_p(\tau)\,d\tau \qquad (2.14)$$

2.3) *PERFECTLY PLASTIC BEHAVIOUR WITH P-Δ EFFECT AND DUCTILITY CONTROL*

2.3.1) Basic statements

One considers the shear type frame in Fig. 2.3 under the action of a ground acceleration a(t) and of the vertical loads W given by

$$W = m\,g/2 \qquad (2.15)$$

with g denoting the gravity acceleration. The condition of the rotational equilibrium leads to

$$M = \frac{W\,u}{2} + \frac{F\,H}{4} \qquad (2.16)$$

Substituting u = 0 and M = M$_o$ in eq. (2.16), one gets the total limit shear force in the absence of P-Δ effect

$$F_o = \frac{4M_o}{H} = m\,a_o \qquad (2.17)$$

Alternatively, putting F = 0 and M = M$_o$ in eq. (2.16), one gets the condition for collapse by purely geometrical effect

$$\frac{W\,u}{2} = M_o \tag{2.18}$$

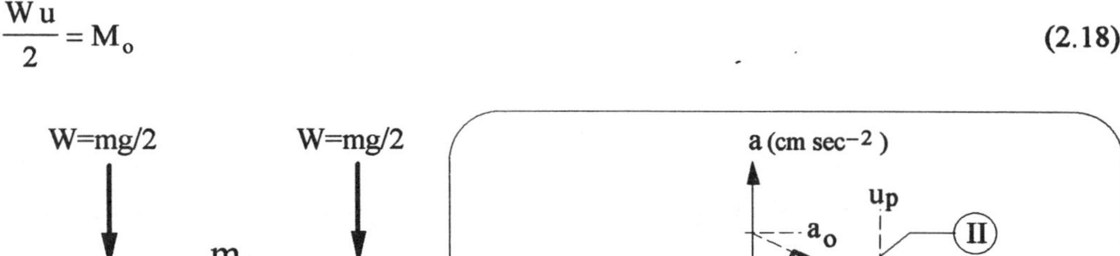

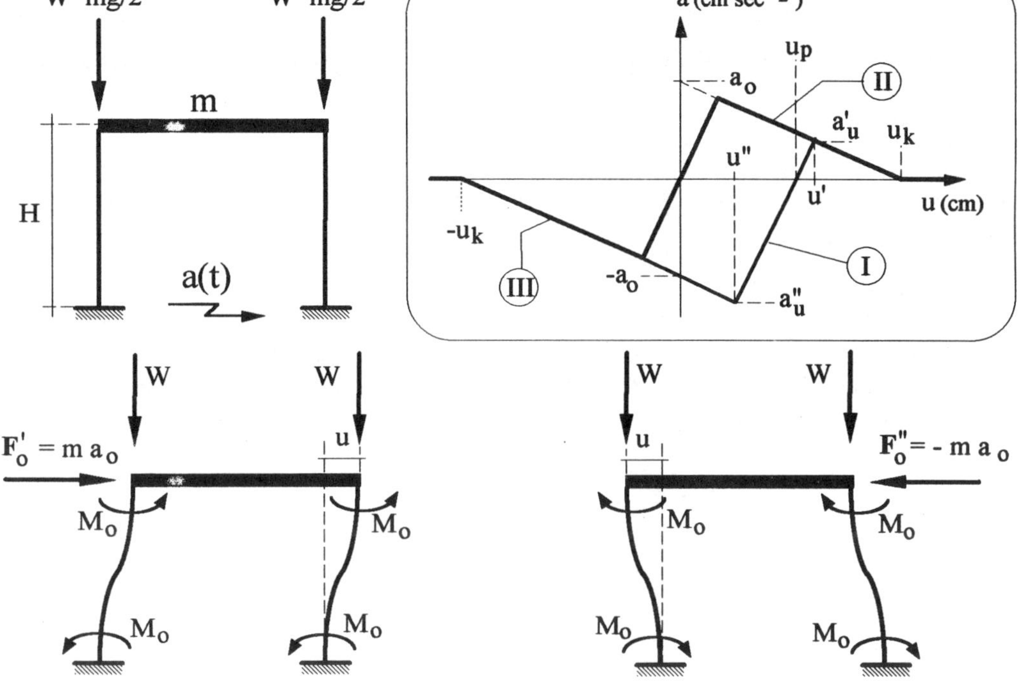

Fig. 2.3 The shear type sdof system under consideration

Substitution of eqs. (2.17) and (2.15) into eq. (2.18) yields

$$m\,g\,u = 4M_o = F_o\,H \tag{2.19}$$

whence, introducing $\vartheta = g/H$, one defines the *collapse displacement*

$$u_k = F_o / (\vartheta\,m) \tag{2.20}$$

corresponding to the limit value F_o of the shear force F at which the residual lateral strength of the frame is zero.

The virgin strength of the system can be expressed in term of acceleration by $a_o = F_o/m$, whence

$$u_k = \frac{a_o}{\vartheta} . \tag{2.21}$$

The characteristic diagram of the structure is shown in Fig.2.3; where a'_u and a''_u represent respectively the elastic limit for $u - u_p > 0$ and for $u - u_p < 0$ once the assigned elastic thresholds $a'_o = a_o > 0$ and/or $a''_o = -a_o < 0$ are violated, u_p being the instantaneous value of the plastic displacement.

The equations of the lines I, II and III shown in Fig. 2.3 are

$$\begin{aligned}
\text{line I:} \quad & a_o(u) = \omega_0^2(u - u_p) \\
\text{line II:} \quad & a_o(u) = a'_o - \vartheta u \\
\text{line III:} \quad & a_o(u) = a''_o - \vartheta u
\end{aligned} \tag{2.22}$$

where the elastic component of the displacement in P-Δ effect has been neglected.
By intersections of the line I with II and II with III one gets the values of a'_u and a''_u and of the corresponding displacements:

$$a'_u = \omega_0^2(u' - u_p) \quad \text{with} \quad u' = \frac{a'_0 + \omega_0^2 u_p}{\omega_0^2 + \vartheta} \tag{2.23}$$

and

$$a''_u = \omega_0^2(u'' - u_p) \quad \text{with} \quad u'' = \frac{a''_0 + \omega_0^2 u_p}{\omega_0^2 + \vartheta} \tag{2.24}$$

2.3.2 The elastic-plastic equation of motion with P-Δ effect

Neglecting the elastic component of the displacement in P-Δ effect, one studies the dynamic behaviour of the system considered by formally the same equation of the motion (2.9) where $f_r(u, \dot{u}, u_p, \dot{u}_p)$ is expressed in the following form:

$$f_r(u, \dot{u}, u_p, \dot{u}_p) = \begin{cases} 2\zeta\omega_0(\dot{u} - \dot{u}_p) + \omega_0^2(u - u_p) & \text{if} \begin{cases} u''(t) < u(t) < u'(t) \\ u(t) = u'(t) \quad \text{and} \quad \dot{u}(t) \leq 0 \\ u(t) = u''(t) \quad \text{and} \quad \dot{u}(t) \geq 0 \end{cases} & (2.25) \\ \\ 2\zeta\omega_0(\dot{u} - \dot{u}_p) + a'_o - \vartheta u' & \text{if} \quad u(t) = u'(t) \quad \text{and} \quad \dot{u}(t) > 0 & (2.26) \\ 2\zeta\omega_0(\dot{u} - \dot{u}_p) + a''_o - \vartheta u'' & \text{if} \quad u(t) = u''(t) \quad \text{and} \quad \dot{u}(t) < 0 & (2.27) \end{cases}$$

where u' and u'' are given respectively by eqs. (2.23) and (2.24), and u_p and $\dot{u}_p$ are given by

$$\dot{u}_p(t) = \begin{cases} \dfrac{(\omega_o^2 + \vartheta)}{\omega_0^2}\dot{u} & \text{if} \quad u(t) = u'(t) \ \text{and} \ \dot{u}(t) > 0 \ \text{or} \ u(t) = u''(t) \ \text{and} \ \dot{u}(t) < 0 \\ 0 & \text{otherwise} \end{cases} \qquad (2.28)$$

$$u_p(t) = \int_0^t \dot{u}_p(\tau)\,d\tau \qquad (2.29)$$

3) BASIC TEST ACCELEROGRAMS

The approach that is presented in this chapter is calibrated with reference to the following accelerographic records obtained on the occasion of the Campano-Lucano Earthquake of Nov. 23, 1980, in Italy, not far from city of Naples. The magnitude of the event was estimated in 6.5, and the epicentral intensity was set at the 7.5 degree of the MKS scale. The event was rather atypical, mainly because of its duration that lasted up to almost two minutes in some sites. Anyway, only the duration of the structural-significant motion will be considered here, that varies from about 52 to more than 86 secs.

Information on the local characters of ground motion is derived from direct inspection of recorded accelerograms, limiting the analysis, for simplicity, to only the NS component. The considered records are summarized in Table 3.I

Recorded ordinates of all accelerograms are converted to cm sec^{-2}, and all earthquakes are preliminarly reduced to the same norm [the *energy*, eq.(5.1)] as the one recorded in Torre del Greco, the most close site to Naples, that is the largest town in the region, so that every record possesses energy $E_0 = 79.46$ cm sec$^{-3/2}$.

Harmonic (Fourier) analysis is performed for every earthquake, thus obtaining for each one an expansion of the type (5.10) with the $a_i(t)$ given as in eq. (5.11). The analysis is carried on to include n = 400 sine and cosine waves for Torre del Greco, Sturno and Calitri earthquakes, while this number is increased up to n= 800 waves for the accelerograms recorded in Brienza and Bagnoli Irpino, that exhibit power spectra scattered on a wider range of frequencies.

A plot of each accelerogram, with the approximation resulting from its Fourier expansion and its power spectrum is quoted in Figs. $3.1.1 \div 3.1.5$.

                                                                A. Baratta and G. Zuccaro

TABLE 3.I

Site	Epicentral Distance (Km)	Local Intensity (MSK)	PGA (g/10)	Duration (sec)	Norm (cm.sec$^{-3/2}$)
TORRE DEL GRECO	80.1	7.0	0.59	52.9	79.46
BRIENZA	41.3	7.0	2.24	78.7	185.19
STURNO	34.8	6.0	2.25	70.7	284.53
CALITRI	21.0	8.5	1.52	86.1	268.25
BAGNOLI	22.3	6.0	1.31	79.1	152.29

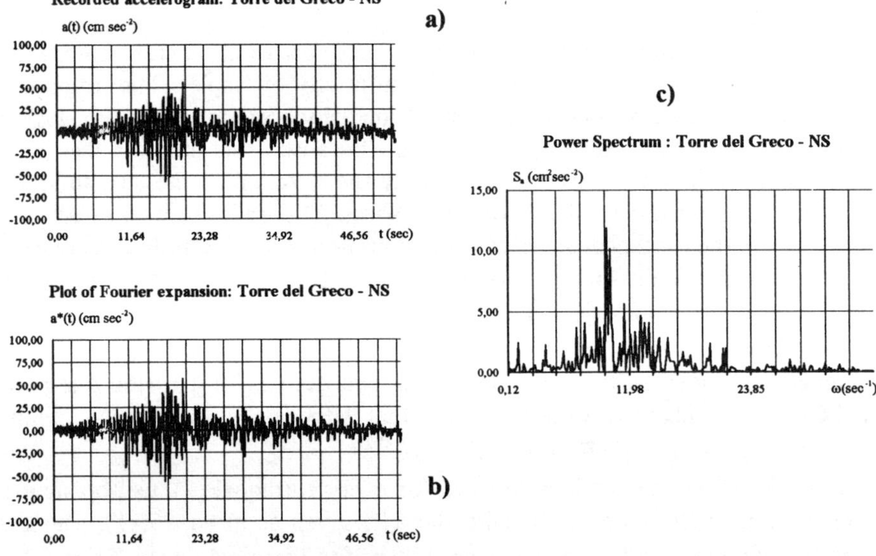

Fig. 3.1.1: TORRE DEL GRECO - Campania Earthquake of 11/23/80
a)Recorded accelerogram
b)Fourier approximation
c)Power spectrum

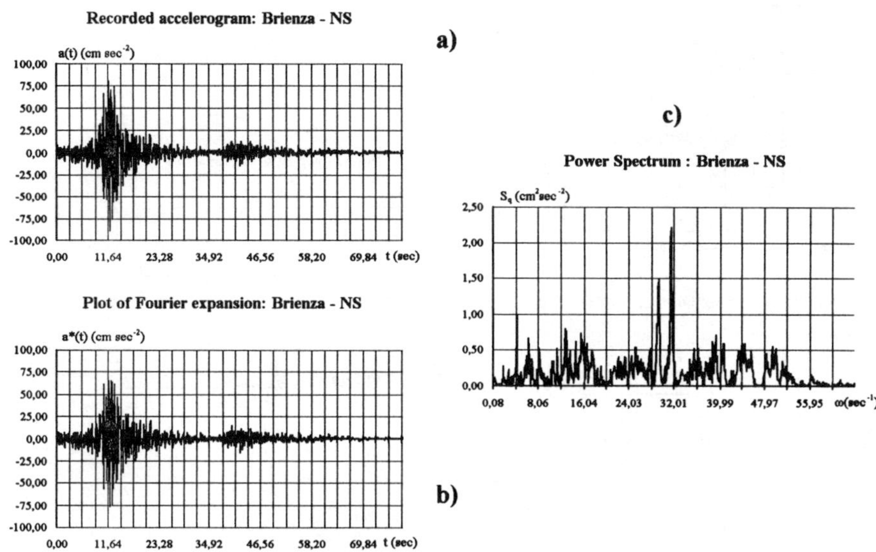

Fig. 3.1.2: BRIENZA - Campania Earthquake of 11/23/80
a)Recorded accelerogram
b)Fourier approximation
c)Power spectrum

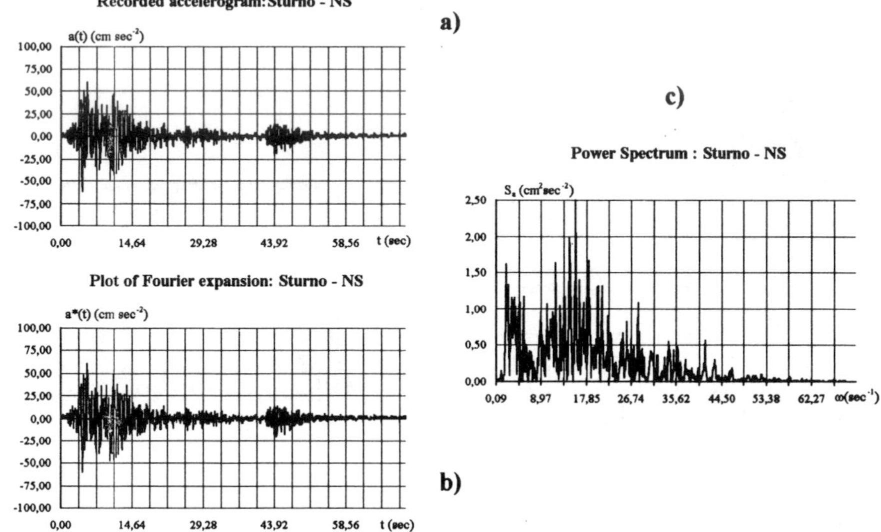

Fig. 3.1.3: STURNO - Campania Earthquake of 11/23/80
a)Recorded accelerogram
b)Fourier approximation
c)Power spectrum

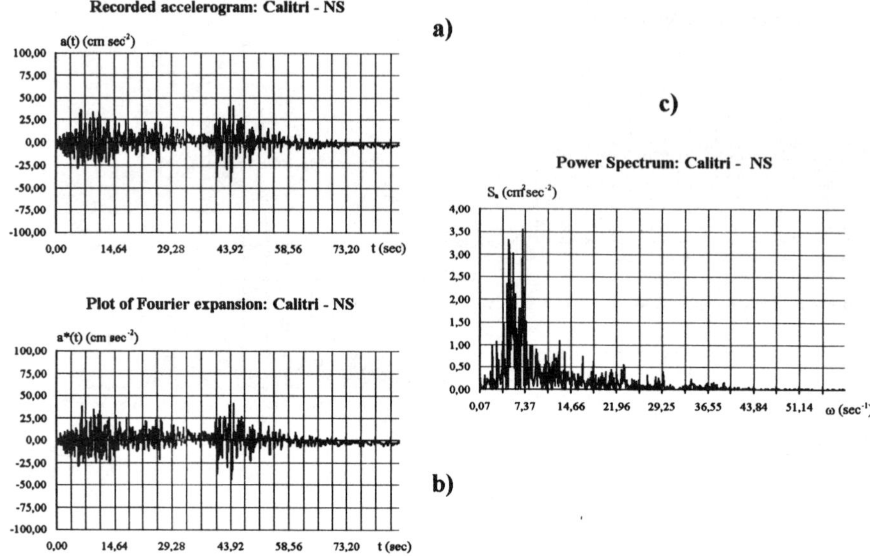

Fig.- 3.1.4:   CALITRI - Campania Earthquake of 11/23/80
               a)Recorded accelerogram
               b)Fourier approximation
               c)Power spectrum

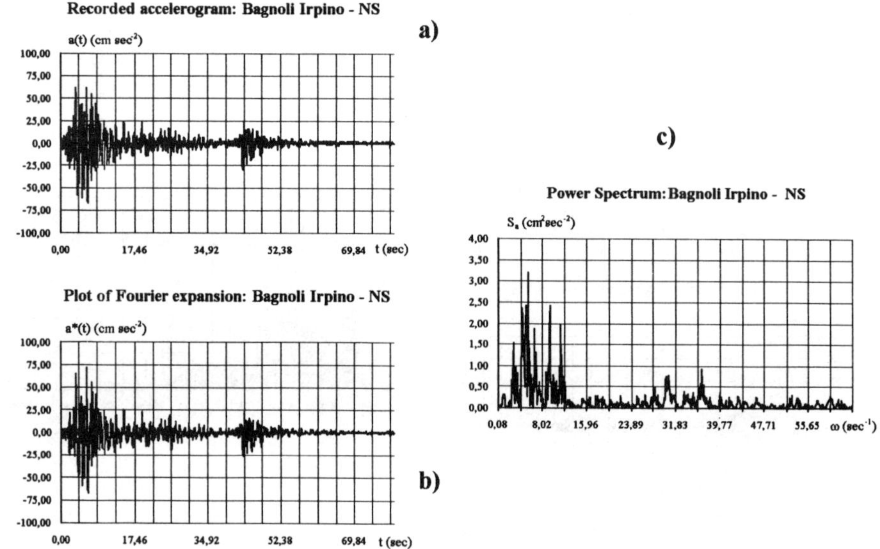

Fig. 3.1.5:   BAGNOLI IRPINO - Campania Earthquake of 11/23/80
               a)Recorded accelerogram
               b)Fourier approximation
               c)Power spectrum

Displacement response spectra for each investigated site and their envelopes are plotted in Fig. 3.2.

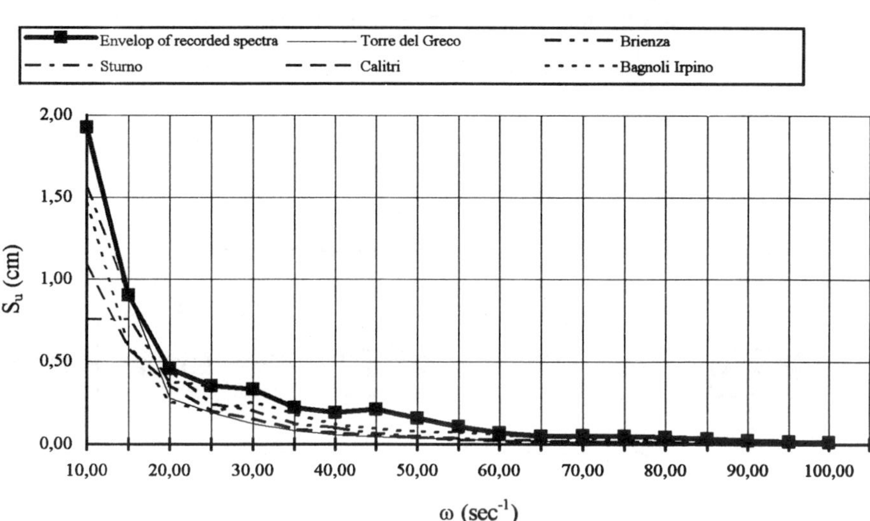

Fig. 3.2: Response spectra for the five test earthquakes and their envelope.

4) OPTIMIZATION PROCESS COMBINED WITH STOCHASTIC GENERATION OF SITE-COMPATIBLE SINTHETIC ACCELEROGRAMS

The technique that is illustrated in this section aims at producing the maximum theoretical values of any required component of the structural response under seismic load at given site. As stressed in the Introduction, the goal is pursued by combining available techniques for synthesis of random accelerograms with optimization procedures capable to maximize some parameters, significant for aseismic design, in the respect of the constraints represented by some *basic values* of the preliminarily identified characteristics of the shaking properties at the site (see Baratta [2,3,4], Baratta and Zuccaro [6,7,8], Zuccaro [5]).

4.1) THE PROCEDURE

It has been already highlighted in the Introduction that a great part of a synthetic seismic generator depends on a large number of random parameters x_i ($i = 1,...,n_x$), correlated to the accelerograms ordinates, plus a finite number of local characters ℓ_j ($j=1,...,n_\ell$) which are well defined site-dependent constants that qualify the algorithm which operates on the x_i's to produce the accelerogram a(t).

In most cases the x_i's are responsible for the *details* of the ground shaking while the ℓ_j's yield the macroseismic properties of the earthquake (i.e. the peak acceleration, the duration, the frequency range). The scheme for earthquake generation is outlined in Fig. 4.1, where reference is made to the procedure proposed by Ruiz and Penzien [9,10], $\overline{x}$ is the vector collecting the random variable's $\tilde{x}_i$'s and $\tilde{\ell}$ is the vector of the local constants ℓ_j.

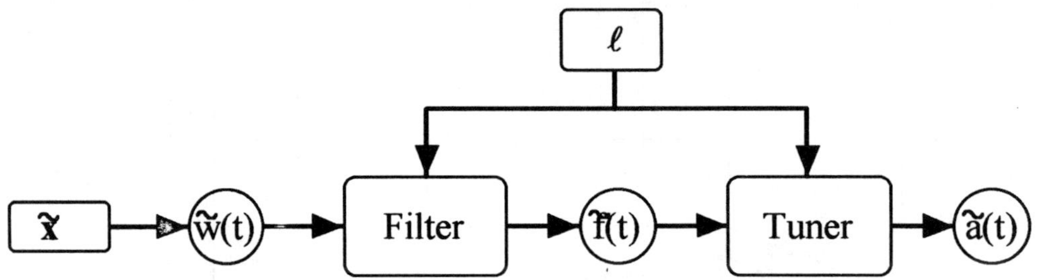

Fig. 4.1: Scheme for earthquake generation

According to such procedure, a vector **x** of numbers is first generated by repeated independent sampling of a random variable $\tilde{x}$ with mean value μ_x and variance σ_x^2. By associating such numerical sequence to a sequence of time instants t_i (i=1,...,n_x) equally spaced by a small time lag Δt, an approximation of a sample function of a white-noise $\tilde{w}(t)$ with mean value μ_x and intensity $I_o = \sigma_x^2 \cdot \Delta t$ is produced. If the random variable $\tilde{x}$ is Gaussian, $\tilde{w}(t)$ is a Gaussian white noise. If $\tilde{x}$ is a standard Gaussian random variable (i.e. $\mu_x = 0$ and $\sigma_x = 1$), then $\tilde{w}(t)$ is a zero-mean unit-variance white noise.

The white noise w(t) is filtered by means of a second-order linear system with damping ratio ξ and undamped circular frequency Ω

$$\begin{cases} \ddot{\eta} + 2\xi\Omega\dot{\eta} + \Omega^2\eta = -w(t) \\ f(t) = -2\xi\Omega\dot{\eta} - \Omega^2\eta \end{cases} \tag{4.1}$$

whose frequency-response-function is

$$H^2(\omega) = \frac{\Omega^4 + 4\xi^2\Omega^2\omega^2}{\left(\Omega^2 - \omega^2\right)^2 + 4\xi^2\Omega^2\omega^2} \tag{4.2}$$

The parameters of the filter are chosen in way to shift the energy of the final generated accelerogram in the frequency ranges that are the most prone to be excited at the site.

TABLE 4.I

Variance and spectral density functions at the four stages of the gneration process

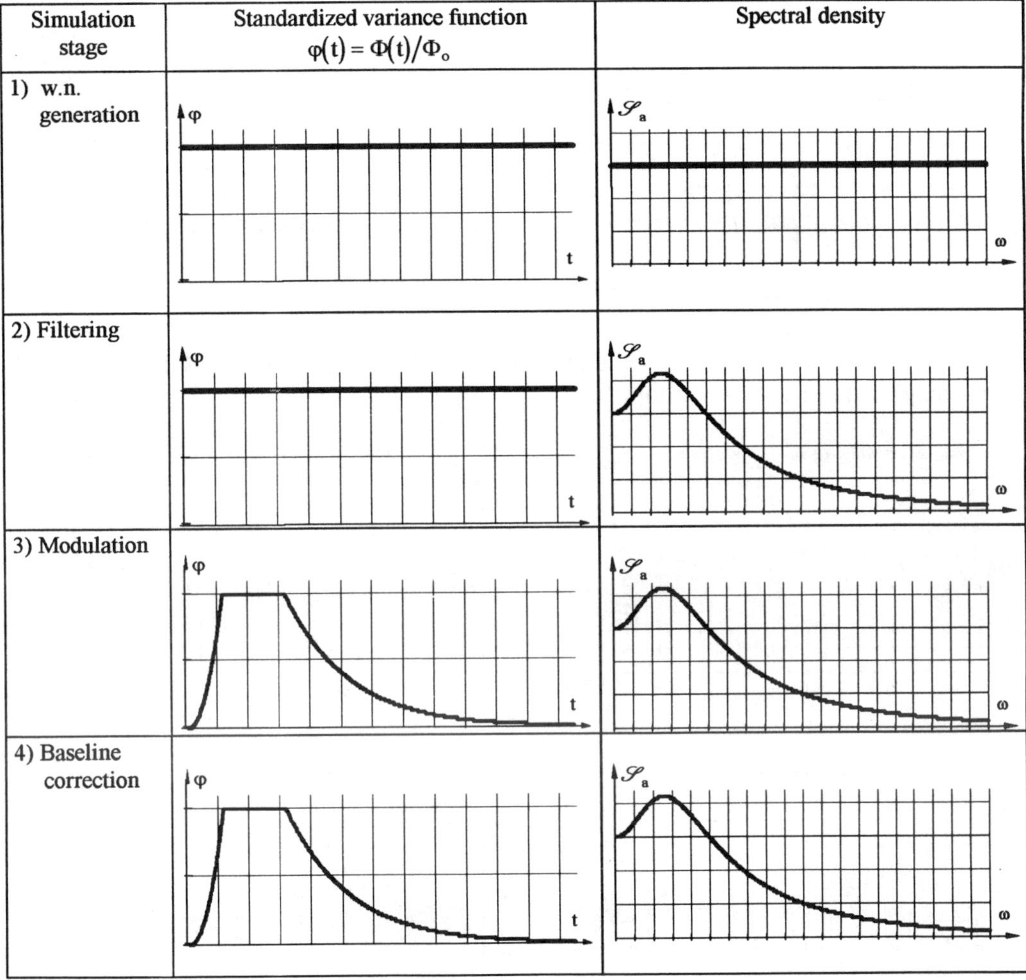

Simulation stage	Standardized variance function $\varphi(t) = \Phi(t)/\Phi_o$	Spectral density
1) w.n. generation		
2) Filtering		
3) Modulation		
4) Baseline correction		

By modulating the resulting coloured noise f(t), one obtains the pseudo-acclerogram s(t)

$$s(t) = f(t)\sqrt{\varphi(t)} \tag{4.3}$$

where $\varphi(t)$, the *modulating function*, aims at reproducing the shape of the energy build-up, strong-motion phase and decay of earthquakes at the site; it is assumed of the type

$$\varphi(t) = \begin{cases} \left(\dfrac{t}{t_i}\right)^2 & \text{if } t < t_i \\ 1 & \text{if } t_i \le t \le t_d \\ \exp\left[-v\left(t - t_d\right)\right] & \text{if } t > t_d \end{cases} \qquad (4.4)$$

Finally, the pseudo-accelerogram is treated as a numerical record, and it is submitted to standard procedure of *baseline correction* yielding the final accelerogram in the form

$$a(t) = s(t) + s_0 + s_1 t + s_2 t^2 \qquad (4.5)$$

where s_0, s_1 and s_2 are sought in a way that ground acceleration, velocity and displacement are vanishing at the end of the motion.

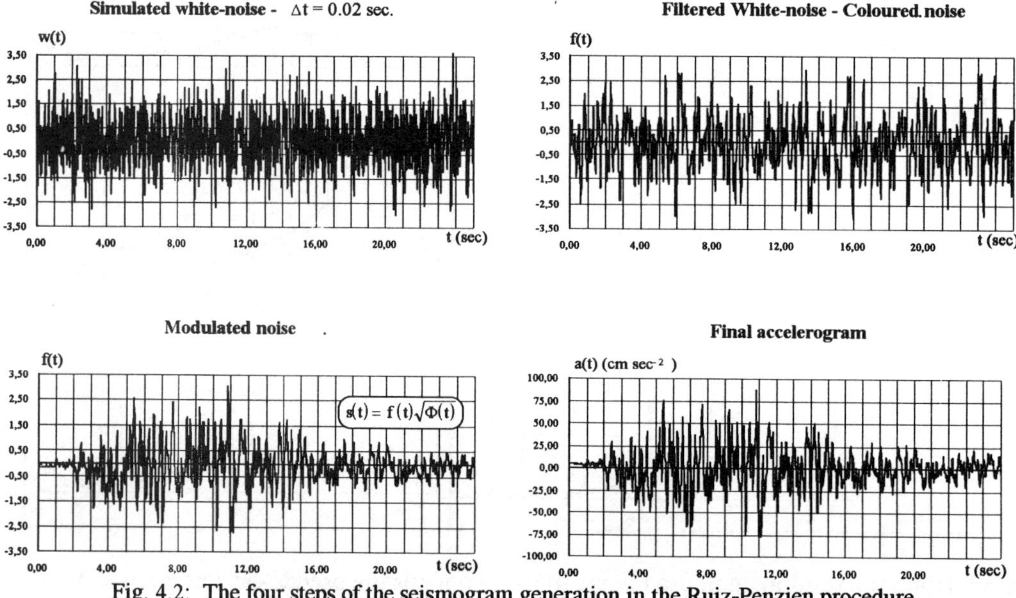

Fig. 4.2: The four steps of the seismogram generation in the Ruiz-Penzien procedure

The various stages of the seismogram generation process are basically illustrated in Table 4.1 and Fig. 4.2.

From the above summary exposition, it is clear that after the vector ℓ of the local characters has been assigned, any particular accelerogram depends on the set of numbers x_i constituting the random vector $\mathbf{x}$. Since the modalities of their generation, such numbers should verify the following relationships, the closer the larger n_x is (remember that n_x is of the order 1,000 or more)

$$\sum_{i=1}^{n} x_i = n\mu_x$$

$$\sum_{i=1}^{n} (x_i - \mu_x)^2 = n\sigma_x^2 \tag{4.6}$$

$$n(a,b) = n[P(b) - P(a)] \ \forall \ (a,b) \in X^2 \tag{4.7}$$

where n(a,b) is a number of components of $\mathbf{x}$ falling in the interval (a,b), X is the range of the values of $\mathbf{x}$, and P(x) is the distribution function of $\tilde{x}$.

Assume now that one is interested, as it is the case here, in the worst situation that can occur for a structure when it is acted on by earthquakes, apart from the decision concerned with gross shaking parameters like PGA or total energy and so on.

Having assumed the above scenario, for a given structure any component R of the structural response will be a function of the sample vector x, conditioned on the assumed value of ℓ

$$R = R\left(\mathbf{x}|\ell\right) \tag{4.8}$$

The problem reduces then to maximize the function R $(\mathbf{x} \mid \ell)$, with the components of $\mathbf{x}$ that obey the above (4.6) - (4.7)

Find $\ R^* = \underset{\mathbf{x}}{\mathrm{Max}} \ R\left(\mathbf{x}|\ell\right)\ $ subjected to equalities

$$\begin{cases} \sum_{i=1}^{n} x_i = n\mu_x \\[2mm] \sum_{i=1}^{n} (x_i - \mu_x)^2 = n\sigma_x^2 \\[2mm] n(a,b) = n[P(b) - P(a)] \ \forall \ (a,b) \in X^2 \end{cases} \tag{4.9}$$

Note that eqs. (4.6) - (4.7) are the conditions for which an admissible test accelerogram a(t) can be associated to $\mathbf{x}$ by the generation process illustrated in the above.

The optimization can be executed by random-search procedures that verify the above constraints almost spontaneously (see Rao [22]); unlike the classical optimization methods which turn out to be inadequate for the constraints described. The procedure has been named by the present writers *Driven Simulation (D.S.)* and can be better understood observing the flow-chart in Fig. 4.3.

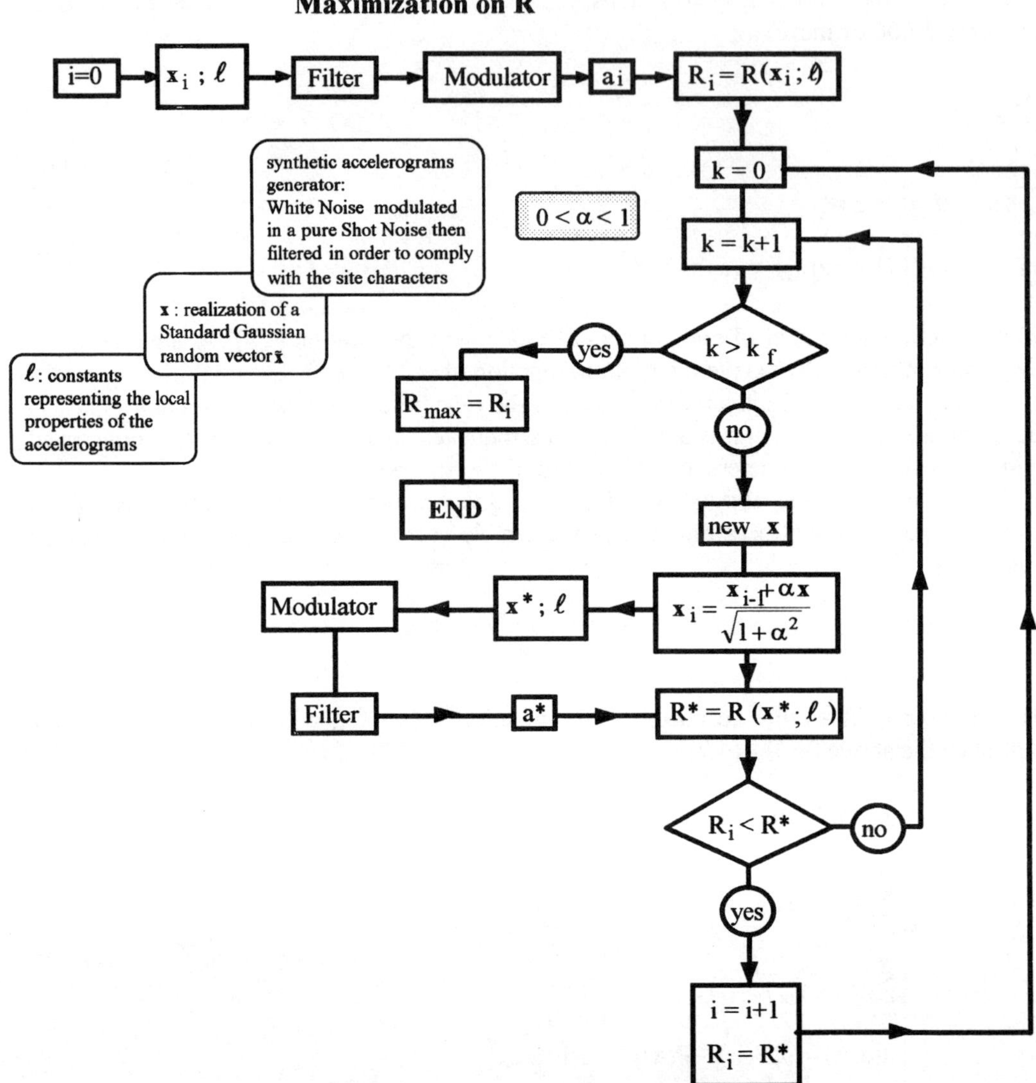

Fig. 4.3: The flow-chart of the D.S. procedure

Let $\mathbf{x_o}$ be the initial vector obtained by a first generation of n_x random numbers from the standard random variable $\tilde{x}$, and calculate a first response value

$$R_o = R(\mathbf{x_o} | \ell) \qquad (4.10)$$

Next, let α be a positive number not larger than unity, and generate a second random vector $\mathbf{x}$ from $\tilde{\mathbf{x}}$ and put

$$\mathbf{x}_1 = \frac{\mathbf{x}_o + \alpha \mathbf{x}}{\sqrt{1+\alpha^2}} \tag{4.11}$$

and calculate

$$R_1 = R(\mathbf{x}_1|\ell) \tag{4.12}$$

If the attempt made by the sample vector $\mathbf{x}$ is not successful, i.e. if $R_1 < R_o$, a new perturbation vector $\mathbf{x}$ is generated and steps (4.11) - (4.12) are repeated until the attempt is successful, i.e. $R_1 > R_o$. In this case $\mathbf{x}_o$ can be subsituted by $\mathbf{x}_1$, R_o by R_1, and the procedure is iterated in way to produce a sequence of vectors $\mathbf{x}_o$, $\mathbf{x}_1,...\mathbf{x}_r,...$ yielding an increasing sequence of response values R_o, $R_1,...R_r,...$, and the procedure ends when, after a reasonable number of trials, say k_f, no vector $\mathbf{x}$ is randomly produced that yields a new successfull $\mathbf{x}_i$.

4.2) THE LOCAL CONSTANTS

Besides local components Ω, ξ, and the parameters of $\varphi(t)$ (the duration of the parabolic build up t_i, the instant t_d at beginning of the decay and the decay constant ν), the duration of the event T and the expected peak ground acceleration a_p have to be estimated. The calibration has been conducted by utilizing the accelerograms recorded at stations around Naples area during the last earthquake (23rd of November 1980, see Sec. 3). The values of the natural frequency Ω and the damping ratio ξ are determined by minimising the error between the Kanai-Tajimi function

$$G(\omega) = \mathscr{S}_o \frac{1 + 4\xi^2 \left(\frac{\omega}{\Omega}\right)^2}{\left[1 - \left(\frac{\omega}{\Omega}\right)^2\right]^2 + 4\xi^2 \left(\frac{\omega}{\Omega}\right)^2} \tag{4.13}$$

and the average of the power spectra measured at different stations during the 1980's event (Fig. 4.4).
The duration t_i of the parabolic build up and the initial instant of the decay t_d are set by optimal fitting of the conventional intensity function $\Phi(t)$, with

$$\Phi(t) = \Phi_o \varphi(t) = \Phi_o \begin{cases} \left(\dfrac{t}{t_i}\right)^2 & \text{if } t < t_i \\ 1 & \text{if } t_i \le t \le t_d \\ \exp\left[-v(t - t_d)\right] & \text{if } t > t_d \end{cases} \qquad (4.14)$$

Average spectral density and Kanai-Tajimi approximation

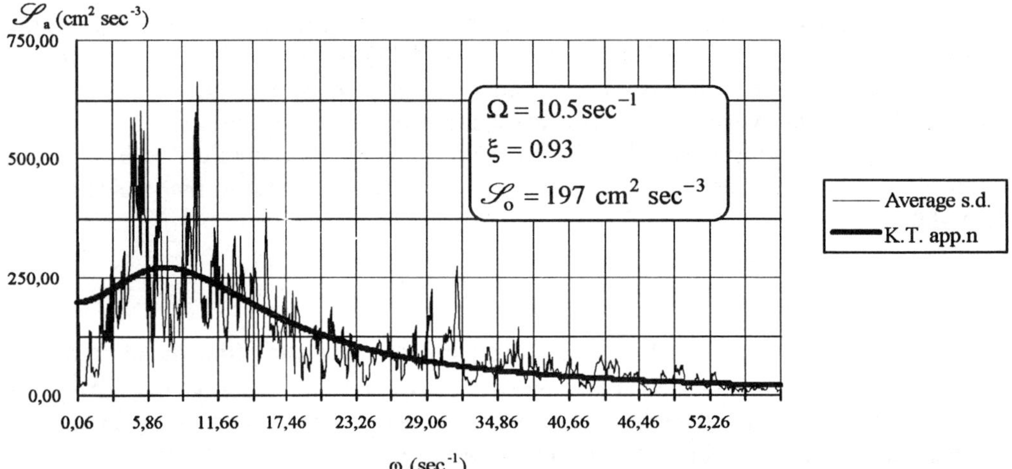

Fig. 4.4: Fitting of the Kanai-Tajimi standard spectrum to average power spectra of recorded earthquakes

to the average variance intensity function $\Phi(t)$ obtained by the available data (Fig. 4.5)
One observes that the lack of instrumental records of earthquakes in Neapolitan area produce a large variability of $\Phi(t)$, however it estimates a satisfatory smooth variance function, a better evaluation of $\Phi(t)$ will be possible once sample records will be available. The duration T of the event to be simulated is assumed equal to the average of the duration in the area of the strong-motion phase.
The peak ground acceleration is assumed in relation to the model of the material behaviour considered.
After the elaboration described above the local constant assumed are: Ω =10.48 sec^{-1}; ξ =0.93; Φ_o= 276 cm^2/sec^4; t_i = 4.2 sec.; t_d = 10.9 sec.; T =25 sec.; v = 0.1.

Average intensity function and conventional approximation

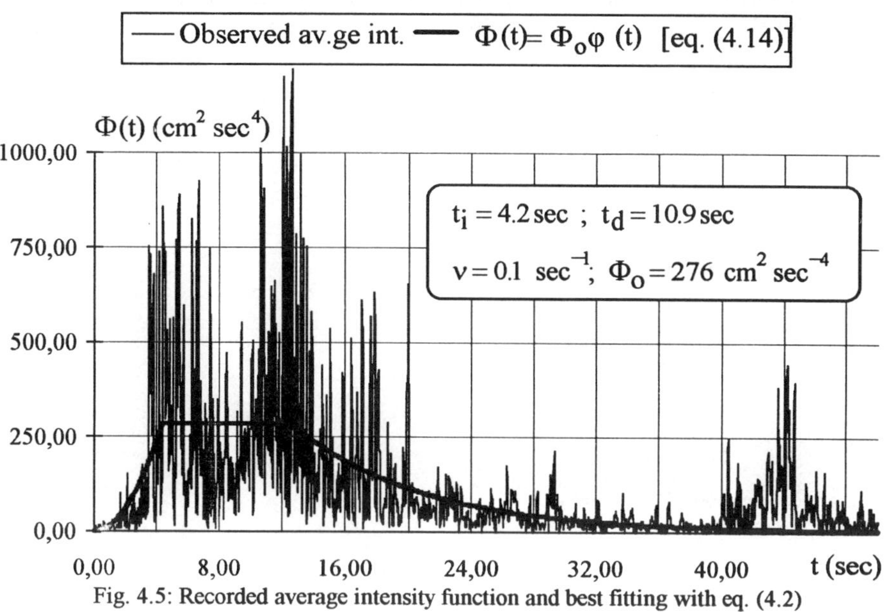

Fig. 4.5: Recorded average intensity function and best fitting with eq. (4.2)

4.3) BASIC LOCAL PARAMETERS ON THE BASIS OF HAZARD ANALYSIS

The approach presented in the paper allows to reduce the uncertainty in checking safety deriving from the accelerogram shape. In order to make the procedure effective, it is thus possible to leave only one uncertain parameter, the peak ground acceleration, that is approached by the following method.

The evaluation of seismic hazard has been performed by application of two distinct procedures: S.I.S.M.A.(Coburn, Spence and Zuccaro, 1986) [23] and SIM-SIS (Baratta and Cacace, 1986) [24]. The Seismic Impact Simulation Model for risk Assessment is based on a systematic approach following the probabilistic theory (Cornell, 1968) [25]. The model assumes that the Hazard at the point derives from earthquakes generated in source zones which have uniform seismicity defined by a linear recurrence relationship (Gutenberg and Richter, 1954) [26]. The site hazard, expressed in terms of probability of exceedance of a particular parameter of ground motion (e.g. intensity or PGA) is determined by the use of source-to-site attenuation relationship derived from past earthquake data. The pattern of arrival times, given an average recurrence rate, is assumed to follow the Poisson law.

The SIM-SIS stands for a SIMulation model of the SeISmic history on the territory. It is based on an arrival Poisson process and on Gutenberg and Richter seismicity law. The territory is overlapped by a very fine grid; one assigns to each cell the probability of epicentral location as a function of the magnitude. The attenuation law is isotropic with random perturbations and a local amplification coefficient. The simulation starts assuming a prefixed window of time (e.g. 100 years), and generating the arrival times of the events and relative magnitudes with independent statistics. Then one localizes epicenters following the correlation with the magnitude and the effects at the site are determined possibly through propagation and amplification procedures.

The results obtained by the application of both models have proved to be essentially coincident. This circumstance encouraged the present writers to further investigate in this direction.

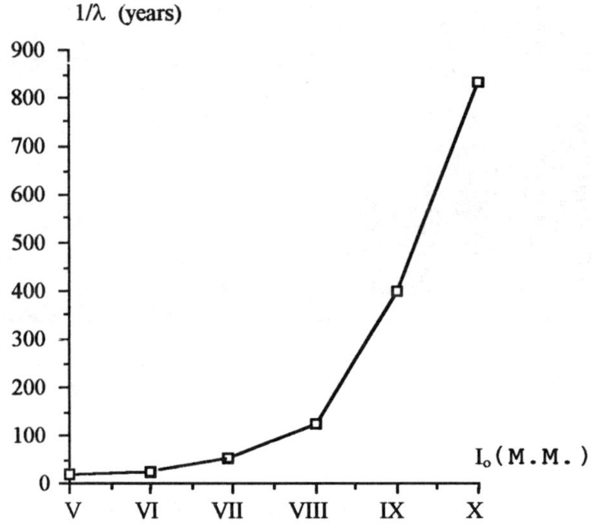

Fig. 4.6 Correlation of epicentral Intensity (MM) against return period, both models.

To estimate the most probable PGA at the site one needs to correlate the damage measure I_o to PGA recorded in the past earthquakes or, alternatively, one can assume the Richter correlation

$$\text{Log } a_p = \frac{I_o}{3} - \frac{1}{2} \tag{4.15}$$

Studies on the correlation between PGA and intensity, aiming at finding a dimensionless measure of the damage at a given level of shaking at ground (Coburn, Sakai, Spence, Pomonis, 1990 [27]), show that the correlation proposed by Margottini (see Coburn & others 1990 for details), based on 1980 earthquake data, yields the best fitting of several sets of damage data from sites all over the world. The PGA-Intensity correlation assumed [27] is

$$\text{Log } PGA = 1.84 + 0.057\psi \tag{4.16}$$

where ψ is a parameter correlated to the intensity, as given below

ψ	-1	2	7.5	10	13	16
I (MM)	V	VI	VII	VIII	IX	X

One observes that the Margottini correlation is more realistic one for Naples area. This is shown in Fig. 4.8 where the correspondence of PGA to return period for both correlation proposed has been plotted. One marks that the trend of PGA to increase with the return period is significantly attenuated in Margottini's approach, in agreement with historical observation.

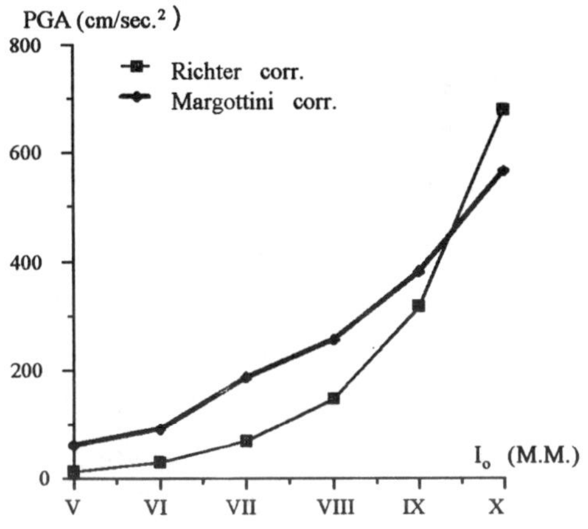

Fig. 4.7 PGA-Intensity correlations

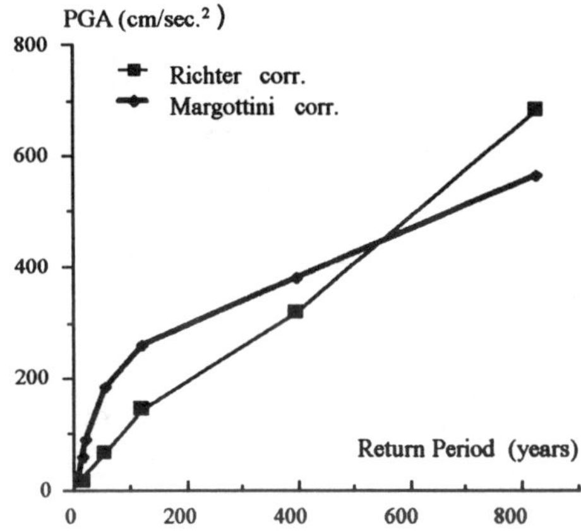

Fig. 4.8 PGA-Return Period relations

4.4) NUMERICAL RESULTS

4.4.1) Elastic behaviour

Any fraction of gravity can be assumed as expected peak ground acceleration just to compute the behaviour of the maximum response spectrum, in that the linearity of the oscillator allows to consider any value of a_p as factor of the spectral ordinates.

The maximum possible values of the acceleration response computed by the procedure proposed (see sec. 4.1) are plotted in Fig. 4.9, and are compared to the envelop of response obtained from 1980 data, both normalized to the same peak acceleration.

In the present case one assumes $a_p = 0.1$ g and a damping ratio of the SDOF system $\zeta = 5\%$.

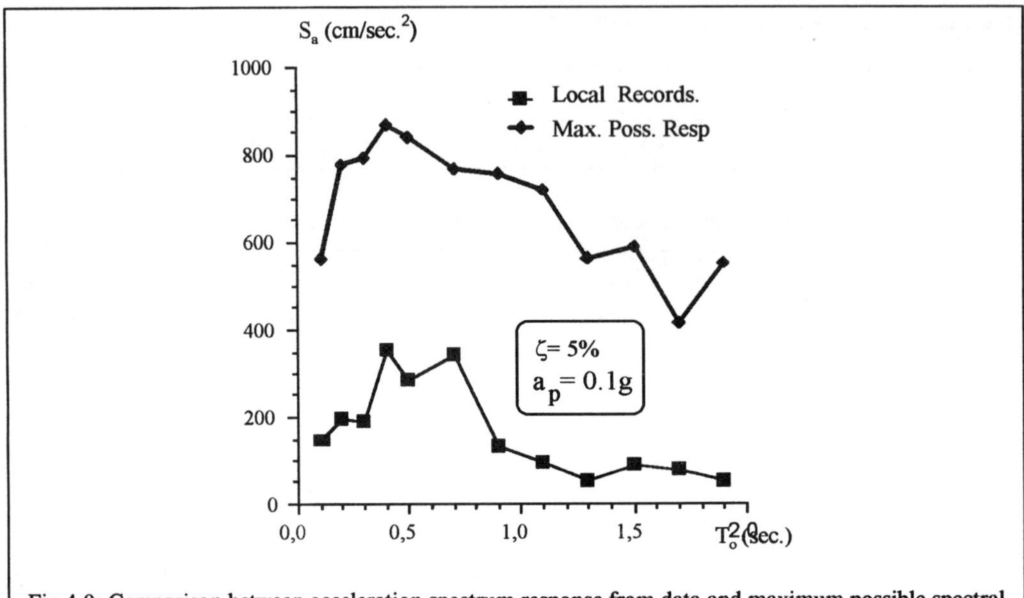

Fig.4.9: Comparison between acceleration spectrum response from data and maximum possible spectral ordinates.

Suitable values for PGA, a_{p_i} , are read by graph of Fig. 4.8, as a function of the time length on which one wants to ensure survival. According to the Margottini correlation, for return periods of 25, 50, 100, 500 years one gets a_p = 95, 180, 230, 425 cm sec $^{-2}$ respectively. The acceleration spectral response is computed by the following relation

$$S_{a_i} = \frac{a_{p_i}}{a_p} \, S_{a_{max}} \tag{4.16}$$

where i = 25, 50, 100, 500 years, a_p=0.1 g and $S_{a_{max}}$ the maximum response shown in Fig. 4.9. The extreme displacement response spectra are plotted in Fig. 4.10, for ζ =5%.

4.4.2) Perfectly plastic behaviour with ductility control

Unlike the elastic behaviour, the STF response in the plastic range cannot be expressed by a factor of the expected peak acceleration a_p at the site. Therefore, the maximum responses in term of displacement for several values of a_p (150, 200, 250 cm/sec^2) and of damping ratio ζ of the structure.(2%, 5%) have been computed.

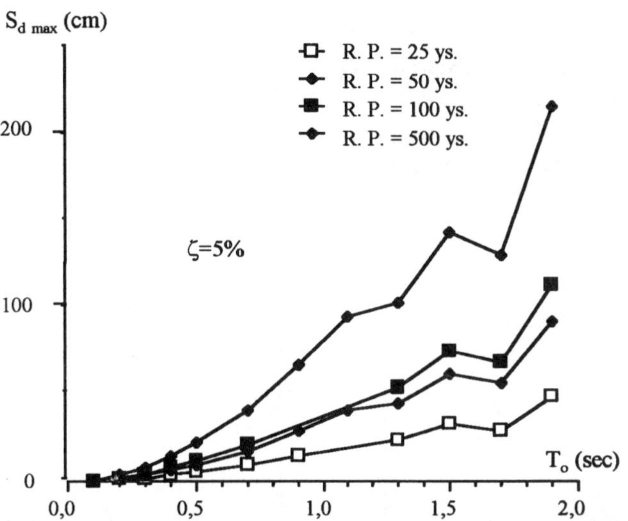

Fig. 4.10 Maximum displacement responses for some Return Periods
design in Naples area

The expected peak acceleration values considered correspond to the expected values for a return period between 25 and 200 years roughly (see Fig. 4.8). A perfectly plastic model of the material behaviour has been assumed (Fig. 4.11) having a limit value of the

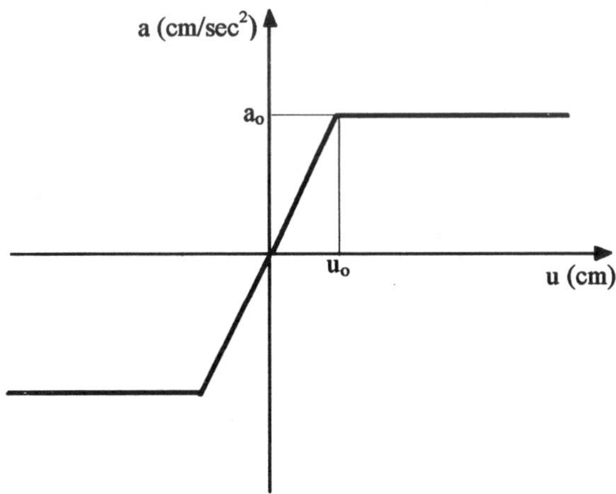

Fig. 4.11 - Perfectly plastic STF

acceleration relevant to the plasticity threshold $a_o = 200$ cm sec$^{-2} \cong 0.2$ g.
In Fig. 4.12 one compares the maximum possible displacement spectrum in the linear elastic range, with the maximum possible displacement spectrum evaluated when a perfectly plastic behaviour is considered, and for a value of the damping ratio $\zeta = 5\%$. One observes that for $a_p < a_o$ the values of the plastic response are not dissimilar from the elastic

ones, while for $a_p > a_o$ the values of the displacement in the case of the perfectly plastic material are, as it was predictable, generally higher of the elastic ones.

Moreover one can observe, through looking at the behaviour of the elastic spectrum, that for values of the natural vibration period around 1.8 sec. the curves always show an inflection before reaching the absolute maximum values of the displacement.

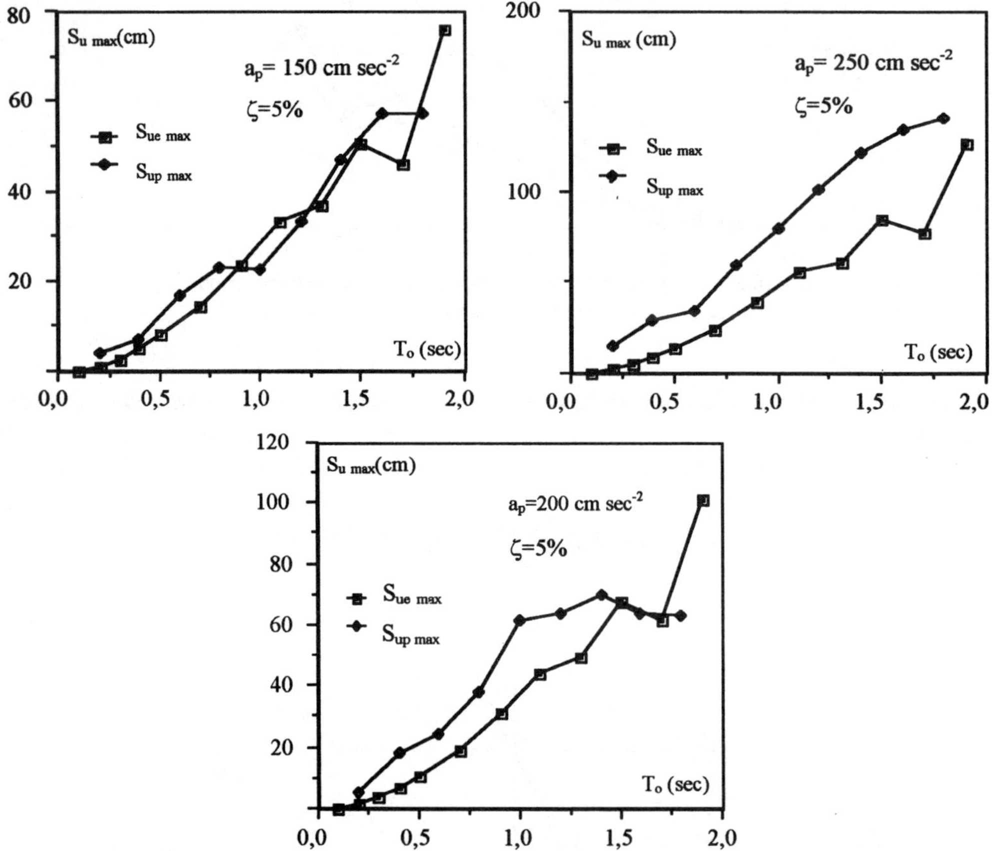

Fig.4.12: Comparison between the maximum possible displacement spectrum for the elastic STF $S_{ue\,max}$ and the maximum possible displacement spectrum for the perfectly plastic material $S_{up\,max}$

In order to study the influence of the ratio between the expected maximum possible peak acceleration a_p and the plastic threshold acceleration a_o on the ductility factor δ, one observes in Fig.4.13 the response spectra of the ductility factor $\delta = u_p/u_o$ for several ratios a_p/a_o (0.75, 1, 1.5) and for two different damping factors $\zeta = 2\%$ and $\zeta = 5\%$.

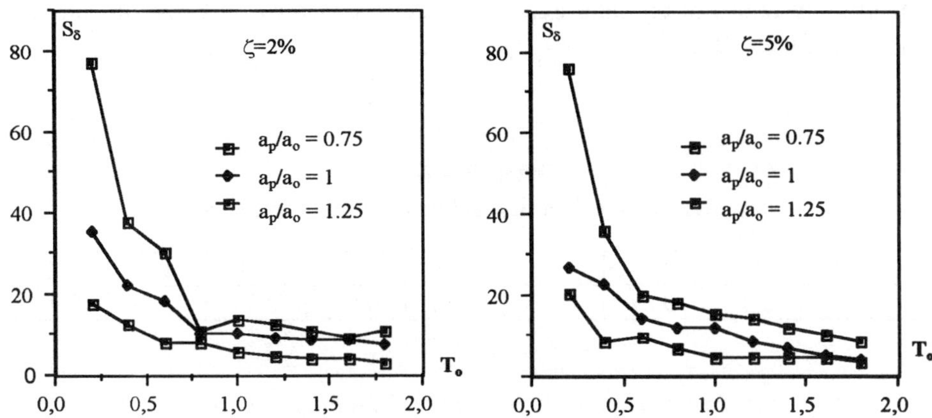

Fig. 4.13: Response spectra of the ductility factor

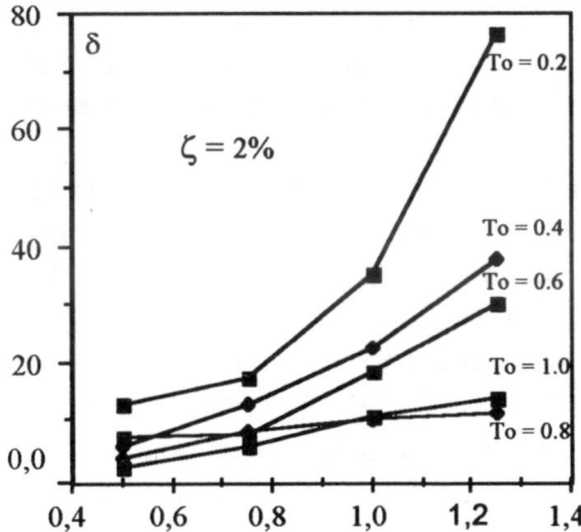

Fig. 4.14: Response of the ductility factor varying a_p/a_o, for $\zeta = 2\,\%$.

In Fig. 4.13 one observes that no evident difference can be detected for the values of δ by varying the damping factor. Moreover, the worst values of the response arise for own periods in the range [0.2, 1.0] sec. and for a damping factor $\zeta = 2\%$. In Fig. 4.14 the ductility factor δ is plotted versus the ratio a_p/a_o for the most critical natural vibration periods of oscillation and for $\zeta = 2\%$.

Tab. 4.II lists the values of the ratio a_p/a_o for a ductility factor $\delta = 4$, usually adopted in plastic analysis, for $\zeta = 2\%$.

TABLE 4.II
a_p/a_o values for a ductility factor $\delta = 4$ and for $\zeta = 2\%$.

T_0	0.2	0.4	0.6	0.8	1.0	1.2	1.4	1.6	1.8
a_p/a_o	--	0.48	0.62	0.53	0.64	0.95	0.77	0.75	0.83

4.4.3) Perfectly plastic behaviour with P-Δ effect and ductility control

After the assumption of a elastic-plastic behaviour of the system, the maximum response is found in terms of displacement, rather than a factor of the expected peak acceleration at the site a_p, like in the previous section. Therefore the maximum possible values of the displacements for different values of a_p (150, 200, 250 cm sec^{-2}) and of the damping coefficient ζ (2%, 5%) are found. The values of the peak accelerations assumed in the analysis represent the maximum expected ground motion for return periods of 25 ÷ 200 years at the site of Naples, such values are based on the available data in the region (Baratta et Zuccaro, 1992) [6]. One assumes, with reference to the symbols of the Fig. 2.3, an elastic threshold $a_o = \pm 200$ cm sec^{-2} and a critical displacement $u_k=66.6$ cm. for a value of $\vartheta = 3$ cm^2 sec^{-2}.

In Fig. 4.15 one reports the comparison between the maximum possible spectrum of the response of the system supporting vertical loads and the maximum response spectrum presented in the previous section, where no destabilizing effect was considered and a perfectly plastic model was assumed.

Figs. 4.15 a), b), and c) show, respectively, the behaviour of the spectra for the values of a_p=150, a_p=200 and a_p=250 cm sec^{-2} and for a damping factor equal to 5 %. Fig. 4.16 shows the same comparison for the damping factor equal to 2 %. The spectra of the Figs. 4.15 and 4.16 are non-dimensional and normalized to the value of the critical displacement u_k (see Sec.2).

Therefore, the values of the response equal to unity represent the collapse of the structure. One observes that the maximum values of the response when the phenomenon of destabilization is considered, are generally greater then the values of the system without vertical loads. The natural period $T_o= 0.4$ sec. appears to be not very sensitive to the stability problem having found, for it, similar values of the maximum response in both cases. Moreover, one observes a moderate influence of the damping factors on the maximum values of the response, while these show more sensitivity to variations in a_p.

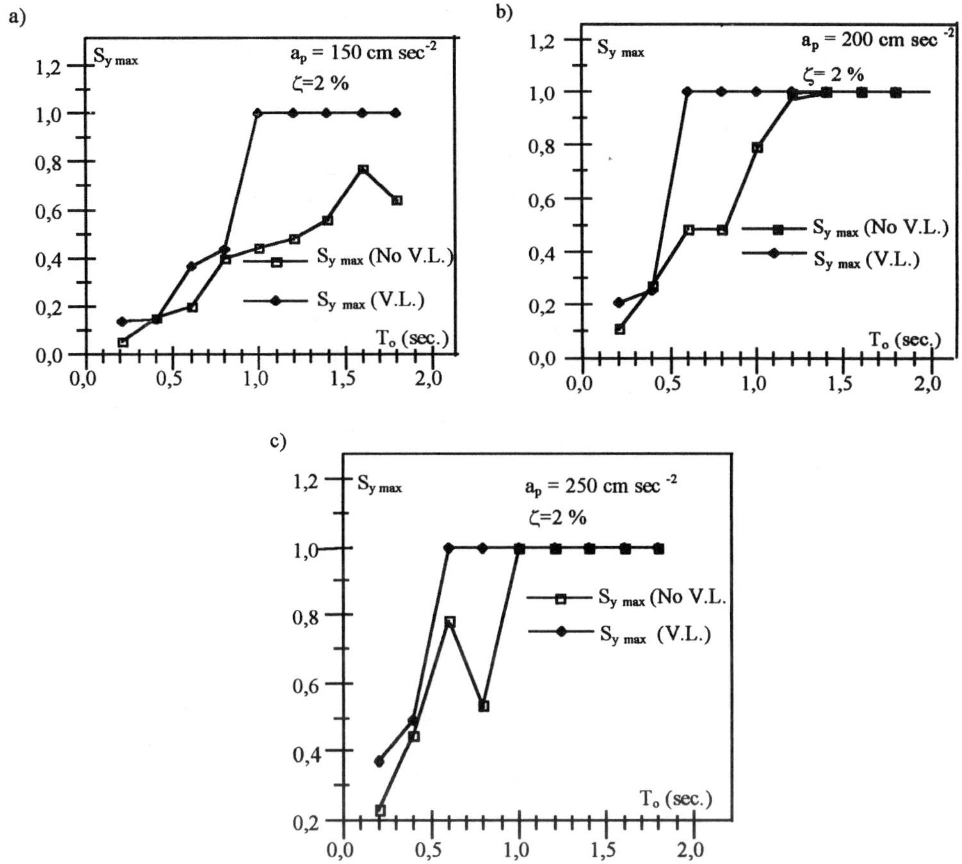

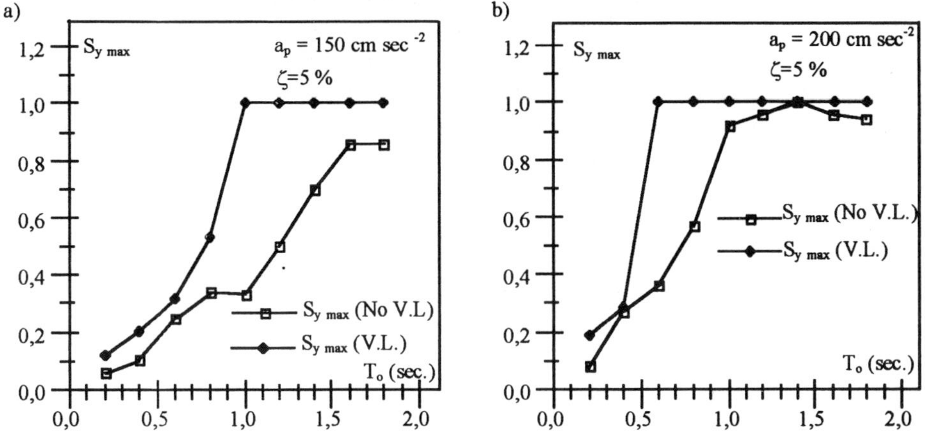

Fig. 4.15:   Comparison between the maximum spectrum of displacement obtained without the Vertical Loads and with Vertical Loads, both in the elastic-plastic phase and for a damping factor $\zeta= 2\%$

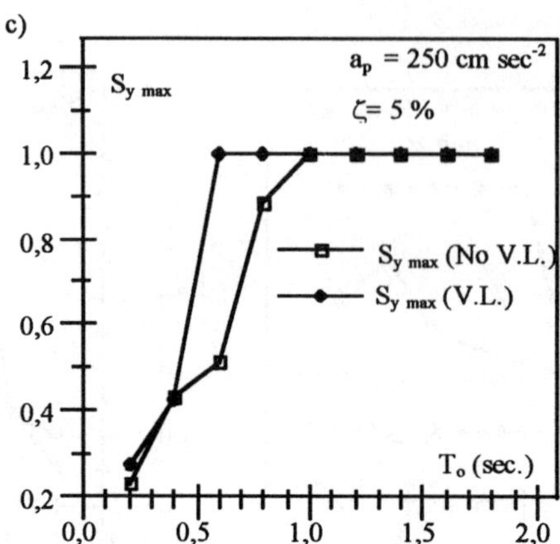

Fig. 4.16: Comparison between the maximum spectra of displacement obtained without the Vertical
Loads and with Vertical Loads, both in the elastic-plastic phase and for a damping factor $\zeta = 5\ \%$.

4.5) DISCUSSION

Considering the results obtained for the elastic behaviour, one can remark that the maximum possible values of the structural response seem to be usable for operative purposes. In fact, if one considers the PGA and the maximum value of the spectral acceleration response obtained by the 1980 earthquake set of data, the amplification factor $A_f = S_a/a_p$ is 2.5 times smaller than the analogous factor obtained by the ratio between the maximum possible values of the acceleration response of the structure and a_p (Fig. 4.9).

This result can be considered acceptable if one takes into account that the response estimated is the maximum possible and the high degree of reliability that can be attained. One observes that beginning from the maximum possible values of the response reported above, any reduction philosophy can be adopted to define the acceleration to assume for High Reliability Aseismic Design (HRAD).

Of course, the choice to introduce any reduction factor depends on the expense percentage the community should be ready to invest to adopt HRAD to structures of particular interest. Analyses on this subject (Zuccaro 1991 [5]) showed that for structures with $T_o > 1.5$ sec (e.g. steel frame) the HRAD is not prohibitive, while for structures with $T_o < 1.5$ sec (e.g. reinforced concrete frame) the adoption of special aseismic structural type is needed.

The introduction of the assumption that the structure, during the excitation, could exceed the elastic limit has allowed to reduce the theoretical threshold of the maximum strength a structure can undergo. Let us consider, to this end, the values of Tab.4.II, which vary between 0.48 and 0.95. Assuming a mean coefficient $a_p/a_o = 0.72$ one gets a design acceleration a_o as 1.4 times the expected peak acceleration a_p at the site.

If one considers $a_p = 0.25$ g (return period 200 roughly years), one obtains $a_o = 0.35$ g; the values of a'_o (the structure strength), obtained performing a limit state analysis according to the code and with a ductility coefficient δ =4 as in Tab. 4.II, is a'_o = 0.1 g for a site of I° seismic category in Italy. Thus, considering a safety coefficient equal to 2, a'_o becomes equal to 0.2 g; in this case the amplification factor A_f = a_o/a'_o = 1.7 results reduced with respect to the elastic case. By analysing the values of Tab. 4.II one can infer that varying the natural vibration period of the structure T_o, the value A_f is always within the range [1,2], demonstrating a considerable reduction of the theoretical bound calculated in the case of elastic response.

From the analysis of the results obtained by investigation of the influence of the stability phenomenon on the maximum possible response of a sdof system one can make following conclusions:
1) the maximum possible response of a SDOF system supporting vertical load seems to be not significantly influenced by the damping factor ζ;
2) the maximum values of the spectra obtained considering the stability phenomenon are greater than those produced without the vertical loads; moreover, the structure reaches the collapse for relatively small values of a_p and any ζ, and for any natural period above 0.8 sec. Hence the structures with longer natural period (e.g. steel structures) show to be more affected by P-Δ problems;
3) since the maximum possible response turns out to be coinsiderably influenced by the maximum expected peak acceleration at the site a_p, it is suggested to pay an attention in the calibration of this parameter through sophisticated hazard modelling before performing high-reliability aseismic tests.

5) CONVEX MODELING FOR TREATMENT OF UNCERTAINTY IN SEISMIC ACTION

In recent years, a proposal has been advanced by Elishakoff & Ben-Haim [21], to treat uncertainty by optimization, that is, practically speaking, to find the worst combination of uncertain parameters, based on the statement that only available, well-established data, should be processed.
A first formal setting of this approach was given by the authors, under the assumption that "uncertainty", i.e. the range of variation of uncertain items, could be included into a convex set, and that the performance of engineering systems could be modeled by a convex functional of the same items. This allows to treat problems in the frame of convex analysis and to get practical results through some well-established properties of convex optimization. Here the treatment is limited to consider the structure response in the linear elastic range.

5.1) THE DRENICK'S APPROACH

The basic idea, that was first set up by Drenick [11], is to find the base accelerogram function yielding the maximum possible value of the response u(t). Following Drenick, assume that a(t) has a finite duration, say T, and a bounded energy $\mathscr{E}_o$

$$\int_0^T a^2(t)\, dt = \|a\|^2 = \mathscr{E}_o^2 \qquad (5.1)$$

Let $I(\mathscr{E}_o)$ be the set collecting all functions defined in $(0,T)$ and possessing the given energy $\mathscr{E}_o^2$. As proved by Drenick

$$U_d = \max_{\substack{t \in (0,T) \\ a(t) \in I(\mathscr{E}_o)}} u(t) = \|u\|_T = \mathscr{E}_o H_T \qquad (5.2)$$

where

$$H_T^2 = \max_{t \in (0,T)} \int_0^T h^2(t-s)\, ds = \int_0^T h^2(t_o - s)\, ds \qquad (5.3)$$

As $T \to \infty$, the base accelerogram $a_c(t)$ yielding the maximum peak response, named by Drenick in a later paper [12] the "critical excitation", tends to assume a shape coincident with the impulse response function reversed with respect to time, and is given by

$$a_c(t) = \pm \frac{\mathscr{E}_o}{H_\infty} h(-t) \qquad (5.4)$$

and the time t_o at which the maximum response occurs is 0. A plot of the critical acceleration for $T \to \infty$ is shown in Fig.5.1

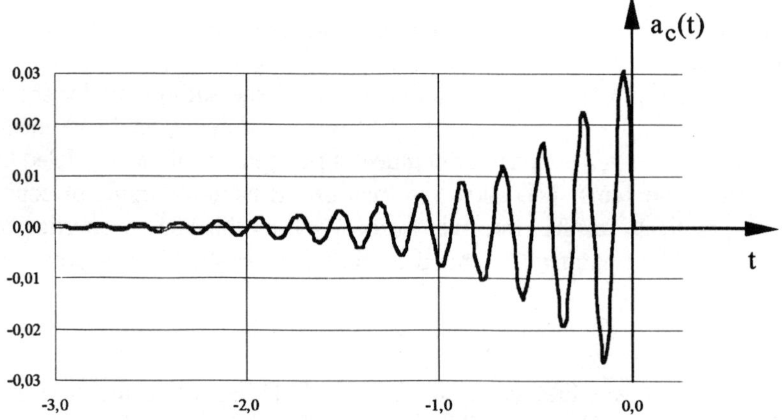

Fig. 5.1: Drenick's critical excitation

The maximum peak response is given by

$$U_{d\infty} = \|u\| = u_c(0) = E_o H_\infty \tag{5.5}$$

where $u_c(t)$ denotes the response to $a_c(t)$. In the same paper, Drenick also proved that the peak response of a structure to the critical excitation as $T \to \infty$ coincides with the standard deviation of its response to a white noise with mean value equal to $\mathscr{E}_o^2$.

Also a suggestion to investigate the critical excitation for spectrum-constrained excitations and a result for band-limited excitation were given by Drenick. Some criticism was argumented in the paper by Drenick himself, mainly about the circumstance that aseismic design based on critical excitation is "far too pessimistic to be practical".

Drenick also investigated the ratio of the maximum response due to critical excitation to the maximum response of actually recorded seismic accelerograms [12]. Response spectra were derived on the basis both of the critical excitation and of a number of accelerograms from a set of recorded earthquakes, and the results were compared. From the comparison of the results, one can conclude that in the range of structures with natural vibration period from 0.5 sec. to 1.2 sec. and damping coefficient in the range 0.02-0.10, the ratio of the critical to experimental response was almost always in the neighborhood of 2.

It should be stressed that by extending the class of possible accelerograms from only the recorded ones to linear combinations of these, the above ratio still decreases in this range to values between 1.3 and 1.6.

5.2) THE SHINOZUKA'S APPROACH

In Shinozuka's approach (1970), one assumes that it is possible to identify a power-spectrum function $\Pi(\omega)$ yielding an upper bound on the ordinates of the power spectrum of any admissible accelerogram. Following Shinozuka's model, one obtains the following bound for the maximum displacement over the whole duration of the excitation

$$U_s = \frac{1}{2\pi} \int\limits_{-\infty}^{+\infty} |H(\omega)| \, \Pi(\omega) \, d\omega \tag{5.6}$$

where

$$H^2(\omega) = \frac{1}{(\omega_0^2 - \omega^2)^2 + 4\zeta^2 \omega_0^2 \omega^2} \tag{5.7}$$

is the frequency response function of the oscillator. It can be easily verified that the bound (5.6) is a generalization of the Drenick's bound (5.2). The difference can be sought in the circumstance that in Shinozuka's approach the upper bound (5.6) can be specialized with respect to some particular shape of the spectrum power of the earthquake that is expected at the site, by suitably modeling the function $\Pi(\omega)$, while in Drenick's approach the distribution of power over the frequency range is not differentiated, and the maximum value of U_s is sought regarding U_s as a functional of $\Pi(\omega)$. It can be proved that the maximum of U_s with varying the function $\Pi(\omega)$ under the condition $\|\Pi(\omega)\| = \text{const.}$, is attained for $\Pi(\omega) \equiv |H(\omega)|$

$$\max_{\Pi(\omega)} \left[\frac{1}{2\pi} \int_{-\infty}^{+\infty} |H(\omega)| \Pi(\omega)\,d\omega \right] = \frac{1}{2\pi} \int_{-\infty}^{+\infty} H^2(\omega)\,d\omega \tag{5.8}$$

Recalling that the impulse response function $h(x)$ and the frequency response $H(\omega)$ function are Fourier transforms of each other

$$\frac{1}{2\pi} \int_{-\infty}^{+\infty} H^2(\omega)\,d\omega = \int_{0}^{+\infty} h^2(x)\,dx \tag{5.9}$$

the coincidence of U_s with U_d when $\Pi(\omega) \equiv |H(\omega)|$ is fully proved. It is of interest to note that $U_s \leq U_d$, and therefore that Shinozuka's approach is sharper than Drenick's one if $\Pi(\omega)$ is properly chosen.

5.3) ELLIPSOIDAL MODELING AND CONVEX OPTIMIZATION

Assume that one is faced with the problem to define the maximum stress which a 2D structure can suffer by earthquake action at a given site, and assume that the structure should not exceed the linear threshold. Finally, assume that the base accelerogram is a member of a linear space $\mathscr{S}$, the set of possible accelerograms, whose basis is yielded by functions (*accelerograms*) a_i ($i=1,...,n,...\infty$), so that any *possible* accelerogram $a(t)$ can be approached by a linear combination of the base functions

$$a(t) = \sum_{i=1}^{n} c_i\, a_i(t) \tag{5.10}$$

where it has been assumed that the average acceleration, i.e. the possible constant term in he expansion, is zero.

One trivial example of such space is the one of generally continuous functions of time, where Fourier expansion yields a denomerable basis, and every accelerogram can find a representation by a linear combination of harmonic functions. In this case, assuming that T is the duration of the oscillation

$$a_i(t) = A_o \sin(\omega_i t) \qquad\qquad i \leq m$$
$$a_i(t) = A_o \cos(\omega_i t) \qquad\qquad m < i \leq n \tag{5.11}$$

with n even, m = n/2, A_o a constant and

$$\omega_i = \frac{2\pi i}{T} \quad (i = 1, \ldots, m) \tag{5.12}$$

the problem reduces to set analysis to give appropriate values to the coefficients c_i.

Note that the response $u_i(t)$ to any of the base functions $a_i(t)$, with initial conditions $u_i(0) = 0; \ \dot{u}_i(0) = 0$ is given by

$$u_i(t) = \frac{e^{-\zeta\omega_o t}\left\{\omega_i\left[\omega_i^2 - \omega_o^2\left(1 - 2\zeta^2\right)\right]\sin\omega_d t + 2\zeta\omega_o\omega_i\cos\omega_d t\right\} + \left(\omega_o^2 - \omega_i^2\right)\sin\omega_i t - 2\zeta\omega_o\omega_i\cos\omega_i t}{\left(\omega_o^2 - \omega_i^2\right)^2 + 4\zeta^2\omega_o^2\omega_i^2}$$

$$u_{i+m}(t) = \frac{-e^{-\zeta\omega_o t}\left[\dfrac{\zeta\omega_o\left(\omega_o^2 + \omega_i^2\right)}{\omega_d}\sin\omega_d t + \left(\omega_o^2 - \omega_i^2\right)\cos\omega_d t\right] + \left(\omega_o^2 - \omega_i^2\right)\cos\omega_i t + 2\zeta\omega_o\omega_i\sin\omega_i t}{\left(\omega_o^2 - \omega_i^2\right)^2 + 4\zeta^2\omega_o^2\omega_i^2} \tag{5.13}$$

$(i = 1, \ldots, m) \ ; \ \omega_d = \omega_o\sqrt{1 - \zeta^2}$

Now, by the linearity of the oscillator, the response to a(t) reads

$$u(t) = \sum_{i=1}^{n} c_i\, u_i(t) \tag{5.14}$$

If nothing else is specified, it is clear that the structure is exposed to undergo uncontrollably high stress.

5.3.1) Basic test accelerograms and central spectrum

The approach that is presented in this section is tested with reference to the five accelerographic records presented in Sec. 3. Assume that one knows that earthquakes, in general, should exhibit a mean power spectrum whose shape is constrained to fit the following expression

$$K(\omega|\Omega_a,\xi_a,\Omega_b,\xi_b) = q\, f\left(\omega|\Omega_a,\xi_a,\Omega_b,\xi_b\right) \tag{5.15}$$

with

$$f\left(\omega|\Omega_a,\xi_a,\Omega_b,\xi_b\right) = \left[G\left(\omega|\Omega_a,\xi_a\right) + G\left(\omega|\Omega_b,\xi_b\right)\right] \tag{5.16}$$

$$G(\omega|\Omega,\xi) = \frac{\Omega^2 + 4\xi^2\omega^2}{\left(\Omega^2 - \omega^2\right)^2 + 4\xi^2\omega^2} \tag{5.17}$$

where Ω and ξ are parameters governing the *shape* of the spectrum, and q is a normalizing factor. These parameters are assumed to have assigned values for any given site under examination, so that eq. (5.15) is intended to have a well defined expression for the problem at hand. Note that eq.(5.15) is nothing else than the sum of two functions of the Kanai-Tajimi type. Moreover, the superposition is introduced in order to have a better approximation for spectra of recorded earthquakes exhibiting more than one dominant frequency.

After this shape is assumed as the *central spectrum*, it is to be expected that accelerograms to be considered in the analysis should not differ from the central spectrum by more than a given amount. Let c_o be the vector of combination coefficients fitting the central spectrum, i.e, with reference to a basis of the type (5.11)

$$c_{oi}^2 = c_{o,i+m}^2 = K\left(\omega_i|\Omega_a,\xi_a,\Omega_b,\xi_b,\right)\Delta\omega \qquad (i \le m) \tag{5.18}$$

Let $\bar{c}_i^{(j)}$ be the i-th coefficient of the Fourier expansion of the j-th accelerogram ($i =1,...,n$; $j=1,...,5$), normalized to unit modulus ($|\bar{c}^{(j)}| = 1$). The central distribution of coefficients, collected in the vector c_0, are obtained for each of the five sites analyzed, through the following steps:

1) For every site find the best values of the parameters $q^{(j)}$, $\Omega_a^{(j)}$, $\xi_a^{(j)}$, $\Omega_b^{(j)}$, $\xi_b^{(j)}$ that give the optimal fit of the ordinates of the Kanai-Tajimi eq. (5.15) with the values of the spectrum from the recorded accelerograms on the corresponding pulsations, with $q^{(j)}$, the normalizing factor in eq. (5.15), such that after calculating

$$s_{oij}^2 = q^{(j)} f\left(\omega_i\middle|\Omega_a^{(j)},\xi_a^{(j)},\Omega_b^{(j)},\xi_b^{(j)}\right) \tag{5.19}$$

it results

$$\sum_{i=1}^{n} s_{oij}^2 = 1 \tag{5.20}$$

2) Put

$$\gamma_{ij} = \frac{\bar{c}_{i+n}^{(j)}}{\bar{c}_i^{(j)}} \tag{5.21}$$

Finally, calculate the ordinates of the nominal target spectra at the considered sites

$$c_{oi}^{(j)} = \frac{S_{oij}}{\sqrt{\left(1+\gamma_{ij}^2\right)}}\, sgn\left(\bar{c}_i^{(j)}\right)$$
$$c_{o,i+n}^{(j)} = \gamma_{ij}\, c_{oi}^{(j)}$$

$$(5.22)$$

3) Calculate for every site the squared scatter of the coefficients obtained by direct
   processing the accelerogram and those calculated by the target spectrum

$$\rho_j^2 = \sum_{i=1}^{n}\left(c_i^{(j)} - c_{oi}^{(j)}\right)^2$$

$$(5.23)$$

In the following table, the final values of ρ_j^2 are quoted, showing that the inequality $\rho^2 \leq 4$
is largely verified.

TABLE 5.I

Site	ρ^2	$\Omega_a\ sec^{-1}$	ξ_a	$\Omega_b\ sec^{-1}$	ξ_b
TORRE DEL GRECO	0.261	9.920	0.109	14.550	0.171
BRIENZA	0.219	32.980	0.478	37.670	0.403
STURNO	0.186	17.940	0.347	3.770	0.399
CALITRI	0.358	0.070	0.059	6.350	0.211
BAGNOLI	0.257	5.150	0.142	28.640	0.762

Assuming that the central spectrum is site-dependent, but that independence holds for the
squared scatter, it is possible to state that the above results represent a definite specific
character of possible ground motion at the site.

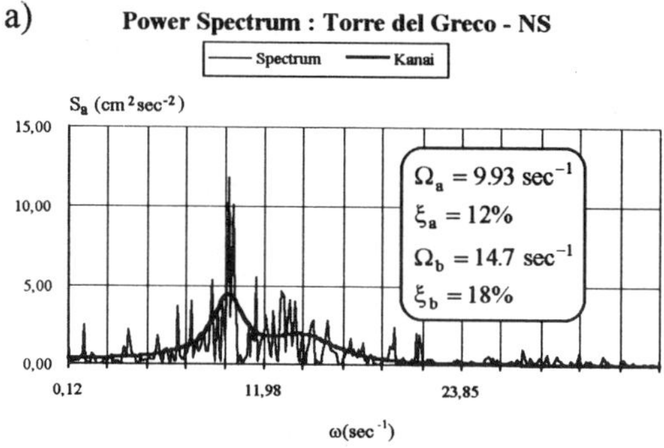

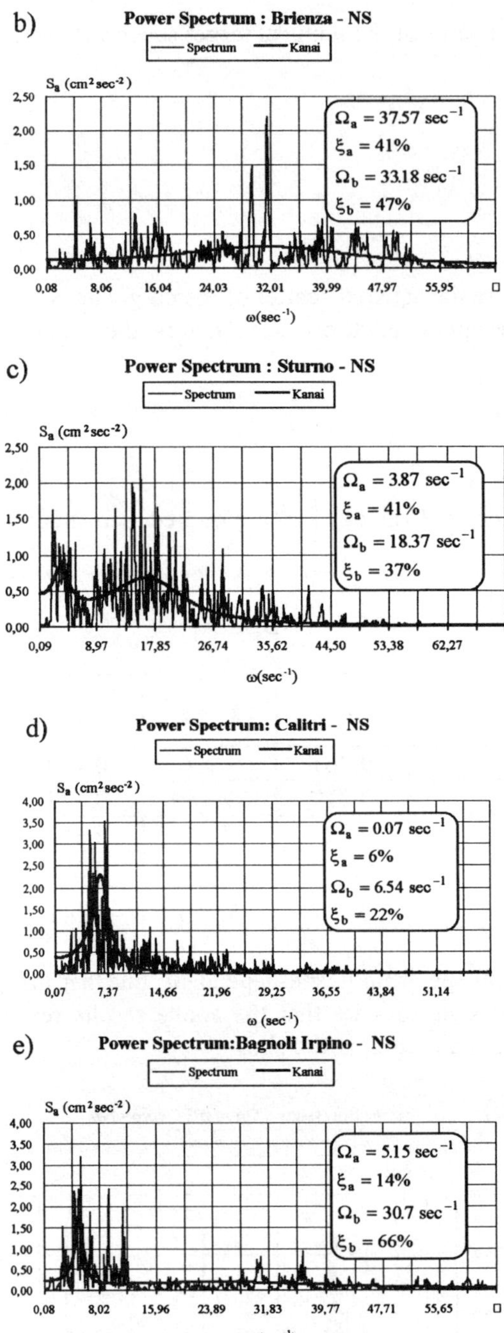

Fig. 5.2: Processed spectra from recorded accelerograms and fitted central spectra.
a)Torre del Greco; b)Brienza; c)Sturno; d) Calitri; e) Bagnoli Irpino

These results are graphically illustrated in Fig. 5.2

5.3.2) Accelerograms as forcing functions with assigned energy
   Let assume that the duration T of the quake is a-priori known, and that it is possible - from geophysics, for instance - to assume a given bound for the energy of the function, like in Drenick's problem

$$\int_o^t a^2(t)\,dt \le \mathscr{E}_o^2 \tag{5.24}$$

If the base functions $a_i(t)$ are orthogonal and all scaled to the same energy, every function of the type (5.10)-(5.11) will be respective of the constraint (5.1), provided that

$$\sum_{i=1}^n c_i^2 \le 1 \quad \text{and} \quad A_o^2 = \frac{2}{T}\mathscr{E}_o^2 \tag{5.25}$$

Any possible function that verifies (5.24) is called an *admissible accelerogram*. Thus, admissible functions can be mapped in the Euclidean space of n-dimensional numerical vectors **c**. Admissible accelerograms are not external to the unit sphere S_n, centered in 0.
The diagram in (0,T) of the maximum and minimum displacement of the structure is required for a(t) varying in S_n. To this end, consider $u_i(t)$, the response function under $a_i(t)$. The response under any a(t) expressed by eq.(5.10), by the system's linearity, is yielded in a straightforward way by eq.(5.14) re-written under vector form denoting vectors and matrices by boldface characters and transposition by superscript "T"

$$u(t)=\sum_{i=1}^n c_i\, u_i(t) = \mathbf{c}^T\mathbf{u}(t) \tag{5.26}$$

where **c** and **u**(t) are the n-dimensional column vectors representing, respectively, all the coefficients c_i and all the displacement values at instant t when the oscillator is acted on by $a_i(t)$.
   Thus, the problem is set as follows for any fixed instant t in (0,T)

$$\begin{aligned}
&\text{Find} \quad \mathbf{c} \in R^n, \text{ for given } t\\
&\text{subject to}\\
&\quad \mathbf{u}^T(t)\,\mathbf{c} = \max \qquad (\text{or min})\\
&\quad \mathbf{c}^T\,\mathbf{c} \le 1
\end{aligned} \tag{5.27}$$

It is well known that in this case - a linear function to be optimized on a convex domain - the solution point is on the boundary of S_n. Thus, problem (5.27) is equivalent to

Find $c \in R^n$, for given t
subject to
$$\mathbf{u}^T(t)\mathbf{c} = \max \qquad (\text{or min})$$
$$\mathbf{c}^T\mathbf{c} = 1$$
(5.28)

The problem is easily solved for any t in (0,T). Indeed, put

$$g(\mathbf{c}) = \mathbf{u}^T\mathbf{c} ; \quad f(\mathbf{c}) = \mathbf{c}^T\mathbf{c} - 1 \tag{5.29}$$

At the solution point

$$\mathbf{grad}\, f = -k\, \mathbf{grad}\, g \qquad \left(\text{for } \mathbf{u}^T\mathbf{c} = \max\right) \tag{5.30}$$

$$\mathbf{grad}\, f = k\, \mathbf{grad}\, g \qquad \left(\text{for } \mathbf{u}^T\mathbf{c} = \min\right) \tag{5.31}$$

whence

$$\mathbf{c} = \pm\gamma(t)\,\mathbf{u}(t) \tag{5.32}$$

with

$$\gamma^2(t) = 1/\left(\mathbf{u}^T\mathbf{u}\right) \tag{5.33}$$

The maximum displacement at any instant t in (0,T) is finally given by

$$u_{\max}(t) = -u_{\min}(t) = \sqrt{\left(\mathbf{u}^T(t)\mathbf{u}(t)\right)} \tag{5.34}$$

A numerical example has been carried assuming that admissible functions are generally continuous, and taking all a_i's to be harmonic functions, as in the Fourier expansion [see eqs. (5.11)÷(5.12)]
In Fig. 5.3, the bound (5.34) is plotted for different values of m = 25÷200, for t ranging in the interval 0÷15 secs. This shows how the refinement of the source functional space makes the bounds to increase progressively, and a good convergence is attained for m = 100.

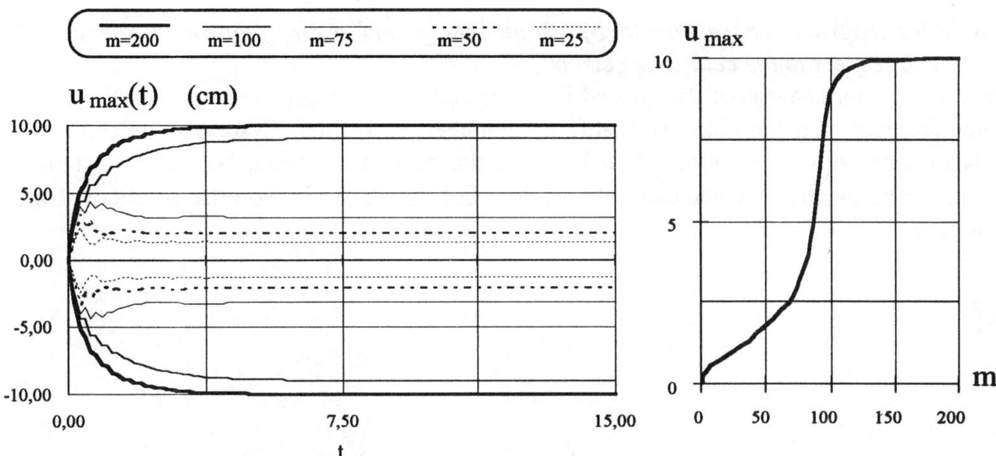

Fig. .5.3: Influence of n, the number of harmonics, on the bound (5.34) (n = 2m)

As it can be seen in Fig. 5.4, where a sample maximizing function for t = 30 secs. and m = 100 is plotted, these bounds are of the type of Drenick's critical excitation, and it is highly questionable that such functions can be accepted as credible earthquakes. It is possible to investigate how this bound is influenced by an improved information.

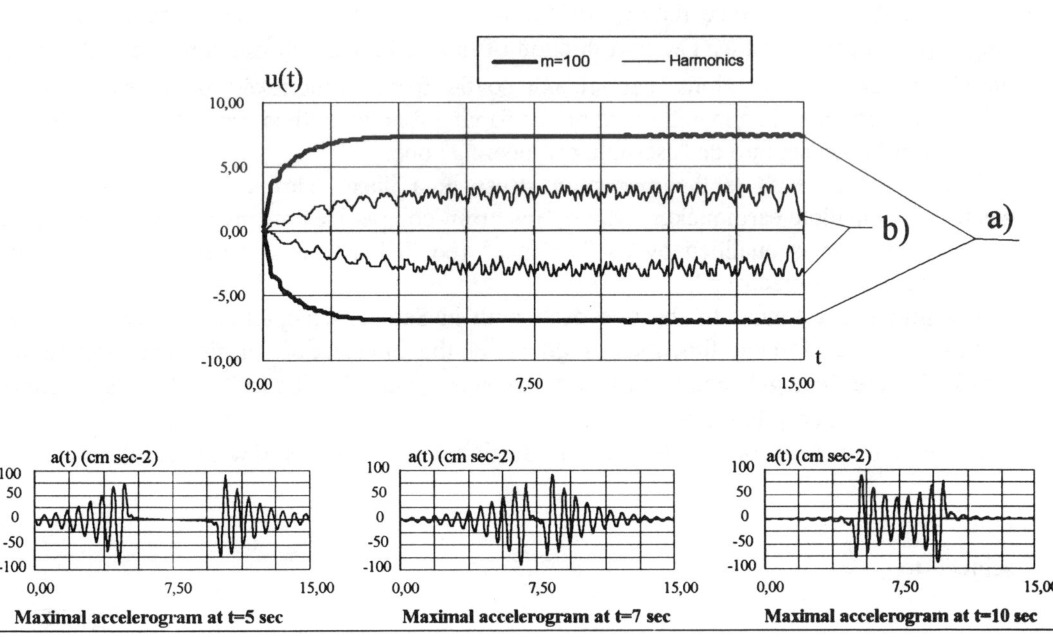

Fig. 5.4: Upper and lower bound and sample maximal accelerogram for t = 30 secs:
a) Bounds (5.34), b) u_{imax} of any harmonic component.

5.3.3) Accelerograms characterized by given energy and a "neighborhood constraint"
with respect to the central spectrum

Assume that after analysis of the ground firmness and/or of previous recorded earthquakes, or anything else, it is possible to qualify earthquakes, and their desultory variability, as accelerograms with given norm (eq.5.1), as in the previous section, but with "distance" from c_o not larger than a given amount, say "ρ". So, the constraints on the coefficients are of the type

$$\sum_{i=1}^{n} c_i^2 = 1 \tag{5.35}$$

$$\sum_{i=1}^{n} \left(c_i - c_{oi}\right)^2 \le \rho^2 \tag{5.36}$$

which is couple with the condition

$$\sum_{i=1}^{n} c_{oi}^2 = 1 \tag{5.37}$$

expressing the circumstance that c_o yields an accelerogram of given norm, realized by assigning $\mathscr{E}_o$ a proper value. The introduction of this additional information means that one intends to deal with functions that are not so far from earthquake-type accelerograms, rather than with generic generally continuous forcing functions, thus translating the original "mechanical" problem into an "aseismic engineering" one.

The expected result is the maximum stress in a linear sdof structure acted-on by spectrum-compatible earthquakes, where "spectrum-compatible" means that at a given site seismic excitation is qualified by its "distance" (eq. 5.36) from a given origin function named the "central spectrum".

The problem is similar to the one dealt with in Sec. 3, except the fact that now the admissible set of forcing functions is given by the intersection of the unit sphere S_n [eq.(5.35), the "energy" constraint] and the side-sphere S_ρ [eq.(5.36), the "spectrum-compatibility" sphere], Fig. 5.5.

Following the lines set forth in the previous section, the problem is now set as follows:

Find $c \in R^n$

subject to

$$c^T c = 1 \tag{5.38}$$

$$\left(c - c_o\right)^T \left(c - c_o\right) \le \rho^2$$

$$u^T(t) c = \max \left(\text{resp. min}\right)$$

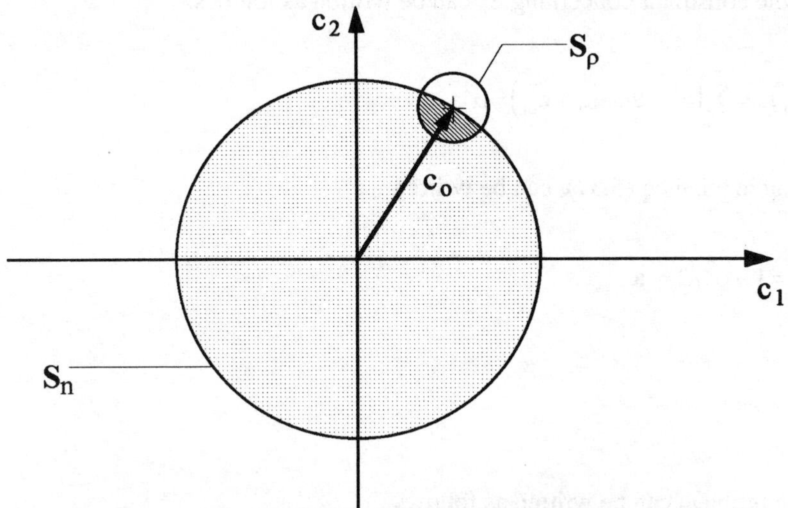

Fig. 5.5: The set of admissible coefficients

where $\rho^2 \leq 4$, otherwise S_ρ includes S_n, and the problem shifts to the formulation (5.28).

Problem (5.38) can be solved in two steps.

Step #1:   Solve the simpler problem (5.28), and find c_1. If c_1 is included in the set S_ρ, then it represents also the solution of the problem (5.38). If c_1 is on S_n but it is outside S_ρ then no internal point to S_ρ exists where the optimal condition (5.30) [respectively (5.31)] holds, and the solution must be searched on the boundary of the intersection of S_n and S_ρ, as described in step 2.

Step #2:   Now the problem transforms into a modified one

Find $c \in R^n$:

subject to:

$$c^T c = 1 \tag{5.39}$$

$$(c - c_o)^T (c - c_o) = \rho^2$$

$$u^T(t) c = \max \quad (\text{resp. min})$$

Note that the constraint concerning S_ρ can be written as follows:

$$\sum_{i=1}^{n}(c_i - c_{oi})^2 = \sum_{i=1}^{n}(c_i^2 - 2c_i\,c_{oi} + c_{oi}^2) = \rho^2 \qquad (5.40)$$

that, bearing in mind eq.(5.37), can be written

$$\sum_{i=1}^{n}(c_i\,c_{oi}) = 1 - \rho^2/2 = a \qquad (5.41)$$

with

$$|a| \leq 1 \qquad (5.42)$$

That is, the problem can be written as follows:

Find $\quad c \in R^n$:
subject to

$$c^T c = 1 \qquad (5.43)$$
$$c_o^T c = a$$
$$u^T(t)\,c = \max\,(\text{resp. min})$$

where the constraints are all under equality form. In order to solve the latter problem, consider the Lagrangian function

$$\mathscr{L}(c, r_1, r_2 | t) = u^T(t)\,c + r_1(c^T c - 1) + r_2(c_o^T c - a) \qquad (5.44)$$

The gradient of the Lagrangian with respect to c must vanish in solution

$$\mathbf{grad}_c\,\mathscr{L} = u(t) + 2\,r_1\,c + r_2\,c_o = 0 \qquad (5.45)$$

whence

$$c = c(t) = -\frac{1}{2\,r_1}\left[r_2\,c_o + u(t)\right] \qquad (5.46)$$

Introducing the expression (5.46) into the equation of S_n one gets

$$\mathbf{c}^T\mathbf{c} = \frac{1}{4\,r_1^2}\left(\mathbf{u}^T\mathbf{u} + r_2^2\,\mathbf{c}_o^T\mathbf{c}_o + 2r_2\mathbf{u}^T\mathbf{c}_o\right) = 1 \tag{5.47}$$

with the positions

$$\mathbf{u}^T(t)\,\mathbf{c}_o = b \quad ; \quad \mathbf{u}^T(t)\,\mathbf{u}(t) = 4U^2(t) \tag{5.48}$$

Recalling that in solution $\mathbf{c}^T\mathbf{c} = \mathbf{c}_o^T\mathbf{c}_o = 1$, one gets

$$4r_1^2 - r_2^2 - 2r_2 b - 4U^2(t) = 0 \tag{5.49}$$

Introducing now eq. (5.46) into the equation of S_n one gets:

$$\mathbf{c}_o^T\mathbf{c} = -\frac{1}{2r_1}\left(r_2 + b\right) = a \tag{5.50}$$

whence

$$r_2 = -\left(2\,r_1\,a + b\right) \tag{5.51}$$

Introducing the latter expression into eq. (5.49) one arrives at the equation for r_1

$$4\left(1 - a^2\right)r_1^2 + b^2 - 4U^2 = 0 \tag{5.52}$$

whence

$$r_1 = \pm\frac{1}{2}\sqrt{\frac{4U^2 - b^2}{1 - a^2}} \tag{5.53}$$

Note that the expression under the root operator in the above equation is essentially positive. Indeed, by elementary use of Cauchy-Schwartz's inequality

$$b^2 = \left[\mathbf{c}_o^T\mathbf{u}(t)\right]^2 \leq \left[\mathbf{c}_o^T\mathbf{c}_o\right]\left[\mathbf{u}^T(t)\mathbf{u}(t)\right] = \mathbf{u}^T(t)\mathbf{u}(t) = 4U^2 \tag{5.54}$$

After r_1 has been calculated, r_2 is given by eq. (5.51). Let calculate the second derivative

$$\frac{\partial^2 \mathscr{L}\left(\mathbf{c}, r_1, r_2, \middle| t\right)}{\partial c_i\, \partial c_j} = \begin{cases} 2\,r_1 & \text{if } i = j \\ 0 & \text{otherwise} \end{cases} \tag{5.55}$$

One verifies that the matrix of second derivatives of the Lagrangian with respect to the basic variables c_i is diagonal, and positive (respectively, negative) definite if r_1 is assumed from eq. (5.53) with the positive (respectively, negative) sign, and the solution corresponds to the conditioned minimum (respectively, maximum) of the objective function $\mathbf{u}^T \mathbf{c}$.

In conclusion, the solution for the problem of the *maximum* response is yielded by:

$$r_1 = -\frac{1}{2}\sqrt{\frac{4\,U^2 - b^2}{1 - a^2}}$$
$$r_2 = -\left(2 r_1\, a + b\right) \tag{5.56}$$
$$\mathbf{c} = \mathbf{c}(t) = -\frac{1}{2\,r_1}\left[r_2\, \mathbf{c}_o + \mathbf{u}(t)\right]$$

The solution for the *minimum* response is given by changing the sign of r_1

$$r_1 = +\frac{1}{2}\sqrt{\frac{4\,U^2 - b^2}{1 - a^2}}$$
$$r_2 = -\left(2 r_1\, a + b\right) \tag{5.57}$$
$$\mathbf{c} = \mathbf{c}(t) = -\frac{1}{2\,r_1}\left[r_2\, \mathbf{c}_o + \mathbf{u}(t)\right]$$

After introducing this value in the results discussed in above, one gets the bounds plotted in Fig. 5.6, for $\omega_0 = 30$ sec^{-1}, where it is also possible to get a comparison with the previous, spectrum-free bound, showing that the new bound is approximately a half of the one constrained only by the intensity upper bound.

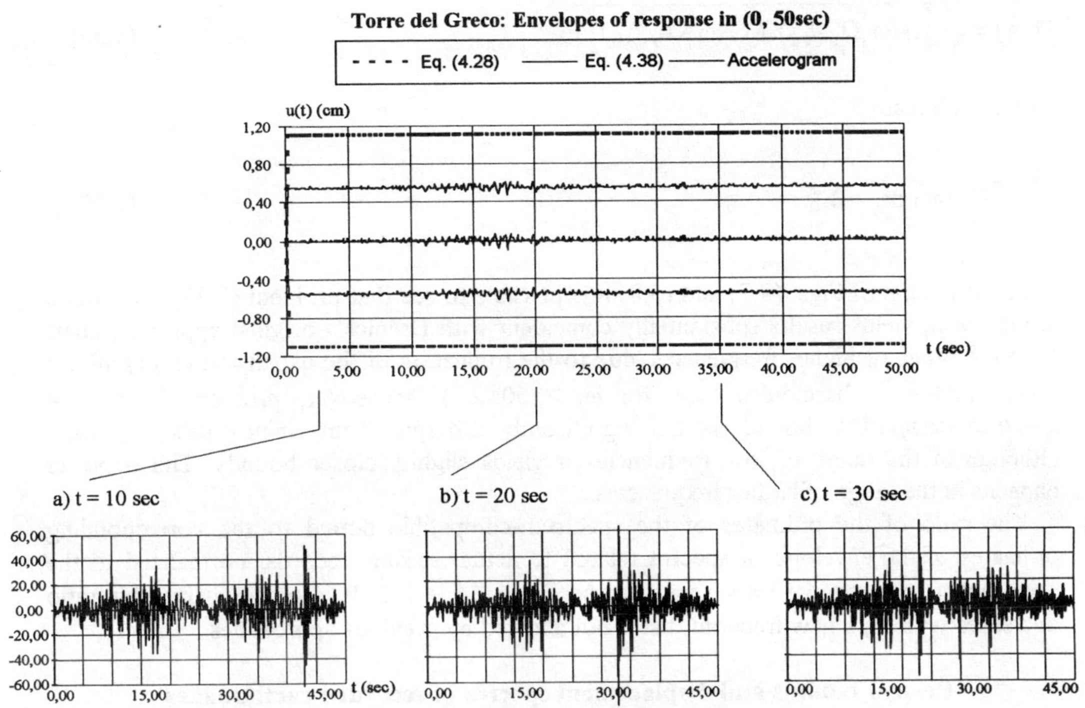

Fig. 5.6:   Upper and lower bounds by problems (5.28) and (5.38)
a) Sample maximal accelerogram at t=10 secs
b) Sample maximal accelerogram at t=20 secs
c) Sample maximal accelerogram at t=30 secs

In Fig. 5.7, the response displacement spectrum calculated by the present procedure is finally plotted, for a value of the damping coefficient $\zeta = 5\%$, and the comparison with the envelope of the same spectra calculated with reference to the accelerograms processed, shows that the present results, although better founded and less sensitive to uncertain parameters than recorded accelerograms, are not too conservative, but yield a percentage increase in the design forces.

In Fig. 5.8 the same spectra obtained via the procedures developped in secs. 3 and 4, are compared to the response spectrum yielded by the Drenick's approach and Shinozuka's approach. As expected, Drenick's bound is practically coincident with the results yielded by problem (5.28). In Shinozuka's approach, one assumes that eq. (5.15), with the parameters presented in Table 5.I yields an upper bound on the ordinates of the power spectrum of any admissible accelerogram. Following Shinozuka's model, one can apply the bound (5.6) for the maximum displacement over the entire duration of the excitation giving $\Pi(\omega)$ the functional form introduced by eq. (5.15)

$$\Pi(\omega) = \sqrt{q\left[G(\omega\,|\,\Omega_a,\xi_a) + G(\omega\,|\,\Omega_b,\xi_b)\right]} \qquad\qquad (5.58)$$

with q such that

$$\frac{1}{2\pi}\int_{-\infty}^{+\infty}\Pi^2(\omega)\,d\omega = E_0^2 \qquad\qquad (5.59)$$

From inspection of Figs. (5.7) and (5.8) it is possible to see that problem (5.28), the bound in the *norm*, yields results substantially coincident with Drenick's original approach, apart from the range of higher frequencies, due to the truncation of the expansion (5.11) after a finite number of harmonics (say, for $\omega_i > 50\text{sec}^{-1}$). Moreover, problem (5.38), the *spectrum-compatible* bound, is not significantly different from Shinozuka's approach, although in the range of low frequencies it yields slightly closer bounds. The opposite happens in the range of higher frequencies.

The ratio of the ordinates of the spectrum-compatible bound to the corresponding ordinates of the envelope of spectra related to actual seismic records, normalized to the same norm as Torre del Greco vary in the range 2.5 ÷ 2.0. It is obvious that this ratio decreases with ρ^2, a new freedom that is not allowed by previous approaches.

Present bounds and displacement spectra of recorded earthquakes

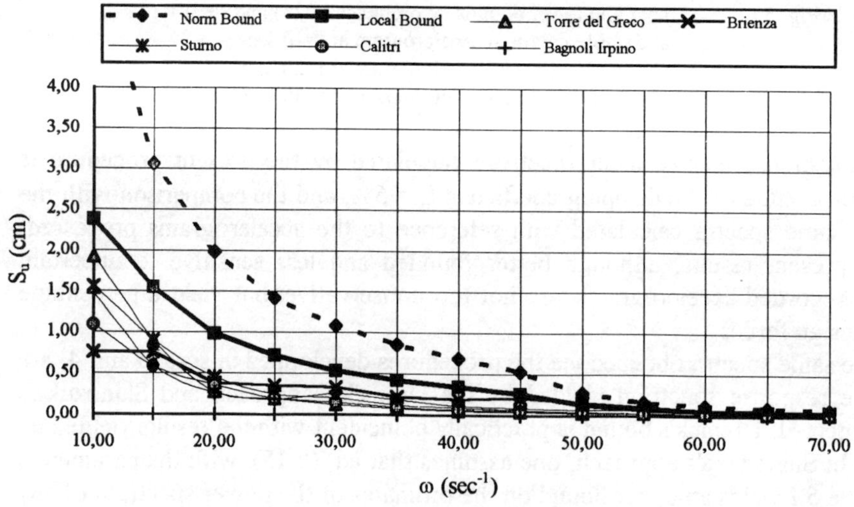

Fig. 5.7:   Response spectra of displacement:
Comparison of bounds by problems (5.28) and (5.38) with the envelope of spectra from recorded accelerograms normalized to the same norm as Torre del Greco

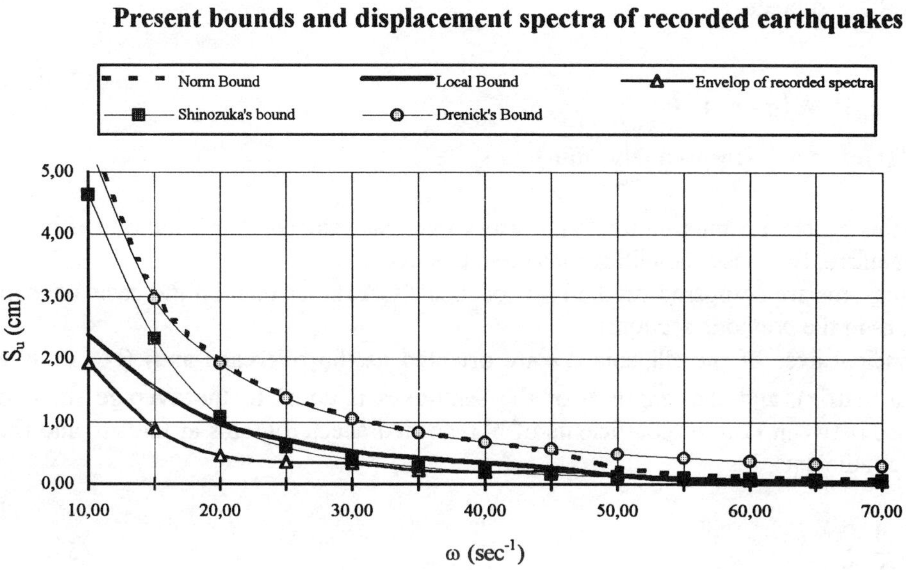

Present bounds and displacement spectra of recorded earthquakes

Fig.5.8:   Response spectra of displacement:
          Comparison of bounds by problems (5.28) and (5.38) with Drenick's and Shinozuka's spectra.

5.3.4) Accelerograms characterized by a "similarity constraint" of the power spectrum to the central spectrum

Assume now that the central spectrum at a given site has been identified [for instance in the form of eq. (5.15)÷(5.16)] and that the distribution of the accelerogram power over the frequency range is given by the coefficients c_{oi} (i = 1,...,n). Assume that the distance between any possible accelerogram at the site is differentiated as it is in eq. (5.36) but that it may be further specialized with respect to any value of frequency; in other words, the constraint on the accelerogram spectrum is specified in the form

$$\left(\mathbf{c} - \mathbf{c_o}\right)^T \mathbf{W}\left(\mathbf{c} - \mathbf{c_o}\right) \le \theta^2 \tag{5.60}$$

where $\mathbf{W}$ is a suitable matrix of coefficients and θ^2 a suitable size factor. If $\mathbf{W}$ is symmetric and positive definite eq. (5.60) identifies an ellipsoid $\mathscr{E}_\theta$ in the n-dimensional space of Fourier coefficients. The problem is set in form analogous to eq. (5.39)

Find $\quad \mathbf{c} \in \mathbf{R}^n$, subject to

$$\mathbf{c}^T \mathbf{c} = 1$$
$$\left(\mathbf{c} - \mathbf{c}_o\right)^T \mathbf{W}\left(\mathbf{c} - \mathbf{c}_o\right) = \theta^2 \qquad (5.61)$$
$$\mathbf{u}^T(t)\mathbf{c} = \max \quad (\text{respectively, min})$$

However, this problem cannot be treated in way to get easy closed-form solutions as in eq. (5.57). Therefore, two basic simplifications are introduced:
1) Accelerograms are compared on the basis of their PGA rather than on the basis of their energy, as in the previous sections;
2) The principal axes of the ellipsoid $\mathscr{E}_\theta$ are oriented as the reference axes (i.e. $\mathbf{W}$ is a diagonal matrix), and the length ℓ_i of the semi-axes is equal to the average squared difference between Fourier coefficients of N recorded accelerograms at the site and the central coefficients, i.e.

$$\ell_i^2 = \frac{1}{N}\sum_{j=1}^{N}\left(c_i^{(j)} - c_{oi}\right)^2 \qquad (5.62)$$

Considering that at any site one has only *one* recorded accelerograms, whose Fourier coefficients are $\overline{c}_i$, the matrix $\mathbf{W}$ can be set as follows

$$W_{ij} = \frac{1}{\left(\overline{c}_i - c_{oi}\right)^2}\delta_{ij} \qquad (5.63)$$

where δ_{ij} denotes the Kronecker's delta. In Fig. (5.9) the location of the admissible domain is illustrated.
According to point 1), the size of the ellipsoid is related to the PGA of the accelerogram. Since the accelerogram is expressed in the form (5.10), one can predict the maximum base acceleration at time t by solving the following problem

Find $\quad \mathbf{c} \in \mathbf{R}^n$, subject to:

$$\left(\mathbf{c} - \mathbf{c}_o\right)^T \mathbf{W}\left(\mathbf{c} - \mathbf{c}_o\right) = \theta^2 \qquad (5.64)$$
$$a_{\max}(t) = \mathbf{a}^T(t)\mathbf{c} = \max \quad \left(\text{respectively, } a_{\min}(t) = \mathbf{a}^T(t)\mathbf{c} = \min\right)$$

where $\mathbf{a}(t)$ denotes the n-dimensional vector whose components are $a_i(t)$. The relevant Lagrangian function is

$$\mathcal{L}(\mathbf{c},r) = \mathbf{a}^T(t)\,\mathbf{c} + r\left[(\mathbf{c}-\mathbf{c}_o)^T\,\mathbf{W}\,(\mathbf{c}-\mathbf{c}_o) - \theta^2\right] \qquad (5.65)$$

Keeping into account eq. (5.63) can be explicitly re-written in the form

$$\mathcal{L}(c_1,\ldots,c_n,r) = \sum_{i=1}^{n}\left\{c_i a_i(t) + r\left[W_{ii}(c_i - c_{oi})^2 - \theta^2\right]\right\} \qquad (5.66)$$

The stationarity conditions

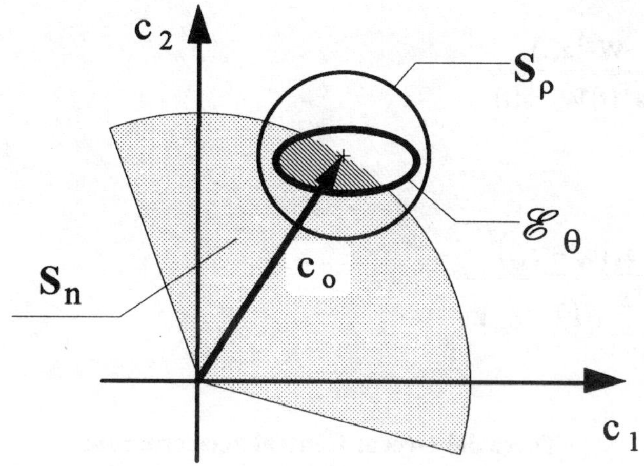

Fig.5.9: Ellipsoidal model for occurrence of earthquakes at the site

$$\frac{\partial \mathcal{L}}{\partial c_k} = a_k(t) + 2rW_{kk}(c_k - c_{ok}) = 0 \qquad (5.67)$$

yield

$$c_k - c_{ok} = -\frac{1}{2rW_{kk}}\,a_k(t) \qquad (5.68)$$

or in vector form

$$\mathbf{c} - \mathbf{c}_o = -\frac{1}{2r}\,\mathbf{W}^{-1}\mathbf{a}(t) \qquad (5.69)$$

The proper value for "r" is obtained after substitution in the ellipsoid's equation

$$\left(\mathbf{c}-\mathbf{c}_o\right)^T\mathbf{W}\left(\mathbf{c}-\mathbf{c}_o\right) = \frac{1}{4r^2}\,\mathbf{a}^T(t)\left[\mathbf{W}^{-1}\right]^T\mathbf{W}\mathbf{W}^{-1}\mathbf{a}(t) = \frac{1}{4r^2}\,\mathbf{a}^T(t)\left[\mathbf{W}^{-1}\right]^T\mathbf{a}(t) = \theta^2 \quad (5.70)$$

whence

$$\frac{1}{2r} = \pm\frac{\theta}{\sqrt{\mathbf{a}^T(t)\left[\mathbf{W}^{-1}\right]^T\mathbf{a}(t)}} \quad ; \quad r = \pm\frac{1}{2\theta}\sqrt{\mathbf{a}^T(t)\left[\mathbf{W}^{-1}\right]^T\mathbf{a}(t)} \quad (5.71)$$

After substitution in (5.69) one gets

$$\left.\begin{matrix}\mathbf{c'}\\[4pt]\mathbf{c''}\end{matrix}\right\} = \mathbf{c}_o \mp \theta\,\frac{\mathbf{W}^{-1}\mathbf{a}(t)}{\sqrt{\mathbf{a}^T(t)\mathbf{W}^{-1}\mathbf{a}(t)}} \quad (5.72)$$

or, explicitly,

$$\left.\begin{matrix}c'_k\\[4pt]c''_k\end{matrix}\right\} = c_{ok} \mp \theta\,\frac{a_k\left(\bar{c}_k - c_{ok}\right)^2}{\sqrt{\displaystyle\sum_{i=1}^{n}a_i^2\left(\bar{c}_i - c_{oi}\right)^2}} \quad (5.73)$$

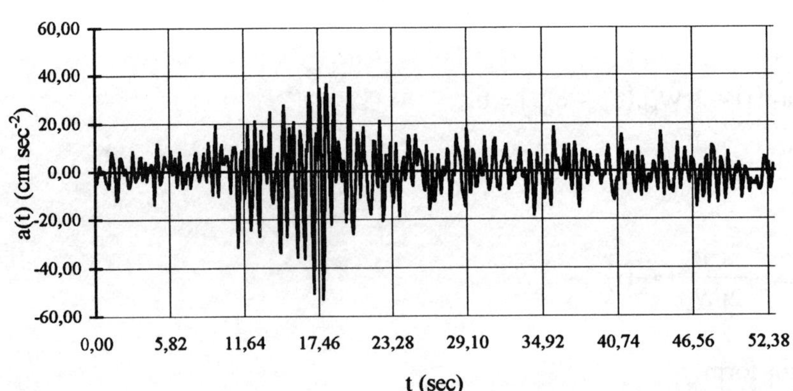

Fig. 5.10: Central accelerogram produced by average Kanai-type spectrum in Torre del Greco

Finally,

$$a_{max}(t) = \sum_{k=1}^{n} c'_k a_k(t) = \sum_{k=1}^{n} c_{ok} a_k(t) - \frac{\sum_{k=1}^{n} a_k^2 \left(\bar{c}_k - c_{ok}\right)^2}{\sqrt{\sum_{i=1}^{n} a_i^2 \left(\bar{c}_i - c_{oi}\right)^2}} = a_o(t) - \sqrt{\sum_{k=1}^{n} a_k^2 \left(\bar{c}_k - c_{ok}\right)^2}$$

$$a_{min}(t) = \sum_{k=1}^{n} c''_k a_k(t) = \sum_{k=1}^{n} c_{ok} a_k(t) + \frac{\sum_{k=1}^{n} a_k^2 \left(\bar{c}_k - c_{ok}\right)^2}{\sqrt{\sum_{i=1}^{n} a_i^2 \left(\bar{c}_i - c_{oi}\right)^2}} = a_o(t) + \sqrt{\sum_{k=1}^{n} a_k^2 \left(\bar{c}_k - c_{ok}\right)^2}$$

$$(5.74)$$

In eq. (5.74) $a_o(t)$ is the central accelerogram at he site of Torre del Greco, i.e. the accelerogram that is generated by the central spectrum (Fig. 5.10).
The functions $a_{max}(t)$ and $a_{min}(t)$ are plotted in Fig. (5.11)

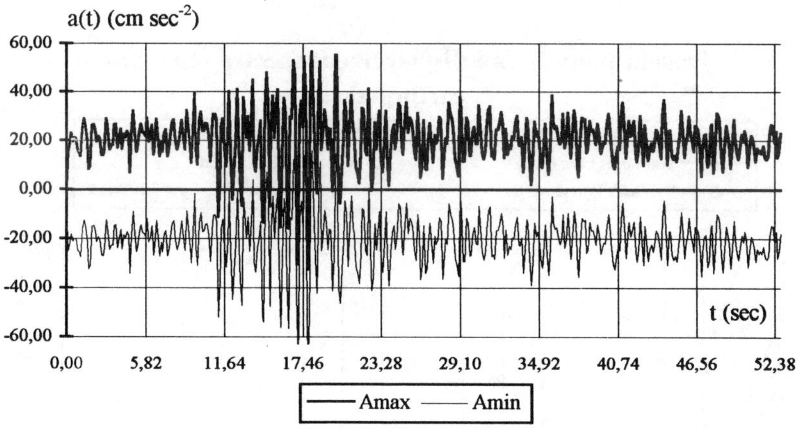

Fig. 5.11: Maximum and minimum possible accelerations in Torre del Greco; $\theta^2 = 1$

The maximum possible peak acceleration can be found as follows

$$a_p(\theta) = \max_{t \in (0,T)} \left\{ \max\left[a_{max}(t), -a_{min}(t)\right] \right\}$$

$$(5.75)$$

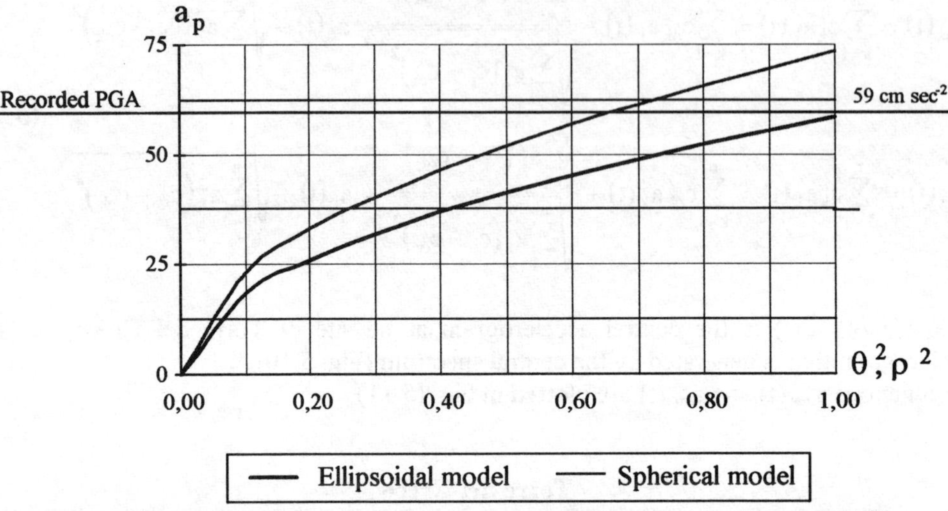

Fig. 5.12: Peak acceleration as a function of the size of the sphere S_ρ and of ellipsoid $\mathscr{E}_\theta$

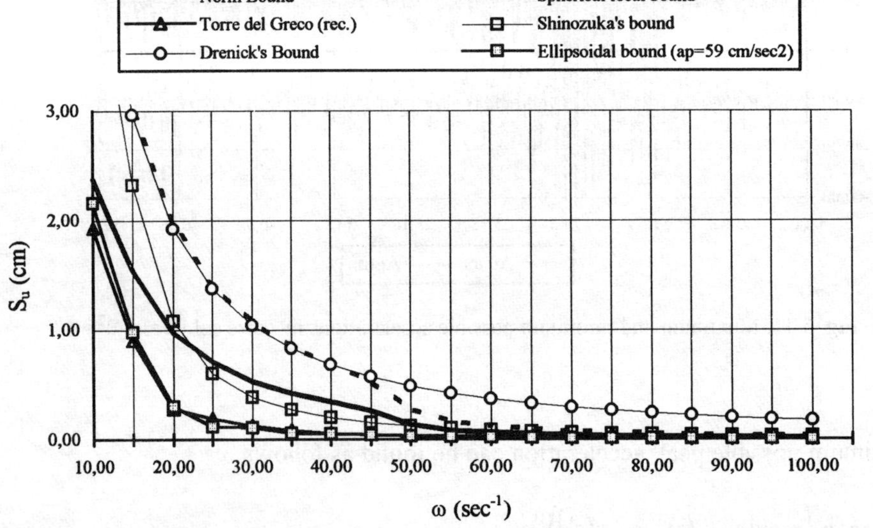

Fig. 5.13: Recorded response spectrum in Torre del Greco and investigated upper bounds.
All bounds are normalized to the same peak acceleration (59cm·sec^{-2})

In Fig. 5.12 the proceeding of a_p with θ^2, as from eq. (5.75), is plotted, whence one can verify that the same PGA as the recorded accelerogram can be realized by the ellipsoidal domain for $\theta^2 = 1$, while the same is obtained through the spherical domain for $\rho^2 = 0.65$. Finally, in Fig. 5.13 the displacement spectrum for the recorded quake in Torre del Greco is plotted in comparison of all possible bounds investigated in this section.

5.4) DISCUSSION

In this section, an approach to analyze the response of a linear structure acted on by seismic shaking, and to calculate upper bounds not dependent on the details of the excitation, has been proposed, based on the solution of an optimal convex problem. Starting from a linear function space, with a given basis, the optimal combination can be found via simple analytical tools. The required data are: i) the assumed energy of the quake (sec. 5.3.1); ii) an estimate of the shape of the central power spectrum at the site (sec. 5.3.2); iii) an upper bound on the maximum "distance" of the power spectrum of realizable quakes at the site from the central one (sec. 5.3.3); iv) a set of upper bounds on the possible difference in the power displayed by the forcing function at every frequency (sec.5.3.4). In all the investigated procedures the optimization problem has been set up in way to allow simple closed-form solutions.

The results are not substantially different from most procedures available in the literature so far, unless the available information is so detailed as at level iv). In particular, a direct comparison has been attempted of the *spherical* model with Shinozuka's approach, yielding very close results. It should be noted that in Shinozuka's approach, however, the admissible spectra are bounded from above by the "central" spectrum, while the present approach includes earthquakes whose spectra are allowed to cross and to overcome the same central spectrum; note, moreover, that in Shinozuka's approach the bound involves only the *stationary* part of the response, while in the present approach, like in Drenick's one, the bound involves also the transient response starting with homogeneous initial conditions.
A decisive improvement is attained by the *ellipsoidal* model (sec. 5.3.4), where it can be seen that the bound results quite sharp, despite the freedom that has been left to the power spectrum shape.

6) CONCLUSIONS

In this chapter, the idea that the most uncertainty involved by earthquake action can be contrasted by a worst-scenario philosophy and evaluated by suitably arranged optimization algorithms, has been pursued and illustrated. The basic idea was first set up and consistently developed by Drenick and Shinozuka, who suggested different techniques to bound the seismic response of a structure, irrespective of the shape of the overcoming accelerogram, by working in the domain of time and of frequency respectively (see Sec.s 5.2.1 and 5.2.2).

The critical point of this approach is that the set of feasible accelerograms should not contain spurious time-histories, which are not compatible with the earthquake features at a given site.

Later on, one of the present writers tried to develop a procedure to obtain the least favourable excitation searching in the set that could be produced by a generator of artificial earthquake-like accelerograms. The model chosen for the generator followed the procedure set up by Ruiz and Penzien in 1969, and the whole class of admissible seismograms so defined was constrained to a given value of the maximum peak ground acceleration (PGA). The generator's parameters were calibrated on the basis of 4 accelerograms recorded at Tolmezzo (Udine, Italy) on the seismic event of May, 1976 (quakes of 6th, 7th, 8th, 9th May, 1979), so that the class of admissible accelerograms was compatible with these samples. The comparison of the least favourable responses of the *undamped* structure to the envelope of responses due to the 4 recorded accelerograms, also yielded values from about 2 in the range of own structure's period around 0.5sec to 7 for more deformable structures (Fig. 5.14).

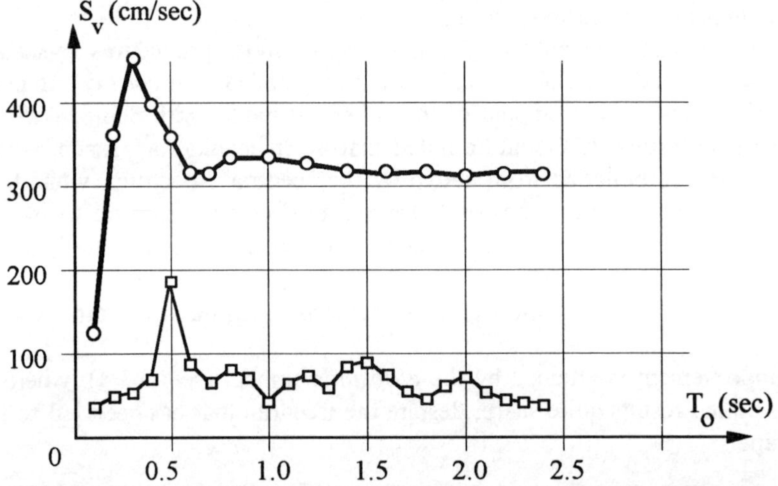

Fig. 5.14:   Comparison of envelope of response spectra from recorded accelerograms at Tolmezzo (Udine, Italy, May 1976) with the spectrum of maximum possible velocities (Baratta, 1980)

The result seems to confirm that at least the Drenick's critical response, if not the critical excitation, should be in the neighborhood of some realizable function during an earthquake, at least if structures resist in the elastic range. Analogous results were found by Baratta and Zuccaro by working on the same seismic model and with the same procedure. In this case, carried on for 5% damping, the range around $T_o = 0.5$, where the ratio is around 2-2.5 is considerably enlarged, probably due to the influence of damping (see Sec 4.4.1).

Again, Drenick's results may be well confirmed: *Earthquake-type functions at a given site may give responses that are comparable to those yielded by critical excitation*. It is also considered that, by introducing the least favourable earthquake in a full probabilistic analysis of seismic hazard, the severity of this approach is rather mitigated by the combination with uncertainties deriving from regional seismicity (see Sec.3 and Sec.4.3). Consideration of anelastic excursions of the structure furtherly reduces the gap between worst excitation and recorded ones (see Sec. 4.4.2 and Sec. 4.4.3).

From the above considerations, one can conclude that the original idea of the critical excitation is worth to be pursued, and deserves some more attention and developments.

A new technique to deal with uncertainty by setting up models suitable for convex (namely, ellipsoidal) optimization was proposed in the early 90's by Elishakoff and Ben-Haim.

This approach lends itself very well to deal with the problem of high-reliability aseismic design, especially when structures are treated as linearly elastic systems. The linearity of the equations allows, in fact, to arrive at closed-form, simple solutions for optimal problems, and, after a seismicity model has been properly formulated, the procedure becomes very straightforward and the bounds rather sharp. Moreover, different types of macro-seismic information can be adopted; the energy of the accelerogram as well as its PGA can be adopted to predict the intensity of the earthquake (see Secs. 5.3.2 and 5.3.4), while its power spectrum can be expected to be in a spherical neighborhood of some *central site-dependent spectrum* or more specifically in a suitably oriented and sized ellipsoidic neighborhood of it (see Secs. 5.3.3 and 5.3.4). In all cases, the more specialized is the information, the sharper is the bound with respect to concretely experimented earthquakes.

ACKNOWLEDGEMENT
Research support by grants of the Italian Ministry of the University and the Scientific and Technological Research (MURST). This Research has been partly carried during a stay of the Authors at the Florida Atlantic University in Sep.-Oct. 1992, supported by MURST.

7) REFERENCES

1. Baratta A., Margiotta P. (1982): Some Remarks on Structural response under earthquakes, *Proc. VII European Conference on Earthquake Engineering*, Athens, Greece, pp. 134-142

2. Baratta A. (1979): Earthquake Simulation in Aseismic *Engineering, Proc. Int. Symp. "Symulation of Systems"*, Sorrento, Italy, pp.517-536, North-Holland P.C.

3. Baratta A. (1980): Analisi dei Dati e Simulazione per la Sicurezza Antisismica delle Costruzioni, *Proc. of the Nat. Conf. "L' Ingegneria Sismica in Italia"*, Udine (Italy), pp. 1-13. (in Italian)

4. Baratta A. (1981): Optimized Earthquake Simulation for High-Reliability Aseismic Design, *Proc. 3rd Int. Seminar on "Reliability of Nuclear Power Plants"*, Paris (France), pp. 299-300..

5. Zuccaro G. (1991): Accelerogramma di Progetto per Verifiche ad Alta Affidabilita': Analisi Costi-Benefici., *Proc. 5th National Congress L'ingegneria Sismica in Italia*, Palermo, Italy, pp. 1504-1517

6. Baratta A. , Zuccaro G. (1992): High-Reliability Aseismic Design, *Proc. Tenth World Conference on Earthquake Engineering*, Madrid, Spain, pp. 3739-3744.

7. Baratta A. , Zuccaro G. (1993): Previsione di Risposta Nonlineare per Progettazione Antisismica ad Alta Affidabilita', *Atti VI Convegno Nazionale: L' Ingegneria Sismica in Italia*, Perugia, Italy, pp. 1193-1199. (in Italian)

8. Baratta A. , Zuccaro G. (1994): Worst Response of Structures under Not Sharply Defined Ground Motion, *Proc. 10th European Conference on Earthuake Engineering*, Vienna, pp. 1179-1183.

9. Ruiz P., Penzien J. (1969): Probabilistic Study of the Behaviour of Structures During Earthquakes, *Internal Report, Earthquake Engineering Research Center, College of Engrg.*, University of California, Berkeley..

10. Ruiz P., Penzien J. (1971): Stochastic Seismic Response of Structures, *Proc. ASCE*, vol. 97, N. EM2, pp.441-456.

11. Drenick R.F. (1970): Model-Free Design of Aseismic Structures, *Proc. ASCE, Journ. of Engineering Mechanics Division*, Vol. 96,N. EM4, pp. 483-493,1970.

12. Drenick R.F. (1973): Aseismic Design by Way of Critical Excitation, *Proc. ASCE, Journ of Engineering Mechanics Division*, Vol. 99, N. EM4, pp. 649-667.

13. Drenick R.F. (1977): On the class of the Non-Robust Problems in Stochastic Dynamics, *in "Stochastic Problems in Dynamics"*, B.L. Clarkson ed., Pitman, London, pp. 237-255.

14. Drenick R.F., Yun C.B: (1979): Reliability of Seismic Resistance Predictions, *Journal of the Structural Division*, Vol. 105, pp.1879-1891.

15. Shinozuka M. (1970): Maximum Structural Response to Seismic Excitations, *Journal of Engineering Mechanics Division*, Vol. 96, 729-738, 1970.

16. Elishakoff I., Pletner B. (1991): Analysis of Base Excitation as an Uncertain Function with Specified Bounds on It and Its Derivatives, in *Structural Vibration and Acoustics (T.C.Huang et al, eds.) DE-Vol. 34, ASME*, pp.177-184.

17. Zuccaro G., Elishakoff I., Baratta A. and Shinozuka M. (in print): Ellipsoidal Modeling of Earthquakes

18. Baratta A., Elishakoff I., Zuccaro G. and Shinozuka M. (in print): A Convex Model for Bounds on The Seismic Response of Linear Structures

19. Ben-Haim Y., Elishakoff I. (1990): *Convex Modeling of Uncertainty in Applied Mechanics*, Elsevier Science Publishers, Amsterdam.

20. Schweppe F.C. (1973): *Uncertain Dynamical Systems*, Prentice Hall, Englewood Cliff, NJ.

21. Elishakoff I., Ben-Haim Y. (1990): Dynamics of a Thin Cylindricxal Shell Under Impact with Limited Deterministic Information on Its Initial Imperfection, *Structural Safety*, Vol. 8, pp. 103-112.

22. Rao S.,S. (1978): *Optimization: Theory and Applications*, Wiley E.L., New Delhi

23. Coburn A.W., Spence R. and Zuccaro G., (1986): Seismic Risk to Population in Campania, a Model for Regional Planning, *Final Report, Joint Research University of Cambridge, "The Martin Centre" and University of Naples "Research Centre LUPT"*

24. Baratta A., Cacace F., (1986): Modello Probabilistico di Pericolosita' Finalizzato all'Analisi di Rischio Sismico, *Proc. 1ˢᵗ International Conference Towards the New Planning*, vol. I°, pp.113-140, Naples, Italy.

25. Cornell C., A., (1968): Engineering seismic risk analysis, *Bulletin of the Seismological Society of America*, **58**, 5, pp 1583-1606.

26. Gutenberg B., Richter C., F., (1954): *Seismicity of the Earth and Associated Phenomena*, Princetown University Press.

27. Coburn A.W., Sakay R.J.S., Spence R. and Pomonis A. (1990): A Parameterless Scale of Seismic Intensity for Use in Vulnerability Assessment, *Final Report. Cambridge Architectural research Ltd.*

APPENDIX

ESSENTIAL SYMBOLS

$\tilde{x}; \tilde{\mathbf{x}}$: random variable (r.v.) & random vector (r.v.)

x_i : vector component

ℓ : vector of local shaking characters

μ_x : expected value of random variables $\tilde{x}$

σ_x : standard deviation of random variables $\tilde{x}$

n : vector or space dimension

N : number of points in n-dimension space

R : structural response component

t : time

i, j, k : indices

α : real number in [0,1]

WN : White noise

$\tilde{w}(t)$: Gaussian white noise

RSG : Random standard Gaussian variable

Ω : pulsation parameter of Kanai Tajimi average spectrum

ξ : damping parameter of Kanai Tajimi average spectrum

$\varphi(t)$: modulating function

$\Phi(t)$: earthquake variance intensity function

Φ_o : maximum value of $\Phi(t)$

t_d : standard time of decay branch in $\Phi(t)$

t_i : end time of the build-up branch in $\Phi(t)$

T : duration of event

a_o : nominal lateral strength of shear-type frame (STF) expressed as acceleration

a_p, PGA : peak ground acceleration

ν : decay exponential rate

I_o : earthquake macroseismic intensity

S_x : response spectrum of the parameter x

T_o : natural vibration period of Single Degree Of Freedom (SDOF) structure

ω_o : natural vibration frequency of SDOF structure

ζ : damping ratio of SDOF structure

$\lambda(I_o)$: earthquake occurrence rate at intensity I_o

T_r, RP : return period of seismic event

ψ : non-dimensional Coburn-Spence intensity parameter [27]

δ : ductility factor of STF

δ_{ij} : Kronecker's delta

A_f : amplification factor

W   : self-weight of transverse beam in the STF
m   : the mass of transverse beam in the STF
u   : horizontal displacement of transverse beam in the STF
$u_g(t)$: ground displacement during the shaking
u_p : horizontal displacement over the elastic limit
M_o : limit bending moment in the columns
H   : height of the column
g   : gravity acceleration
F   : the total shear force in the column
F_o : limit value of the total shear force
$\vartheta = g/H$: critical parameter of the P-Δ STF
u_k : collapse horizontal displacement of STF
a'_a, a''_a : virgin lateral strength for right-ward and left-ward motion in the P-Δ STF
      (expressed as acceleration)
a'_u, a''_u : lateral strength for right-ward and left-ward motion in the P-Δ STF starting
      with displacement u (expressed as accelerations)
$f_r(u, \dot{u}, u_p, \dot{u}_p)$: restoring force in the P-Δ STF
$y = u/u_k$: horizontal displacement in the P-Δ STF normalized to the critical
      displacement
$a(t), \ddot{u}_g(t)$: seismic accelerogram
$h(x)$: impulse response function of the STF
$\mathscr{E}_o, \|a\|$: the square root of the energy of the accelerogram $a(t)$
$\wp(\mathscr{E}_o)$: the set of functions having energy E_o^2
H_T : the square norm of the impulse response function in the time interval $[0, t]$
H_∞ : the square norm of the impulse response function in the time interval $[0, \infty]$
$a_c(t)$: critical excitation according to Drenick
$u_c(t)$: STF responce under the critical excitation $a_c(t)$
c   : vector of Fourier coefficients
a(t) : vector of basic accelerogram
f_c, f_g : generic functions of Fourier coefficients
ω_i : pulsation of the i-th term in the Fourier expansion (FE)
ω : the current pulsation
$K(\omega | \Omega_a, \xi_a, \Omega_b, \xi_b)$: generalized Kanai-Tajimi central spectrum
θ : size parameter of the feasible ellipsoid $\mathcal{E}$
ρ : radius of the sphere $\mathcal{S}$
Q   : normalizing factor
$\mathbf{c}_o$: vector of Fourier coefficient of the central accelerogram at a site
r_i : Lagrange multipliers
$\bar{c}_{ij}$: i-th coefficient in the Fourier Expansion of the j-th accelerogram $a_j(t)$

CHAPTER 2

**WHAT ARE THE RANDOM AND FUZZY SETS AND HOW TO USE THEM
FOR UNCERTAINTY MODELLING IN ENGINEERING SYSTEMS?**

A. Bernardini
University of Padua, Padua, Italy

ABSTRACT

Probabilistic and set-based models of the uncertainty are combined reviewing the idea of random set. Moreover fuzzy sets are defined as a particular case, i.e. consonant random sets. Inclusion and extension, through a deterministic function, of random sets or relations give powerful methods to evaluate probabilistic bounds of the response of deterministic systems with uncertain parameters or uncertain input variables.

Combination of independently given information about the same system is discussed both in the ambit of set-based and probability-based model of uncertainty. The limits entailed by the proposed solutions are highlighted together with their overcoming suggested by evidence theory and fuzzy logic.

Random set theory suggests robust and simple procedures to take into account the uncertainties of the involved variables and parameters in the field of structural engineering, and more generally in civil and mechanical engineering. Numerical applications to reliability evaluations and structural mono-objective optimisation are presented to demonstrate their usefulness.

Lastly the applications of fuzzy logic to decision making, pattern recognition, fuzzy controller, fuzzy expert systems, multi-objective optimisation are briefly discussed. The procedures are clarified through some simple numerical examples.

1. WHY DO WE NEED RANDOM AND FUZZY SETS ?

1.1 INTRODUCTION

The uncertainty about a variable or a parameter, on which the time varying behaviour of a complex system (a machine, a structural frame, an environmental system, a human being, etc..) depends, generally occurs in two alternative patterns [1]:

a- *Conflict* or strife between disjoint alternative values; for example in the case of dice play, 6 alternative and conflicting outcome values are possible. Here conflict raises from the fact that the occurrence of one value excludes all the other values and contrasts with the a priori forecast. The modelling of this type of uncertainty was the task of the old theory of chances and of the modern theory of probability.

b - *Non-specificity* relative to a set of alternative values, each of which is to be considered equally possible. For example, a measure obtained by means of a low sensitivity instrument yields an interval of possible values; should further investigation get a precise value, it would not conflict with the a priori forecast, provided it lies inside the interval. This type of uncertainty is the object of interval analysis [2] or more generally of the classical set theory [3].

Nevertheless, it is easy to understand that both types of uncertainty may frequently occur at the same time, even when information is collected through usual statistical techniques, e.g. responses to multi-choice or incomplete questionnaires [4].

In point of fact, also the statistical treatment of measures obtained by means of low sensitivity instruments (and each instrument can be considered as such) leads to information which cannot be represented by an histogram of frequencies relative to prefixed and disjoint intervals of the measured variable. If this is done routinely, it is performed at the cost of some approximation, which cannot be acceptable in any case.

The discussion so far presented allows us to introduce the theories of random sets and of fuzzy sets which appear as the natural extension of the theories previously hinted at, in that they combine the theory of probability and the set theory, as will be shown in §§ 1, 2 and in §4, in which some applications concerning the field of structural engineering in particular are developed. Nevertheless, in   §3 the limits of these theories will be demonstrated when trying to combine pieces of uncertain information separately given on the same space of variables. This is a fundamental step in the development of a more general model of approximate reasoning,   today of great importance as a support to computer-aided procedure in the field of dynamic systems control, pattern recognition and clustering, implementation of expert systems, multi-objectives optimisation. These topics will be briefly presented in §5.

In the following of this §1, the formal definitions of random sets and the procedures to analyse their probabilistic content will be presented. Moreover, we intend to demonstrate how the structure of fuzzy sets can be derived as a particular case of random set. We will here concentrate on finite sets; extension to non finite sets is possible with only minor changes (and yet essential from a mathematical point of view) in the formulations (e.g. max and min substituted by sup and inf; see [5]), which, nevertheless, are not so important in our discussion about the fundamental tenets of the theory and in their applications to engineering problems.

1.2 RANDOM SETS

1.2.1 Fuzzy Measures and Extension Principle

In the analysis of the response of uncertain systems, and more generally in information and control theory, the word *extension* is very frequently used, to which however two different meanings are attached.

According to a first meaning, information of any kind (deterministic, probabilistic, qualitative or linguistic...) about one or more system variables can *extend* information to an other variable related to the first one(s) by a given *relation of compatibility*. For example, in the field of classical set modelling of uncertainty, with reference to a variable x in the set X, the relation of compatibility can be given through a set-valued mapping G from the singletons $\{x\}$ of the set X and the set of the subsets of the range Y of a variable y (the power set of Y, $\mathbf{P}\{Y\}$) :

$$G : X \rightarrow \mathbf{P}\{Y\} \tag{1.1}$$

The deterministic information given through a restriction of the actual values of the variable x to a subset $A \subseteq X$, can be extended to a corresponding restriction of y to the subset

$$G (A) = \bigcup G (x) \mid \{x\} \in A \tag{1.2}$$

Moreover,

$$\forall \ A , B \subseteq X, \quad A \subseteq B \ \Rightarrow \ G(A) \subseteq G(B) \tag{1.3}$$

More generally suppose now that the information (restriction) is given on X through an *ordering* measure g defined on the set of the subsets of X:

$$g : \mathbf{P}\{X\} \rightarrow [0, 1] \mid g(\varnothing) = 0 ; \qquad g(X) = 1 ;$$
$$\forall \ A, B \subseteq X, \ g(A) \leq g(B) \ \Leftrightarrow \ A \subseteq B \tag{1.4}$$

A measure satisfying (1.4) is called a *fuzzy measure*, and clearly the *characteristic function* χ of the subsets A, B of the set X or a probabilistic measure Pr of the events A, B on the space X are particular cases of such a general class of measures.

Relation (1.2) can now be *extended* (and here the word assumes its second meaning) to the case of fuzzy measures in the following way. The information given through a fuzzy measure g on X, can be extended to a class of fuzzy measures on each subset C of the set

Y; these measures, according to the ordering hypothesis given in (1.4), must satisfy the condition:

$$g(C) \geq g(\{ \; x \in X \mid G(x) \subseteq C \; \}) \tag{1.5a}$$

It is "reasonable" ([6] Shafer, 1987), but not always necessary, to assume the minimum measure

$$g(C) = g(\{ \; x \in X \mid G(x) \subseteq C \; \}) \tag{1.5b}$$

as representing the extension on Y of the given fuzzy measures on X .

1.2.2 Probabilistic Measures

The probability measure is a fuzzy measure associated to subsets (i.e. events) satisfying the additivity law:

$$\forall \; A, B \mid A \cap B = \varnothing , \quad g(A \cup B) = g(A) + g(B) \tag{1.6}$$

Let the compatibility relation be a point-valued mapping $f: X \to Y$. In this case, with reference to random variables with probability distribution $P(x)$, the relation (1.5b) gives, for every singleton $y \in Y$:

$$g(y) = Pr(\{y\}) = \Sigma \; (P(x) \mid f(x) = y) \tag{1.7}$$

a well known equation in the theory of the functions of random variables.

1.2.3 Random Sets as Dempster-Shafer Structures

The extension (1.5b) of a probability measure through a multi-valued mapping is not a probability measure.

Example 1.1

Let $X = X_1 \times X_2$, with $X_1 = X_2 = \{ 0, 1 \}$; $Y = [0, 2.5]$; $C_1 = [0.5, 2]$; $C_2 = [2, 2.5]$
The joint probability distribution on X and the set-valued mapping from X to Y are given as follows:

$$P(0, 0) = 0.2 \quad ; \quad P(1, 0) = 0.3 \quad ; \quad P(0, 1) = 0.4 \quad ; \quad P(1, 1) = 0.1 \; ;$$

$$G(0, 0) = [0, 1.1] ; \; G(1, 0) = [0.5, 1.5] ; \; G(0, 1) = [0.9, 2] ; \; G(1, 1) = [1.5, 2.5)$$

Consequently: $\quad g(C_1) = 0.3 + 0.4 = 0.7; \quad\quad\quad g(C_2) = 0 ;$
$$g(C_1 \cup C_2) = g([0.5, 2.5]) = 0.3 + 0.4 + 0.1 = 0.8 \neq g(C_1) + g(C_2)$$

Let the compatibility relation be a multi-valued mapping G between a finite space Z, where a probability distribution P(z) is defined, and a finite set of subsets of X, $A_i = G(z_i)$. This compatibility relation extends the probability measure (defined by P(z)) to the subsets A_i

$$m(A_i) = P(z_i) \qquad (1.8) \; ; \qquad \Sigma_i \; m(A_i) = 1 \qquad (1.9)$$

This mathematical structure has been named a *random set* on X : $\Im(A_i , m(A_i))$, with *focal elements* A_i and a *basic probabilistic assignment* given by the *weights* $m(A_i)$ satisfying the normalisation condition (1.9). The notion of random set was apparently first introduced by Robbins [7] and for the modern treatment, reference is made to the monograph by Matheron [8]. Summary of recent development on random sets is given by Peng, Wang and Kandel in [9].

The weight $m(A_i)$ expresses the extent to which all available and relevant evidence supports the claim that a particular element of X belongs to the set A_i alone (i.e. exactly to set A_i) and it does not imply any additional claims regarding subsets of A_i ; if there is some additional evidence supporting the claim that the element belongs to a subset B of A_i, it must be explicitly expressed by another value m(B). The main difference between a probability distribution function and a basic assignment is that the first one is defined on X, whereas the second is defined on the power set of X, **P**{X}. As a consequence, we obtain the following properties of m :

1) it is not required that m(X)=1 ;
2) it is not required that $m(A) \leq m(B)$ when $A \subset B$;
3) no relationship between m(A) and $m(A^c)$ is required (A^c is the complementary set of A).

Each focal element A, in fact, must be treated as an object *"per se"* ; if $A \subset B$ and $m(A) \leq m(B)$, it means that the object A is less probable than the object B. It is to be noted that :

a) if m(X)=1 , there is a unique focal element and this focal element is X itself (maximum ignorance) ;
b) conversely, if there is a unique focal element $A \subset U$, then m(A)=1 and m(X)=0.
c) if there are two or more focal elements, then m(X)<1.

It should be stressed that the definition of random set refers to distinct non-empty subsets of X. If these distinct non-empty subsets are singletons (the single elements, thus non-overlapping, of X) and each of them has a probability assignment, then we have a probability distribution on X. Note that during real world information processing however, the non-empty subsets may be overlapping.

When a random set structure $\Im(A_i, m(A_i))$ is known for a space X , it is not possible to evaluate the probability for every singleton of X or any subset C of X. But it is possible to evaluate upper and lower bounds: precisely, as pointed out by Dempster [10]:

$$\Sigma\, m(\,A_i\,)\mid A_i \subseteq C \;\leq\; \Pr(\,C\,) \;\leq\; \Sigma\, m(\,A_i\,)\mid A_i \cap C \neq \varnothing \tag{1.10}$$

or, indicating with the words *Belief* and *Plausibility* the bounds, as proposed by Shafer in his fundamental book on Evidence Theory[11]:

$$\Pr(\,C\,) \;\geq\; \Sigma\, m(\,A_i\,)\mid A_i \subseteq C \qquad = \mathrm{Bel}\,(\,C\,) = 1 - \mathrm{Pl}\,(\,C^c\,) \tag{1.11a}$$

$$\Pr(\,C\,) \;\leq\; \Sigma\, m(\,A_i\,)\mid A_i \cap C \neq \varnothing \;=\; \mathrm{Pl}\,(\,C\,) = 1 - \mathrm{Bel}\,(\,C^c\,) \tag{1.11b}$$

It is easy to demonstrate that Bel and Pl are fuzzy measures. Particularly Bel is the extension according the minimum condition (1.5b). Moreover each one of the set functions m , Bel and Pl define completely the two others [11].

Example 1.2

With reference to Example 1, let $Z = X_1 \times X_2$, $X = Y$,

$$A_{ij} = G(\,x_i,\, x_j\,)\,,\; i,j = 1,\, 2\,; \qquad m\,(A_{ij}) = P(x_i,\, x_j)\,,$$
$$C = [0.5\,,\, 2.5]\,.$$

Then: Bel $(C) = g\,(C) = 0.3 + 0.4 + 0.1 = 0.8$;   Pl $(C) = 0.2 + 0.3 + 0.4 + 0.1 = 1.0$

The class of fuzzy measures satisfying the bounds (1.10) can be considered as the class of probability measures on the space X compatible with the given probability assignment on the focal elements, as pointed out by Dubois and Prade [12]. Each probability measure of this class corresponds to a particular distribution of the weight $m(A_i)$ on the singletons of A_i . In the limit cases that the weights $m(A_i)$ are concentrated on the lower and respectively upper bounds of the focal subsets A_i, the cumulative probability distributions functions are given by the relations:

$$F_{low}\,(x) = \mathrm{Pl}\,(\{\; x' \in X \mid x' \leq x \;\}) \tag{1.12a}$$

$$F_{upp}\,(x) = \mathrm{Bel}\,(\,\{x' \in X \mid x' \leq x \}) = 1 - \mathrm{Pl}\,(\,\{x' \in X \mid x' > x\}\,) \tag{1.12b}$$

These probability distributions can be used to obtain upper and lower bounds of the Expectations of any real-valued function defined over X, e.g. E[x] [10]. A third interesting probability measure on X (*White Probability* WhPr in the following) can be obtained assuming the hypothesis that the weights $m(A_i)$ are uniformly distributed on the singletons of A_i , with finite cardinality $|A_i|$:

$$F_{whp}\,(x) = \sum_{x' \leq x}\; \sum_{x' \in A_i}\; m(A_i)\,/\,|A_i| \tag{1.12c}$$

The corresponding Expectations are bounded by the values obtained through the relations (1.12a,b).

These measures seem very useful in the treatment of incomplete statistical information about a set of possible states of a system, or a finite interval of a physical variable.

When the reported data about the state variables are not complete, the statistical information can be considered relevant to the subset of the system states compatible with the data. The frequencies of the observed subsets define a probabilistic assignment on such focal elements.

Example 1.3 [4, 13]

Let's consider the case of a poll taken for candidates in the elections. Often, the interviewees do not specify the single favourite candidate, but rather, express their views about two or more acceptable candidates ; in several cases they express their views about an unacceptable candidate only. Under these circumstances, the estimate of the available support of a given candidate cannot be calculated, rather, it can be bounded from below and from above. The minimum possible support a candidate enjoys equals the proportion to which that single candidate was identified as a preferable one ; the maximum possible support of a candidate equals the sum of proportions, in which the specific candidate appears singularly or with other fellow candidates, as a preferable one. Let the results of a poll conducted in a fictitious election be summarised as follows ; with m defining the support proportion to a candidate or to a group of candidates (subset of the set of candidates $X = \{ 1,2,3,4 \}$) :

$$m(\{1\})=0.1 ; \quad m(\{2\})=0.12 ; \quad m(\{3\})=0.01 ; \quad m(\{4\})=0.06 ; \quad m(\{1,2\})=0.52 ; \quad m(\{2,3\})=0.19.$$

Here we have only 7 focal elements, because no support has been given to the other groups of candidates. Then the supports (denoted by S) enjoyed by the single candidates, i.e. the probabilities of election, are bounded as follows :

$$m(\{1\}) = 0.1 \ < S_1 < m(\{1\})+ m(\{1,2\}) = 0.62 ;$$
$$m(\{2\}) = 0.12 < S_2 < m(\{2\})+ m(\{1,2\})+ m(\{2,3\}) = 0.83;$$
$$m(\{3\}) = 0.01 < S_3 < m(\{3\})+ m(\{2,3\}) = 0.2;$$
$$m(\{4\}) = 0.06 < S_4 < m(\{4\}) = 0.06 .$$

Minimum S_1 is obtained by assigning the total value of $m(\{1,2\})$ to the other candidate, numbered 2, while the maximum S_1 is obtained by assigning the entire $m(\{1,2\})$ to the first candidate. Analogous procedure is performed for calculating min S_j and max S_j for $j > 1$.

It is very important here underline that, in their definition, the meaning of the introduced measures remains in the field of the probability theory.

1.2.4 Consonant Random Sets

Let us consider the particular case of a random set whose n focal elements are nested, i.e. can be ordered in such a way that

$$A_i \subseteq A_{i+1} \quad (i = 1, 2, 3, ... \ n\text{-}1) \tag{1.13}$$

A random set satisfying (1.13) is called *consonant*, according to a definition introduced by Shafer [11]. Consonant random sets satisfy the decomposability properties [11] :

$$\forall\, A, B \subseteq X, \ Bel\,(\,A \cap B\,) \ = \ \min\,(\,Bel\,(A),\, Bel\,(\,B\,)\,) \tag{1.14a}$$

$$\forall\, A, B \subseteq X, \ \ Pl\,(\,A \cup B\,) \ = \ \max\,(\,Pl\,(A),\ Pl\,(\,B\,)\,) \tag{1.14b}$$

The logical meaning of this type of information can be understood considering that the frequency distributions is centred around the subset A_1. In fact in this case

$$Bel\,(\,A_1\,) \ = \ m\,(\,A_1\,) \ ; \qquad Pl\,(\,A_1\,) = 1 \tag{1.15}$$

Remembering the distinction given in § 1 between conflict and non-specificity, it is possible to observe that probabilistic and consonant measures are two extreme cases : in the first case the uncertainty is totally conflicting, while in the second case the non-specificity prevails. In the second case, if it happens a posteriori that the actual value belongs to A_1, this value is compatible with the positive probabilistic assignment attributed a priori to all the focal elements. Moreover, the classical set modelling, which is a totally non-specific representation of the uncertainty, can be considered a consonant measure with a single focal element.

For every singleton $\{x\}$ of X , the Plausibility can be evaluated through the function

$$\mu\,(\,x\,) = \ Pl\,(\,\{x\}\,) = \ \ \Sigma\ m(\,A_i\,)\,|\ \{x\} \in A_i \tag{1.16}$$

By taking into account relations (1.14) and (1.11), it is very important to note that this point-valued function (*contour function* according to Shafer [11]) defines implicitly the Belief / Plausibility measures of every subset of X. From this point of view, the consonant measures have the same property of the probabilistic measures: both can be completely described by a point-valued function (the probability distribution or the contour function), while a non-consonant random set requires a more complex description through a set-valued function.

1.3 FUZZY SETS

In a context totally separated from the research work of Dempster and Shafer, the idea of fuzzy set was developed by L. Zadeh in the Sixties [14] as an extension of classical set theory. He suggested that the membership to a subset A of a universal set X could not always be a crisp property, absolutely verified or not-verified, but also, in some cases, partially verified. So the boundaries between A and A^c should be considered as separated by a fuzzy zone, where the *Law of excluded middle* is no more valid and {x} partially belongs both to A and to A^c .

Formally the extension was obtained through the modification of the *characteristic function* of a subset A of a universe set X

$$\chi \ : \ X \to \{0, 1\} \tag{1.17}$$

to a mapping (a point-valued function named *membership function*):

$$\mu \ : \ X \to [\, 0, 1] \tag{1.18}$$

Therefore, we are facing a logic (fuzzy logic) in which propositions exist whose degree of truth goes from 0 (false) to 1 (true), every gradation being permitted in between. This logic seems to match closely human thinking when a true-false judgement has to be given about complex propositions relative to the real world, as we will see in the following in §3 and §5.

According to (1.18) the *cardinality* of a fuzzy subset A of a (finite) set X is defined by the relation:

$$|A| = \Sigma \, \mu(x) \mid x \in X \tag{1.19}$$

The *height* of a fuzzy subset A is the largest membership grade obtained by any element in that set.

$$h(A) = \max \mu(x) \mid x \in X \tag{1.20}$$

A fuzzy set is called *normal* when its height is equal to 1, else *subnormal*.

One of the main aspects in the proposal by Zadeh is the extension to fuzzy sets of the operations of union, intersection and complementation of classical set theory. The standard definitions suggested are as follows:

intersection:	$\mu_{A \cap B}(x) = \min(\mu_A(x), \mu_B(x))$	(1.21a)
union :	$\mu_{A \cup B}(x) = \max(\mu_A(x), \mu_B(x))$	(1.21b)
complementation:	$\mu_{Ac}(x) = 1 - \mu_A(x)$	(1.21c)

But many alternative rules have been later proposed by other authors, as we will discuss in the following chapters. Here, on the contrary, we focus our interest on the connection between concept of fuzzy set and concept of random set, as uncertain measures of a variable.

This connection can clearly be envisaged using the dual representation of a fuzzy set through their α-cuts $^\alpha A$. They are classical subsets of X defined, for any selected value α of membership, by the relations

$$^\alpha A = \{x \mid \mu(x) \geq \alpha\} \tag{1.22}$$

This representation allows to extend to fuzzy sets the notion of *convexity*: a fuzzy set is said to be convex if their α-cuts are convex, for any value of α.

When a fuzzy set is implicitly given through the sequence of its α-cuts $^{\alpha}A$, its membership function can be reconstructed through the following decomposition theorem:

$$\mu(x) \;=\; \max_{\alpha} \; \min (\alpha , \chi_{\alpha A} (x)) \tag{1.23}$$

where $\chi_{\alpha A} (x)$ is the characteristic function of the classical subset $^{\alpha}A$.

Example 1.4

Let $X = \{1, 2, 3, 4, 5\}$ and A a fuzzy subset of X defined by the membership grades:

$$\mu_A (1) = \mu_A (5) = 0 \; ; \qquad \mu_A (2) = \mu_A (4) = 0.5 \; ; \qquad \mu_A (3) = 1$$

Then: $^{0}A = X = \{1, 2, 3, 4, 5\}$; $\qquad ^{0.5}A = \{2, 3, 4\} \; ; \; ^{1}A = \{3\}$

From the relation (1.23) we obtain:

$$\mu_A (1) = \max (\min (0, 1), \min (0.5 , 0), \min (1 , 0)) = 0$$
$$\mu_A (2) = \max (\min (0, 1), \min (0.5 , 1), \min (1 , 0)) = 0.5$$
$$\mu_A (3) = \max (\min (0, 1), \min (0.5 , 1), \min (1 , 1)) = 1$$

So it is clear that the α-cuts $^{\alpha}A$ give the non-specificity of uncertainty, as do the focal elements of a random sets, while the associated values of α are related to the strife between the different α-cuts. Moreover, the α-cuts $^{\alpha}A$ are a nested sequence of subsets of the set X, as can be demonstrated observing that, for any couple α_1 , α_2 :

$$\alpha_1 > \alpha_2 \;\Rightarrow\; ^{\alpha 1}A \subseteq {}^{\alpha 2}A \tag{1.24}$$

It is therefore clear that a consonant random set can be associated to any given normal fuzzy set, by assuming its membership function as contour function of the corresponding random sets, their α_i-cuts $^{\alpha_i}A$ (for discrete decreasing values of α: $\alpha_1 = 1, \alpha_2 < \alpha_1 \dots \alpha_{i+1} < \alpha_i \dots, \alpha_{n+1} < \alpha_n , \alpha_{n+1} = 0$) as focal elements and the basic probabilistic assignment:

$$m (^{\alpha_i}A) = \alpha_i - \alpha_{i+1} \tag{1.25}$$

The fuzzy set should be normal in order to respect relation (1.9).

From this point of view, the membership function of a fuzzy subset A of a set X can be better considered as a *possibility* distribution $\pi(x)$ of the single values x , according the definition given by Zadeh himself in 1978 [15], and later extensively developed by other authors, in particular Dubois and Prade [11]. In a comparison between Evidence Theory and Possibility Theory, *Necessity* (Nec(B)) and *Possibility* (Pos(B)) measures of any classical subset B of the universal set X are respectively Belief (Bel(B)) and Plausibility (Pl(B)) of B evaluated through the associated consonant random set.

In fact, according to (1.14b) and (1.11a), for any (finite) subset B:

$$\text{Pos (B)} = \text{Pl (B)} \quad = \quad \max \pi(x) \mid x \in B \tag{1.26a}$$

$$\text{Nec (B)} = \text{Bel (B)} = 1 - \text{Pl (B}^c) = 1 - \max \pi(x) \mid x \in B^c \tag{1.26b}$$

This comparison suggests a possible derivation of the membership function of a fuzzy set from statistics of set-valued data. A general method is discussed by Dubois and Prade in [16] for the derivation of possibility measures on the basis of statistics of a variable's measures given as intervals (a very common case in physical measures).

Moreover, the same comparison suggests the frequentistic content of information summarised by a fuzzy set, whenever it is built up by means of objective statistical data. In Fig. 1.1, for a triangular convex normal fuzzy on R^1 (a *fuzzy number*), the corresponding upper (1.12a) and lower (1.12b) cumulative probability distributions of the associated class of random variables are displayed. In addition, the cumulative probability distribution of the white probabilities (1.12c) is indicated, together with the upper and lower bounds of expectation E[x].

In Fig. 1.2, for a convex fuzzy subset of the finite set of the integers from 1 to 5, the upper and lower bounds of the probability of the singletons are shown together with the distribution of the white probabilities.

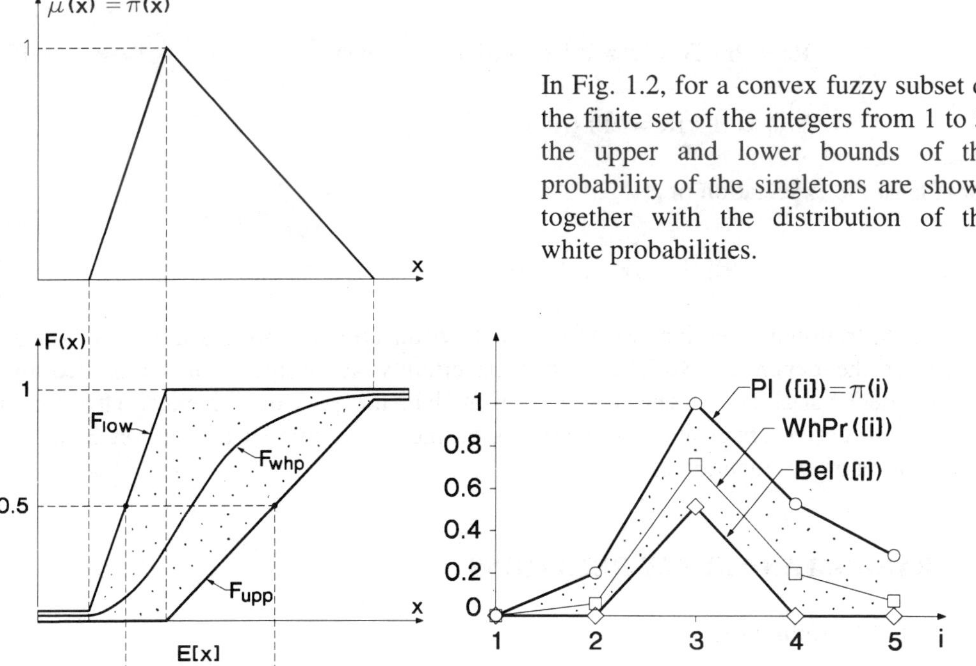

Fig. 1.1 - The frequentistic content of a fuzzy set (a triangular fuzzy number).

Fig. 1.2 - Plausibility, Belief and White Probabilities of the singletons of a consonant random set (fuzzy set).

2.   HOW DO RANDOM AND FUZZY SETS WORK?

2.1. INTRODUCTION

Let's now go back to   the basic problem defined in § 1.2.1 : the extension of a random set measure through a relation of compatibility with a dependent variable. Here, the relation of compatibility is modelled as follows :

$$z = f(\mathbf{x}) = f (x_1, x_2, ...x_n)   ,\mathbf{x} \in \mathbf{X} = X_1 \times X_2 \times ...X_n \tag{2.1}$$

where f is a deterministic point-valued function (defined in an explicit or implicit manner, for example through differential equations and boundary condition , or otherwise through a list of propositions which can assume only the values TRUE or FALSE).

In the ambit of probability theory, when the joint probability distribution $p(\mathbf{x})$ is assigned, we want to solve the fundamental problem of determining the correspondent probability distribution $p(z)$ for the dependent variable z, and therefore the value of the probability of any subset B (event) of the range Z.

For example, z could be the safety margin of a limit state, and therefore we must calculate the Reliability Re or the Probability of Failure P_{fail}:

$$Re = Pr (B = \{z \in Z \mid z > 0\}) = \int_B p(z) \; dz \tag{2.2a}$$

$$P_{fail} = 1 - Re = Pr (B^c) \tag{2.2b}$$

or at least the expectation of z:

$$E[z] = \int_Z p(z) \; z \; dz \tag{2.3}$$

In the following section we will show how uncertainty induces a random set in the space of the dependent variable, when uncertainty about the values assumed by the independent variables is measured as indicated in the previous chapter. Therefore it is possible to derive upper and lower bounds on the probability of any subset B, or on the expectation of any function of **x**.

2.2 RANDOM AND FUZZY RELATIONS

2.2.1 Joint random set

Firstly, let us recall that information about many variables determining the behaviour of a system should be given in a joint manner, at least in the most general case. For example, it is well known that the probability density distribution $p(\mathbf{x})$ of random variables cannot be derived directly from the marginal distributions of the single variables, except in the particular case of stochastically independent variables, when:

$$p(\mathbf{x}) = p\,(\,x_1, x_2, \ldots\ldots x_n) = p\,(x_1) \cdot p(x_2)\ \ldots\ldots\ p\,(x_n) \qquad (2.4)$$

It is obvious that even in the case of random sets or fuzzy sets it is not generally possible to describe the probabilistic content of information, if measures of random sets or fuzzy sets are separately assigned to the single variables. Information must be assigned or collected in a joint manner. Therefore we want to extend the definitions of Random Set and Fuzzy Set, given above in Chapter 1, to multi-dimensional spaces ; more precisely to the space $\mathbf{X}$ correspondent to the Cartesian product of sets X_i of the single basic variables, as shown in the relation (2.1).

To simplify the representation (see Fig. 2.1) we refer to two variables and therefore to the two-dimensional space $\mathbf{X} = X_1 x X_2$. In this space a random relation is defined by a sequence A_1 , A_2 , ... A_i ....A_n of focal elements, generally not-disjoint and a basic probabilistic assignment:

$$m : \{\,A_1\,,\,A_2\,,\,\ldots\,A_i\,\ldots.A_n\,\} \rightarrow [0,\,1] \qquad |\ \ \Sigma\ m(A_i) = 1 \qquad (2.5)$$

For every subset $B \subseteq \mathbf{X}$, it is possible to evaluate the values of the set-valued functions Bel(B) and Pl(B) and the bounds of probability, in a manner exactly correspondent to the relations (11a) and (11b) of Chapter 1; and therefore:

$$Pr\,(B)\ \in\ [\,Bel(B),\,Pl(B)] \qquad (2.6)$$

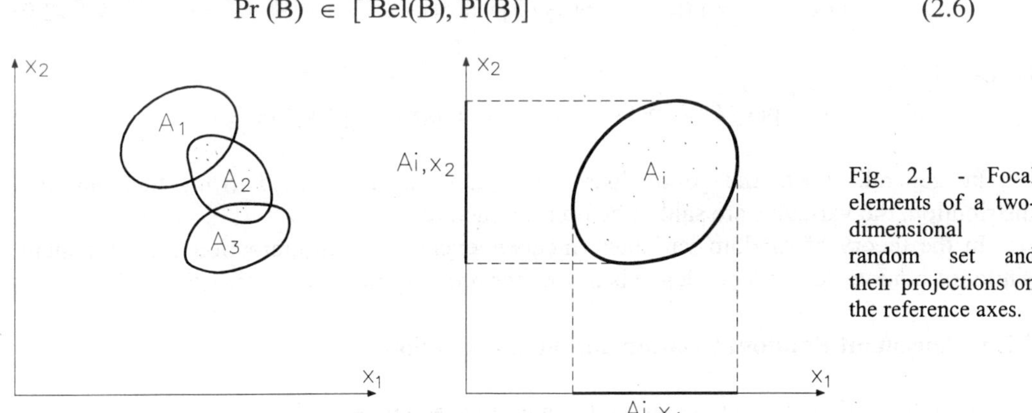

Fig. 2.1 - Focal elements of a two-dimensional random set and their projections on the reference axes.

2.2.2 Marginal Random Sets

For every focal element A_i, the projections $A_{i,x1}$ and $A_{i,x2}$ on the x_1 and x_2 axes can be calculated, as displayed in Fig. 2.1b. In this way, two random sets can be defined, whose focal elements are, respectively, $A_{i,x1}$ e $A_{i,x2}$ and whose basic probabilistic assignment is

$m(A_i)$. If, for two or more different values of index i, the projections are coincident, it is possible to consider one single focal element with basic probabilistic assignment equal to the sum of the values correspondent to such indexes. Therefore in general:

$$m(A_{i,Xj}) = \sum_k m(A_k) \mid A_{k,Xj} = A_{i,Xj} \qquad (2.7)$$

It is evident that the marginal random sets are uniquely determined on the basis of the assigned random relation, while the opposite is not true, for two distinct reasons:
a) because the projections do not determine uniquely the focal elements, unless a rule is known or assumed a priori to derive focal elements from the projections. The Cartesian product is an example of such a rule:

$$A_i = A_{i,X1} \times A_{i,X2} \qquad (2.8)$$

b) because the projections could correspond, according to (2.7), to more focal elements, among which it is necessary to split the marginal basic probabilistic assignment.
In the probabilistic case, it is the second reason that brings about the indeterminateness of the joint probability distribution $p(x_1, x_2)$, when we start from the marginal distributions $p(x_1)$ and $p(x_2)$; on the contrary, the first reason doesn't apply because the focal elements are the singletons $A_i = \{(x_1, x_2)_i\}$. The indeterminateness disappears only when the variables are stochastically independent variables, i.e. when conditional probabilities equal unconditional probabilities:

$$p(x_1 / x_2) = p(x_1) \quad ; \quad p(x_2 / x_1) = p(x_2) \qquad (2.9)$$

hence:

$$p(x_1, x_2) = p(x_2 / x_1)\, p(x_1) = p(x_1 / x_2)\, p(x_2) = p(x_1)\, p(x_2) \qquad (2.10)$$

In general, when the joint distribution can be determined from the marginal distribution, the variables are said to be non-interactive.
In the theory of random variables, independence and non-interaction are equivalent, while for random sets independence is a stronger property than non-interaction [17].

2.2.3 Consonant Random Relation and Fuzzy Relations

If the focal elements can be ordered in a nested sequence, such that

$$A_i \subseteq A_{i+1}, \qquad i = 1, 2, ...n\text{-}1 \qquad (2.11)$$

properties analogous to the one-dimensional case hold (see § 1.2.4).
Information given by the random relation can be fully given through the point valued contour function, i.e. the possibility values $\pi(x_1, x_2)$ of the singletons $\{(x_1, x_2)\}$, that is the

membership function $\mu(x_1, x_2) = \pi(x_1, x_2)$ of a fuzzy relation. The focal elements are in this case the α-cuts of the fuzzy relation for the finite succession $\alpha_1 = 1$, $\alpha_2 < \alpha_1$, ..., $\alpha_{n+1} < \alpha_n$, $\alpha_{n+1} = 0$, with probabilistic assignment $m(^{\alpha_i}A) = \alpha_i - \alpha_{i+1}$.

Then it is easy to observe that the marginal consonant random sets are the fuzzy sets whose membership function is simply defined by the relations (X_1 and X_2 are finite sets; for non finite sets, the "sup" operator should be used instead of "max"):

$$\mu(x_1) = \max_{X_2} \mu(x_1, x_2) \quad ; \quad \mu(x_2) = \max_{X_1} \mu(x_1, x_2) \qquad (2.12)$$

i.e. they are the *projections* [17] of the fuzzy relations on the space of the single variables.

Generally, this marginal distribution does not determine uniquely the joint distribution, i.e. the variables are interactive, and therefore based on data given in a joint manner. However if all the focal elements are Cartesian products of intervals on X_1 and X_2, i.e.

$$A_i = A_{i,X1} \times A_{i,X2} \ , \qquad\qquad i = 1, 2, ...n \qquad (2.13)$$

and the sequence of $A_{i,X1}$ e $A_{i,X2}$ (and consequently of the A_i) are nested, it is possible to demonstrate [17] that

$$\mu(x_1, x_2) = \min\left(\mu(x_1), \mu(x_2)\right) \qquad (2.14)$$

That is: Cartesian consonant random sets are non-interactive fuzzy relations, whose membership function is uniquely determined starting form the marginal fuzzy sets through relation (2.14).

2.3 INCLUSION OF RANDOM AND FUZZY RELATIONS

A deterministic relation between n variables can be given through a crisp subset R of the Cartesian product $\mathbf{X} = X_1 \times X_2 \times ...X_n$, corresponding to the compatible values. When two deterministic relations are given through the subsets R_1 and R_2, R_1 is included by R_2 ($R_1 \subseteq R_2$) if every singleton of R_1 belongs also to R_2. Moreover:

$$R_1 \subseteq R_2 \ \rightarrow \quad (R_1 \cup R_2 = R_2 \ ; \ R_1 \cap R_2 = R_1 \ ) \qquad (2.15)$$

Clearly, when non-specific uncertainty is treated through set theory, inclusion can be used to perform analysis of the response *on the safe side*: as an example, if a limit state is respected when the data for the system variables are in R_2, it is surely respected when the actual data are in $R_1 \subseteq R_2$.

The inclusion operator can be extended to random or fuzzy sets and relations.

For consonant random sets (i.e. normalised fuzzy relations) the extension was suggested by Zadeh through the simple rule:

$$R_1 \subseteq R_2 \quad \Leftrightarrow \quad \forall\, x \in X : \mu_1(x) \le \mu_2(x) \tag{2.16}$$

When the standard definitions suggested by Zadeh (relations 1.21a, b) are assumed as operators of union and intersection, it is possible to observe that:

$$R_1 \cup R_2 = \; \cup \max(\mu_1(x), \mu_2(x))/x \; = \; R_2 \tag{2.17a}$$
$$R_1 \cap R_2 = \; \cup \min(\mu_1(x), \mu_2(x))/x \; = \; R_1 \tag{2.17b}$$

This coincidence is generally lost when other definitions [17] are given as union and intersection operators. As an example, using the *probabilistic* dual operators:

$$R_1 \cup R_2 = \; \cup \; (\mu_1(x) + \mu_2(x) - \mu_1(x) \cdot \mu_2(x))/x \; \supseteq R_2 \tag{2.18a}$$
$$R_1 \cap R_2 = \; \cup \; (\mu_1(x) \cdot \mu_2(x))/x \qquad\qquad \subseteq R_1 \tag{2.18b}$$

In the case of non-consonant structures, the definition of inclusion is quite more complex, requiring the comparison of every pair of focal elements of the two random sets. Precisely, the definition of inclusion of two random sets $\Im_1$, $\Im_2$ on X ([18], [19]) is obtained by means of a generalisation of inclusion of fuzzy sets :

$$\Im_1(A_{i1}, m_1) \subseteq \Im_2(A_{j2}, m_2) \quad \Leftrightarrow \quad m_1(A_{i1}) = \; \Sigma_j \; (w(A_{i1}, A_{j2}) \tag{2.19a}$$
$$m_2(A_{j2}) = \; \Sigma_i \; (w(A_{i1}, A_{j2})$$

where w are the entries of a set-valued assignment matrix satisfying the conditions:

$$w(A_{i1}, A_{j2}) > 0 \;\; \text{if} \;\; A_{i1} \subseteq A_{j2} , \quad \text{else} \quad w(A_{i1}, A_{j2}) = 0 ;$$

$$\tag{2.19b}$$

$$\Sigma_{i,j} (w(A_{i1}, A_{j2}) = 1$$

Example 2.1

	A_{12}	A_{22}	A_{32}	Total $m_1(A_{i1})$
$A_{11} \subseteq (A_{12}, A_{22}, A_{32})$	0.1	0.1	0.3	0.5
$A_{21} \subseteq A_{32}$	0	0	0.5	0.5
Total : $m_1(A_{i2})$	0.1	0.1	0.8	1

In the case of consonant random sets, the relations (2.19a, b) correspond to Zadeh's definition of included fuzzy sets (as can be easily verified in the numerical example when both the subsets A_{i1} and A_{j2} are supposed nested : $A_{11} \subseteq A_{21}$; $A_{12} \subseteq A_{22} \subseteq A_{32}$).

In any case, both for consonant and non-consonant random sets the following theorem can be demonstrated [19]:

$$\mathfrak{I}_1 \subseteq \mathfrak{I}_2 \;\Rightarrow\; \forall\, B \subseteq X : [Bel_1(B), Pl_1(B)] \subseteq [Bel_2(B), Pl_2(B)] \qquad (2.20)$$

i.e. :
$$Bel_1(B) \geq Bel_2(B) ; \;\; Pl_1(B) \leq Pl_2(B) \qquad (2.21)$$
and finally :
$$Bel_2(B) \leq Pr(B) \leq Pl_2(B) \qquad (2.22)$$

Therefore : when a random set can be evaluated by including the random set corresponding to the actual data, it is possible to determine (on the safe side) upper and lower bounds on the probability of every subset (event) in the same space.
Let us observe that, in the case of a probabilistic distribution (a particular non-consonant random set), the inclusion is trivial: the only inclusive probabilistic distribution is the same distribution. In fact, the focal elements of both the first and the second random sets are the disjoint singletons $\{x_i\}$, which have not-empty intersection with themselves only. The basic probability assignment matrix can only assume the pattern shown in Table 1.

	$\{x_1\}$	$\{x_2\}$	$\{x_3\}$	......	$\{x_n\}$	Total
$\{x_1\}$	w_{11}	0	0	......	0	$w_{11} = p_1(x_1)$
$\{x_2\}$	0	w_{22}	0	......	0	$w_{22} = p_1(x_2)$
$\{x_3\}$	0	0	w_{33}	......	0	$w_{33} = p_1(x_3)$
......	......	......	......	......	......	......
$\{x_n\}$	0	0	0	......	w_{nn}	$w_{nn} = p_1(x_n)$
Total	$w_{11} = p_2(x_1)$	$w_{22} = p_2(x_2)$	$w_{33} = p_2(x_3)$	......	$w_{nn} = p_2(x_n)$	1

Tab. 1 - Basic probability assignment matrix for two probability distributions $p_2(x)$ and $p_1(x)$.

Therefore, inclusion, in the sense here defined, cannot be useful in probability theory, because it is strongly related to non-specificity of information, while in probability uncertainty derives only by strife, as discussed in Chapter 1. Moreover:
a) It is possible to include probability distributions by means of non-consonant random sets, whose non-disjoint focal elements include the singletons, and therefore to find upper and lower bounds on the exact probability of every subset B through the Belief and the Plausibility of the same subset B calculated using the focal elements and basic probabilistic assignment of the inclusive random set. An interesting application of these results to the bracketing of Monte-Carlo simulation of a function of random variables is reported by Dubois and Prade [19].
b) It is always possible to include a non-consonant random set by means of a consonant one, i.e. a fuzzy set or relation with membership function $\mu(x)$ and therefore, for every subset B, to bound the interval [Bel(B), Pl(B)] (containing Pr(B)) through the inclusive interval (X finite set; for non-finite sets, the "sup" operator should be used):

$$[\text{Nec(B)} , \text{Pos(B)}] \quad = \quad [1 - \max_{x \in B^c} \mu(x) , \max_{x \in B} \mu(x)] \qquad (2.24)$$

2.4   EXTENSION PRINCIPLE

Let us return again to the basic problem proposed in § 1 : once the deterministic function f (a mathematical model of the system) is assigned and supposing that the uncertainty about the variables can be restricted on **X** by a random relation, generally a non-consonant random relation, we want to find the corresponding restriction extended to the dependent variable z on Z. Clearly, it will be a random set on Z. When its focal elements and basic probability assignment are known, it will be possible to calculate upper and lower bounds to the probability of every subset (event) of Z (e.g. upper and lower bound to the reliability or failure probability defined by relations (2.1)). Firstly we will consider the non-consonant case, and successively the consonant one.

2.4.1 Extension Principle for Random Sets

Let us consider a random relation defined on **X** by n focal elements R_i and basic probability assignment $m_i(R_i)$, $i = 1,2,..n$. Remembering the probabilistic meaning of the basic probability assignment, the probability m_i will be assigned to the subset A_i, image of R_i onto Z through the point-valued mapping f. Supposing that in general more than one focal element R_k can map on the same image A_i , the random set on Z is defined as follows:

$$A_i = f(R_k) = \{z = f(x) \mid x \in R_k\} ; \quad m(A_i) = \Sigma \; m_k \mid f(R_k) = A_i \qquad (2.25)$$

This relation can be easily extended to the case of a set-valued function G mapping from **X** to the set of the subsets of Z (the power set of Z : **P** $\{Z\}$):

$$G : \mathbf{X} \rightarrow \mathbf{P} \{Z\} \qquad (2.26)$$

$$A_i = \quad \cup \; Z_x = G(x) \mid \; x \in R_i \qquad (2.27)$$

Therefore, the focal element A_i is the (generally not convex) union of the subsets Z_x mapped from each point **x** of R_i through the set-valued function G.

Example 2.2

Let us assume: $X_1 = X_2 = \{1, 2, 3\}$; $Z = N$; $G(x) = \text{int} [(x_1 \cdot x_2)^2 - 1 , (x_1 \cdot x_2)^2 + 1]$

$R_1 = \{1, 2, 3\} \times \{1\}$; $m(R_1) = 0.6$

$R_2 = \{2\} \times \{1, 2\}$; $m(R_2) = 0.4$

Hence: $A_1 = [0, 2] \cup [3, 5] \cup [8, 10]$; $A_2 = [3, 5] \cup [15, 17]$

In Fig. 2.2 the contour function on Z (i.e. the Plausibility of the singletons) is displayed.

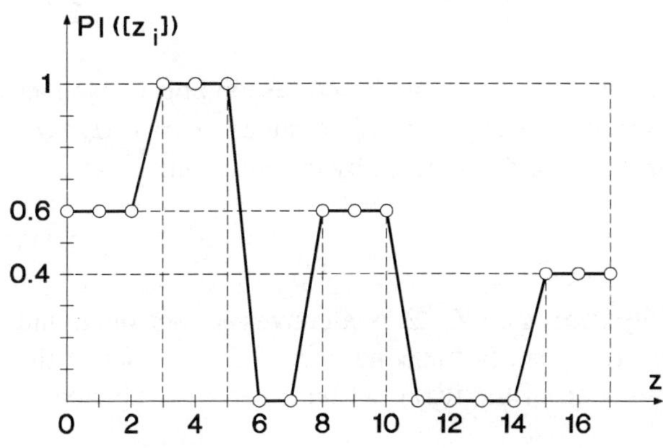

Fig. 2.2 Contour function of the random set induced on Z with the data of the Example 2.2.

The following fundamental theorem can be considered as the extension to the random sets of the theorems on which the set approach to uncertainty (e.g. interval analysis) is based.

Given a continuous point-valued or set-valued function G , and two random relations R_1 , R_2 on **X** mapping to the random sets $G(R_1)$ and $G(R_2)$ respectively:

$$R_1 \subseteq R_2 \quad \Rightarrow \quad G(R_1) \subseteq G(R_2) \tag{2.28}$$

where the inclusion operator has the meaning defined in § 3.

Combining this theorem with the theorem given in § 3, upper (Pl(B)) and lower (Bel(B)) bounds on the safe side of the probability of every subset (event) B on Z can be found through the extension principle, based on a consonant random relation including the actual information.

2.4.2 Extension Principle for Fuzzy Relations

The extension principle leads to particularly simple and robust procedures when the available data correspond to a consonant random set , i.e. a normalised fuzzy relation on X; or at least, on the safe side, when the available data can be included in a consonant structure. The resulting random set on Z is consonant, i.e. it is the fuzzy set whose membership function is given by the following equation:

$$\mu(z) = Pl(\{z\}) = \begin{cases} \max \quad \mu(x) \mid x \in A(z) = G^{-1}(z) \\ \\ 0 \quad \text{if } A(z) = \varnothing \end{cases} \tag{2.29}$$

It is very important to observe that this equation can be numerically evaluated by employing the α-cuts of the normal fuzzy relations R, as defined in § 1.3:

$$^{\alpha}R = \{x \in \mathbf{X} \mid \mu(x) \geq \alpha\} \tag{2.30}$$

As shown above, the α-cuts are the focal elements of the corresponding consonant random relation. Therefore, for a chosen decreasing sequence of α values : $\alpha_1 = 1$, $\alpha_2 < \alpha_1$, ..., $\alpha_{n+1} = 0$, the α-cuts of the induced fuzzy set on Z are given by the equation:

$$^{\alpha}Z = G(^{\alpha}R) \tag{2.31}$$

If f is a continuous point-valued mapping from X to Z, $^{\alpha}Z$ is a convex subset on Z and therefore the fuzzy set is a normal convex fuzzy set (a fuzzy number if Z is the set of the real numbers). Finally, the membership function of the fuzzy set on Z can be evaluated as shown in § 1.3, equation (1.23).

For non finite sets the α-cuts should be substituted by the strong α-cuts [17]:

$$^{\alpha^+}R = \{x \in \mathbf{X} \mid \mu(x) > \alpha\} \tag{2.32}$$

and the sup operator should be used in (1.23), and (2.29).

Example 2.3
Let us consider a "consonant version" of the random set used in numerical example 2.2:

$$X_1 = X_2 = \{1, 2, 3\} \; ; \qquad X = X_1 \times X_2; \quad G(\mathbf{x}) = \text{int} \, [\, (x_1 \cdot x_2)^2 - 1 \, , \, (x_1 \cdot x_2)^2 + 1 \,]$$

$$R_1 = \{1, 2, 3\} \times \{1, 2\} \; ; \qquad m(R_1) = 0.6$$

$$R_2 = \{2\} \times \{1, 2\} \; ; \qquad m(R_2) = 0.4$$

We can immediately calculate the membership function of the normalised fuzzy relation on X corresponding to the consonant random set:

$$\mu(1,1) = m(R_1) = 0.6; \qquad \mu(1,2) = m(R_1) = 0.6; \qquad \mu(2,1) = m(R_1) + m(R_2) = 0.6 + 0.4 = 1;$$

$$\mu(2,2) = m(R_1) + m(R_2) = 0.6 + 0.4 = 1; \qquad \mu(3,1) = m(R_1) = 0.6; \qquad \mu(3,2) = m(R_1) = 0.6;$$

The α-cuts are obviously coincident with the focal elements of the consonant random set :

$$^{1}R = \{x \in \mathbf{X} \mid \mu(x) \geq 1\} \; = \{(2,1), (2,2)\} \qquad\qquad = R_2$$

$$^{0.6}R = \{x \in \mathbf{X} \mid \mu(x) \geq 0.6\} = \{(1,1), (1,2), (2,1), (2,2), (3,1), (3,2)\} \; = R_1$$

The fuzzy set induced by G onto Z is defined by its α-cuts (eq. 2.31) ; for $\alpha = 1$, we get :

$$^{1}Z = G(^{1}R) = [3,5] \cup [15,17]$$

For $\alpha = 0.6$, we get :

$$^{0.6}Z = G(^{0.6}R) = [0,2] \cup [3,5] \cup [3,5] \cup [15,17] \cup [8,10] \cup [35,37] = [0,2] \cup [3,5] \cup [8,10] \cup [15,17] \cup [35,37]$$

This result is coincident with (but much more efficient than) the one obtainable through the application of (2.29) ; for example:

for z=10, we have : $\mu(10) = \max(\mu(3,1), \mu(1,3)) = \max(0.6, 0) = 0.6$,   and, in fact, $z = 10 \in {}^{0.6}Z$;

for z=5 : $\mu(5) = \max(\mu(1,2), \mu(2,1)) = \max(0.6, 1) = 1$,   and, in fact, $z = 5 \in {}^{1}Z$.

2.4.3 Extension Principle for Cartesian Fuzzy Relations . Vertex Method

Particularly simple and frequently used in the applications is the extension of Cartesian consonant random sets through a point-valued function f . In this case, taking into account (2.14), equation (2.29) can be written as follows:

$$\mu\,(z) = Pl\,(\{z\}) = \begin{cases} \sup\ \min\,(\,\mu\,(x_1),\,\mu\,(x_2),....\,\mu\,(x_n)\,)\,)\mid\mathbf{x}\in\ A(z) = f^{-1}\,(z) \\ 0 \quad\text{if}\ \ A(z) = \varnothing \end{cases} \tag{2.33}$$

and equation (2.33) can be reduced to the easier interval analysis problem:

$$^{\alpha+}Z = [\,z_L{}^{\alpha}\,,\,z_R{}^{\alpha}\,] = f\,(^{\alpha+}X_1,\ ^{\alpha+}X_2,\ ...,\ ^{\alpha+}X_n) \tag{2.34}$$

where $^{\alpha+}X_1,\ ^{\alpha+}X_2\,,\ ...,\ ^{\alpha+}X_n$ are the strong α-cuts of the fuzzy sets $X_1\,,\ X_2\,,\ ...,\ X_n$ with membership functions $\mu\,(x_1),\,\mu\,(x_2),....\,\mu\,(x_n)$.

Particularly interesting is the case when the function f is of class C^1 (i.e. continuous with its first partial derivatives) and strictly monotone (increasing or decreasing) with respect to every variables $x_1, x_2, ...x_n$ in their domains $^{\alpha+}X_1,\,^{\alpha+}X_2\,,\ ...,\ ^{\alpha+}X_n$. In this case the Vertex Method [20] can be used to solve (2.34), i.e. $z_L{}^{\alpha}\,,\,z_R{}^{\alpha}$ are respectively the minimum and the maximum value of the function f in the 2^n vertexes of the n-dimensional box $^{\alpha+}X_1$ x $^{\alpha+}X_2$ x ...x $^{\alpha+}X_n$. Moreover, let us denote with $x_j{}^{\alpha}$, j = 1, 2 , ... 2^n the vertices. It is possible to demonstrate [21] that the indexes L and R of the vertexes yielding respectively the minimum and the maximum of f are invariant when the values of α are changed:

$$[\,z_L{}^{\alpha}\,,\,z_R{}^{\alpha}\,] = [\,f\,(x_L{}^{\alpha}) = \min_j x_j{}^{\alpha}\ \,,\,f\,(x_R{}^{\alpha}) = \max_j x_j{}^{\alpha}\,]\ (\,j = 1, 2, ...\,2^n\ ) \tag{2.35}$$

By combining these two theorems, the computational effort requested to evaluate (2.34) is drastically reduced. It is enough :

(i) to evaluate the function f in the 2^n vertexes for a selected value of α (e.g. for $\alpha = 0$),

(ii) determine from the minimum and the maximum values obtained the critical indexes j = L and j= R, and then

(iii) evaluate the function in the vertexes $z_L{}^{\alpha}$ and $z_R{}^{\alpha}$ for every value of α (or a selected list of values of α if α assumes too many or infinite distinct values).

3.   WHY DO WE NEED FUZZY LOGIC?

3.1 INTRODUCTION

The following is a basic problem of any theoretical model of uncertainty and of any consistent theory of information. Suppose: (i) a certain level of information be known or assumed about a system (premise A) AND (ii) further information be obtained independently (i.e. without utilising previous information) (premise B) ; how can we combine the two items of knowledge in order to achieve an updated, *a posteriori* description embracing all available information?

The solution to this problem is an essential one if we want, for example: (i) to construct rule-based expert systems, i.e. based on deductive logic; (ii) to extend to uncertain information the classic deductive reasoning rules such as modus ponens, modus tollens and hypothetical syllogism; (iii) to construct mathematical modelling of approximate reasoning.

In the paragraphs we will examine the to date available solutions pertaining to the ambit of classical set theory of uncertainty (§ 3.2) and probability theory (§ 3.3); we will highlight and discuss their applicability limits(§ 3.4) and will introduce the available answers in the ambit of evidence theory (§ 3.5) and fuzzy set theory (§ 3.6).

3.2 SET THEORY

The answer to the posed problem is quite simple if sought in the field of convex modelling of uncertainty or generally in the field of classic Set Theory. If A AND B correspond to distinct solutions to the two anti-optimisation problems based on *a priori* and further information respectively (i.e. A and B provide conditions which *necessarily* are to be satisfied by their combination), *a posteriori* uncertainty is described by their intersection:

$$C = A \cap B \tag{3.1}$$

Therefore, the combination of the two items of knowledge reduces uncertainty whenever A and B do not coincide.

If, on the other side, A AND B are weaker conditions, i.e. they are just indications of *possibility* which exclude that the true solution *necessarily* belongs to the complementary set, then uncertainty of the combination of non-coincident sets enlarges. In fact, based on De Morgan's Law, we have:

$$C = (A^c \cap B^c)^c = A \cup B \tag{3.2}$$

3.3 PROBABILITY THEORY

In the ambit of probabilistic modelling of uncertainty, update follows from the assiomatic definition of conditioned probability of two events A and B in a probability space:

$$Pr\ (\ A\ /\ B\ ) = \ Pr\ (A \cap B\ )\ /\ Pr\ (B) \tag{3.3}$$

and from the Bayesian techniques. A simple example will clarify how these techniques operate.

Example 3.1

Let us consider the population composed of different portions of concrete usually employed for structural purposes. For this population, joint probability distributions are supposed to be known:
a) between a variable relative to mixture composition $c \in C$ (for example the water/cement ratio, in weight) and a variable relative to mechanical resistance $r \in R$ (for example the cubic characteristic resistance after 28 days);
b) between the result $s \in S$ from a non destructive test (e.g. a sclerometric test s) and the resistance r.
For the sake of simplicity, we assume:

$$C = \{c_1, c_2, c_3, c_4\}\ ;\quad R = \{r_1, r_2, r_3, r_4\}\ ;\qquad S = \{s_1, s_2, s_3, s_4\}$$

and uniform marginal distributions for each variable, i.e. $p(c_i) = p(r_i) = p(s_i) = 0.25$, $i = 1,...,4$.
The joint probability distributions are defined through the conditioned probability distribution shown in Tab.1 as far as the relation between c and r is concerned, and from the following relations as far as the relation between s and r is concerned:

	p(r/c₁)	p(r/c₂)	p(r/c₃)	p(r/c₄)
r_1	0.72	0.12	0.04	0
r_2	0.24	0.60	0.24	0.04
r_3	0.04	0.24	0.60	0.24
r_4	0	0.04	0.12	0.72
	1.00	**1.00**	**1.00**	**1.00**

Tab. 1

$p(r_i/s_i)\ = 0.6;\ \ p(r_{i+1}/s_i) = 0.2\ \ i = 1, 2, 3$
$p(r_i/s_i)\ = 0.8\qquad\qquad\qquad i = 1, 4$
$p(r_{i-1}/s_i) = 0.2\qquad\qquad\qquad i = 2, 3, 4$

The joint probability distributions $p(r, c)$ and $p(r, s)$ can be calculated by means of (3.3):

$$p(r, c) = p(r\ /c)\ p(c)\quad ;\quad p(r, s) = p(r\ /s)\ p(s)$$

as indicated in the following Tables 2.a and 2.b:

	p(r, c₁)	p(r, c₂)	p(r, c₃)	p(r, c₄)	p(r)
r_1	0.18	0.03	0.01	0	**0.25**
r_2	0.06	0.15	0.06	0.01	**0.25**
r_3	0.01	0.06	0.15	0.06	**0.25**
r_4	0	0.01	0.03	0.18	**0.25**
p(c)	**0.25**	**0.25**	**0.25**	**0.25**	**1.00**

Tab. 2a

	p(r, s₁)	p(r, s₂)	p(r, s₃)	p(r, s₄)	p(r)
r_1	0.20	0.05	0	0	**0.25**
r_2	0.05	0.15	0.05	0	**0.25**
r_3	0	0.05	0.15	0.05	**0.25**
r_4	0	0	0.05	0.20	**0.25**
p(s)	**0.25**	**0.25**	**0.25**	**0.25**	**1.00**

Tab. 2b

In their turn, these distributions $p(r, c)$ and $p(r, s)$ can be seen as marginal distributions in the plane (CxR) and (SxR) respectively of a joint probability distribution in the space (CxRxS) defined by the relation:

$$p(r, s, c) = p(r, s)\, p(c/r) = p(r, c)\, p(s/r) = p(r, s)\, p(r, c) / p(r) \qquad (3.4)$$

Let us suppose now that we get a new deterministic information about the exact value of one of the variables, e.g. $c = c^*$. We immediately conclude that:

$$p(r, s / c = c^*) = p(r, s, c^*) / p(c^*) \qquad (3.5)$$

and moreover the updated distributions for r and s are the new marginal ones.

Example 3.2

For $c=c_2$ the joint probability distribution evaluated from the data in Tables 2a and 2b is shown in Table 3.

	$p(r, s_1, c_2)$	$p(r, s_2, c_2)$	$p(r, s_3, c_2)$	$p(r, s_4, c_2)$	$p(r, c_2)$
r_1	0.024	0.006	0	0	**0.030**
r_2	0.030	0.090	0.030	0	**0.150**
r_3	0	0.012	0.036	0.012	**0.060**
r_4	0	0	0.002	0.008	**0.010**
$p(s, c_2)$	**0.054**	**0.108**	**0.068**	**0.020**	**0.25 = p(c_2)**

Tab. 3

Suppose now that we get the deterministic information that $c = c_2$; the marginal distributions, relative to r and s respectively, are modified as indicated in the last row and column in Table 4:

	$p(r, s_1 / c_2)$	$p(r, s_2 / c_2)$	$p(r, s_3 / c_2)$	$p(r, s_4 / c_2)$	$p(r / c_2)$
r_1	0.096	0.024	0	0	**0.120**
r_2	0.120	0.360	0.120	0	**0.600**
r_3	0	0.048	0.144	0.048	**0.240**
r_4	0	0	0.008	0.032	**0.040**
$p(s / c_2)$	**0.216**	**0.432**	**0.272**	**0.080**	**1.00**

Tab. 4

If we get further information $s = s_3$, we obtain: $p(r / s_3, c_2) = p(r, s_3 / c_2) / p(s_3 / c_2)$

	$p(r / s_3, c_2)$
r_1	0
r_2	0.120 / 0.272 = 0.441
r_3	0.144 / 0.272 = 0.530
r_4	0.008 / 0.272 = 0.029
	0.272 / 0.272 = 1.000

Tab. 5

It is possible to observe that the extension of a probabilistic measure through a point-valued deterministic mapping f (relation 1.7) can be considered as a particular application of the Bayes theorem: in both cases, we have to combine probabilistic *a priori* information with deterministic information.

In fact, let a joint probability distribution $p(\mathbf{x}) = p(x_1, x_2..)$ of the variables $x_1, x_2,$ $...x_n$ be given in the space defined by the Cartesian product $\mathbf{X} = X_1 \times X_2 \times ...X_n$. The mapping $y=f(\mathbf{x})$, whose range is Y, *extends* to the space $\mathbf{X} \times Y$ the joint probability distribution:

$$p(\mathbf{x}, y) = \begin{cases} p(\mathbf{x}) & \text{if } x = f^{-1}(y) \\ 0 & \text{else} \end{cases} \tag{3.6}$$

and then the marginal probability on Y reads:

$$p(y) = \Sigma \; p(\mathbf{x}, y) \mid x = f^{-1}(y) \tag{3.7}$$

Example 3.3

Let us suppose now that the relation between c and r in Example 3.1 is deterministic: it can be given by a classical subset of the plane CxR, which can be represented through table 6 that reports the values of the characteristic function χ (r, c).

	$\chi(r, c_1)$	$\chi(r, c_2)$	$\chi(r, c_3)$	$\chi(r, c_4)$
r_1	1	0	0	0
r_2	0	1	0	0
r_3	0	0	1	0
r_4	0	0	0	1
	1	1	1	1

Tab. 6

	$p(r, c_1)$	$p(r, c_2)$	$p(r, c_3)$	$p(r, c_4)$	$p(r)$
r_1	0.25	0	0	0	0.25
r_2	0	0.25	0	0	0.25
r_3	0	0	0.25	0	0.25
r_4	0	0	0	0.25	0.25
$p(s)$	0.25	0.25	0.25	0.25	1.00

Tab. 7

In probabilistic terms, as the marginal distributions are supposed not to be modified and the proposed relation of Table 6 is a one to one mapping (just one term equal to 1 in each row and each column), the hypothesis made is equivalent to assume the joint probability distribution displayed in Table 7.

If the relation between s and r is left unchanged as in Table 2.b, and the deterministic information $c = c_2$ is gained, we obtain immediately:

	$p(r, s_1 / c_2)$	$p(r, s_2 / c_2)$	$p(r, s_3 / c_2)$	$p(r, s_4 / c_2)$	$p(r / c_2)$
r_1	0	0.0	0	0	0
r_2	0.20	0.60	0.20	0	1.00
r_3	0	0	0	0	0
r_4	0	0	0	0	0
$p(s / c_2)$	0.20	0.60	0.20	0	1.00

Tab. 8

The marginal distribution of r in Tab. 8 indicates that $c=c_2$ implies necessarily that $r=r_2$; on the other side, the marginal distribution of s coincides with the *a posteriori* distribution expected for the result of the sclerometric test. In general, if c and r are linked through a one to one mapping $c=f(r)$ and (r, s) are variables whose joint probability distribution $p(r, s)$ is known, we get:

$$p(s/c) = p(r = f^{-1}(c), s)/p(c)$$

If the mapping $c=f(r)$ is not a one to one mapping, the inverse is a multivalued mapping and we obtain, according to 3.7:

$$p(s/c) = \Sigma \; p(r, s) \mid r = f^{-1}(c)/p(c)$$

3.4 LIMITS ENTAILED BY THE PROBABILISTIC SOLUTION

The above discussion together with the results exposed in the previous §§ 1, 2 highlights some limitations and possible extensions of the Bayesian approach emerging in the following cases:

1) the extension is given through a set-valued mapping $G : \mathbf{X} \rightarrow P\{Y\}$; therefore, it associates every $\mathbf{x} \in \mathbf{X}$ with a subset $A \subseteq Y$. In this case, a (generally non-consonant) random set measure is induced on Y and consequently, it is possible to calculate the interval [Bel (B), Pl(B)] containing $Pr(B)$ for each $B \subseteq Y$ (see § 1.2.3);

2) the variables $(x1, x2..)$ are linked by a joint random relation. The extension principle presented earlier (§ 2.4), allows one to determine the corresponding random set measure on Y and then, for each $B \subseteq Y$, the interval [Bel (B), Pl(B)] containing $Pr(B)$. Some particular cases are of interest:

2.a) the random relation on $\mathbf{X}$ is consonant (and then it is completely defined by the possibility distribution $\pi(\mathbf{x}) = \mu(\mathbf{x})$) : the random set induced on Y is consonant too and its possibility distribution $\pi(y) = \mu(y)$ is defined by the relation (2.29);

2.b) the random relation on $\mathbf{X}$ is a Cartesian random relation (i.e. the focal elements are a nested succession of Cartesian products of intervals, each of which pertaining to one variable only) : the variables are non-interactive (§ 2.2.3) and their extension is given by relation (2.33).

In an alternative and more useful manner formula (2.29) can be expressed representing the deterministic relation   by means of a characteristic function χ_B on the space $\mathbf{X} \times \mathbf{Y}$ (as in Example 3.3):

$$\mu\,(y) = \sup_{X} \min\,(\mu_A\,(\mathbf{x}),\, \chi_B\,(\mathbf{x},\, y)) \tag{3.8}$$

Otherwise relation (3.8) can be derived in a form quite similar to the Bayesian procedure: we can write the *cylindrical extension* of $\mu_A\,(\mathbf{x}\,)$ on the space $\mathbf{X} \times \mathbf{Y}$:

$$\mu_A\,(\mathbf{x},\, y) = \mu_A\,(\mathbf{x}\,) \tag{3.9}$$

The combination in this space is expressed by the rule:

$$\mu_C\,(\mathbf{x},\, y) = \min\,(\mu_A\,(\mathbf{x},\, y),\, \chi_B\,(\mathbf{x},\, y\,)\,) \tag{3.10}$$

and then $\mu(y)$ is nothing else but the *projection* (see relation 2.12) of $\mu_C\,(\mathbf{x},\, y)$ on Y (analogous to the Bayesian case of marginal distribution on Y):

$$\mu\,(y) = \sup_{X} \mu_C\,(\mathbf{x},\, y) \tag{3.11}$$

3) The situation is quite more complex when function G or f is not deterministic, because the problem goes well beyond Bayesian formulation. Now we are facing the combination of two kinds of information, being both uncertain and derived by two distinct bodies of evidence: one relative to the independent variables and the other relative to the function which relates independent variables to a dependent variable. More generally, in probabilistic terms, we have to combine information which yields two distinct probability distributions on the same space. This topic leads us to the core of the problems from which both Shafer's Evidence Theory and Zadeh's Fuzzy Set Theory originated. Both theories may be seen as attempts to solve this problem.

3.5 THE DEMPSTER'S RULE OF COMBINATION

Example 3.4

Let us go back to the initial example and suppose that both sclerometric and mix composition tests, performed independently (i.e. in a non joint way), yield evidences about concrete resistance r in the form of the following (non consonant) focal elements and probability assignments:

$$A: \quad m(A_1 = \{r_1, r_2\}) = 0.2 \,;\, m(A_2 = \{r_1, r_2, r_3\}) = 0.7 \,;\, m(A_3 = \{r_2, r_3\}) = 0.1$$
$$B: \quad m(B_1 = \{r_1\}) = 0.05 \quad;\, m(B_2 = \{r_1, r_2\}) = 0.1 \quad ;\, m(B_3 = \{r_2, r_3, r_4\}) = 0.85$$

What conclusions can be drawn about any subset $C \subseteq R$, as far as $Pr(C)$ is concerned ? Or what conclusions can be drawn, at least, as far as the interval $[Bel(C), Pl(C)]$ containing $Pr(C)$ is concerned?

Shafer proposed to utilise the following rule, initially propounded by Dempster [10] :

$$m(C) = \sum_{A_i \cap B_j = C} m(A_i)\, m(B_j) / K \quad ; \quad K = 1 - \sum_{A_i \cap B_j = \varnothing} m(A_i)\, m(B_j) \qquad (3.12)$$

The result is a third random set, whose focal elements (according to eq. (3.1)) are the intersections of the initial focal elements and whose probability assignment is obtained as the product of the corresponding probabilities (following the hypothesis of independence by which they were initially evaluated), normalised in order to take into account that a part of initial evidence (non null products of probabilities) may focus on empty intersections.

In our numerical example we obtain:

$$
\begin{array}{llll}
m(A_1 \cap B_1) = & m(\{r_1\}) & = 0.2 \times 0.05 & = 0.01 \\
m(A_1 \cap B_2) = & m(\{r_1, r_2\}) & = 0.2 \times 0.1 & = 0.02 \\
m(A_1 \cap B_3) = & m(\{r_2\}) & = 0.2 \times 0.85 & = 0.17 \\
m(A_2 \cap B_1) = & m(\{r_1\}) & = 0.7 \times 0.05 & = 0.035 \\
m(A_2 \cap B_2) = & m(\{r_1, r_2\}) & = 0.7 \times 0.1 & = 0.07 \\
m(A_2 \cap B_3) = & m(\{r_2, r_3\}) & = 0.7 \times 0.85 & = 0.595 \\
m(A_3 \cap B_1) = & m(\varnothing) & = 0.1 \times 0.05 & = 0.005 \\
m(A_3 \cap B_2) = & m(\{r_2\}) & = 0.1 \times 0.1 & = 0.01 \\
m(A_3 \cap B_3) = & m(\{r_2, r_3\}) & = 0.1 \times 0.85 & = 0.085 \\
\end{array}
$$

and then : $K = 1 - 0.005 = 0.995$

$$
\begin{aligned}
m(\ C_1 &= \{r_1\}) &&= (0.01 + 0.035) / 0.995 &&= 0.0452 \\
m(\ C_2 &= \{r_1, r_2\}) &&= (0.02 + 0.07) \ / 0.995 &&= 0.0905 \\
m(\ C_3 &= \{r_2\}) &&= (0.17 + 0.01) \ / 0.995 &&= 0.1810 \\
m(\ C_4 &= \{r_2, r_3\}) &&= (0.595 + 0.085) / 0.995 &&= 0.6832
\end{aligned}
$$

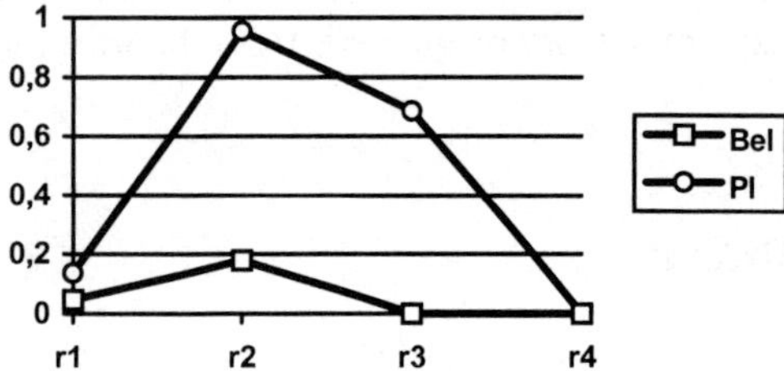

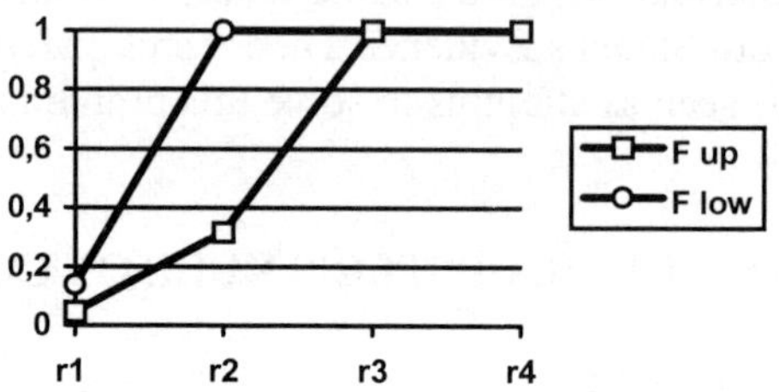

Fig. 3.1a -Plausibility and Belief of the singletons { r_i }   Fig. 3.1b - Upper and lower cumulative distribution functions

Fig. 3.1a and 3.1b portray the corresponding intervals in which the probabilities of the singletons and the cumulative probabilities are included.

Dempster's rule coincides exactly with Bayes' rule when the first body of evidence is probabilistic (the focal elements are all singletons) and the information conveyed by the second body of evidence is deterministic.

We can verify this fact in the above example, by assuming that the initial body of evidence is given by the joint probability distribution $p(c, r, s)$. The focal elements are nothing else but the 64 singletons with probability assignment $m(\ A_{ijk} \{(r_i, s_j, c_k)\ \}\ ) = p\ (r_i, s_j, c_k)$; while deterministic information is defined by the unique focal element:

$$m(\ B = \{r_1, r_2, r_3, r_4\ \} \times \{s_3\ \} \times \{c_2\ \}) = 1$$

The non empty intersections are just the four focal elements, with probability (Tab. 3, column 3):

$$m\ (C_i = \{\ (r_i, s_3, c_2)\ \} \cap B = \{(r_i, s_3, c_2)\ \}\ ) = p(r_i, s_3, c_2)$$

The empty intersections are the remaining 60 focal elements, whose probability summation is obtained as 1 minus the sum of the probabilities of the previous four focal elements. Therefore it yields simply:

$$K = 1 - (\ 1 - 0.0680) = 0.0680$$

Thus, after *a posteriori* information, the probability assignment of the 4 focal elements assumes the values $m\ (\ C_i\ ) = p\ (r_i / s_3, c_2)$ of Table 9, exactly coinciding with those of Tab. 5.

	$m\ (C_i) = p(r /s_3 ,c_2)$
r_1	0
r_2	0.030 / 0.068 = 0.441
r_3	0.036 / 0.068 = 0.530
r_4	0.002 / 0.068 = 0.029
	0.068 / 0.068 = 1.000

Tab. 9

As a comment about Dempster's rule, the following can be observed:

a) its first proposal dates back to the 18th century and precisely to Lambert's speculation about the Theory of Chances [22]. In an independent way, it was propounded by Dempster in 1967 and since then posed as a basis of Shafer's Theory of Evidence, whose research program explicitly set out from a reflection about the limits of Bayes' Theory:
"In the Bayesian theory, the task of telling how our degrees of belief ought to change as new evidence is obtained falls to Bayes' rule of conditioning: we represent the new evidence as a proposition and condition our prior Bayesian belief function on that proposition. Here we find no obvious symmetry in the treatment of new and old evidence. And more importantly, we find that the assimilation of new evidence depends on an astonishing assumption: we must assume that the exact and full effect of that new evidence is to establish a single proposition with certainty. In contrast to Dempster's rule of combination, which can accommodate new evidence that justifies only partial beliefs, Bayes' rule of conditioning requires *that the new evidence be expressible as a certainty"* [11, pp. 25-26);

b) it was not free from criticism deriving especially from the observation that independence of bodies of knowledge does not necessarily imply stochastic independence. Some paradoxes have been worked out by applying Dempster's rule and alternative solutions have been proposed, even if frequently less simple and effective. Readers interested in this issue can find useful reference in [23], where a complete Theory of Imprecise Probability is propounded which, in the Author's mind, overcomes the contradictions of Dempster's rule.

3. 6 THE FUZZY COMPOSITION RULE

Dempster's rule holds even in the case of consonant structures, i.e. when random sets or random relations can be completely defined by means of membership functions (possibility distributions) of fuzzy sets or fuzzy relations. However, the result of the combination of consonant structures is generally a non consonant structure.

Example 3.5

With reference to the previous example 3.4, let us consider the combination of the following consonant structures of data relative to the concrete resistance r , derived from information about mix composition and sclerometric tests respectively:

A : $m(A_1 = \{r_1, r_2, r_3, r_4\}) = 0.4$;
$\quad\quad m(A_2 = \{r_2, r_3\}) = 0.6$
B : $m(B_1 = \{r_2, r_3, r_4\}) = 0.2$;
$\quad\quad m(B_2 = \{r_3, r_4\}) = 0.8$

They correspond to fuzzy sets whose membership functions (possibility distributions) are displayed in Fig. 3.2. The combination is defined by the following focal elements and relative probability assignments, that clearly do not represent a consonant structure; in this case, it is simply K=1.

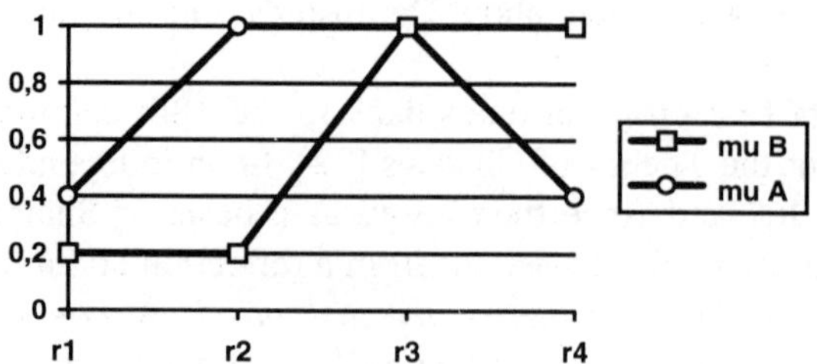

Fig. 3.2 - Possibility distributions of consonant random sets A and B.

$$m(A_1 \cap B_1 = B_1) \quad = \quad 0.4 \times 0.2 \quad = \quad 0.08$$
$$m(A_1 \cap B_2 = B_2) \quad = \quad 0.4 \times 0.8 \quad = \quad 0.32$$
$$m(A_2 \cap B_1 = A_2) \quad = \quad 0.6 \times 0.2 \quad = \quad 0.12$$
$$m(A_1 \cap B_2) = m(\{r_3\}) \quad = \quad 0.6 \times 0.8 \quad = \quad 0.48$$

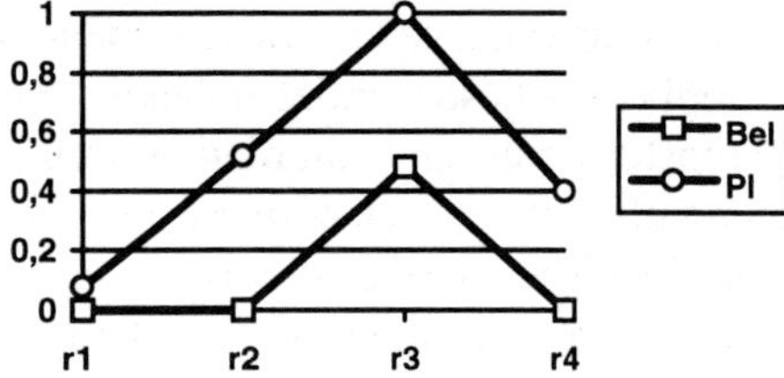

Fig. 3.3a - Plausibility and Belief of the singletons $\{r_i\}$

Fig. 3.3b - Upper and lower cumulative distribution functions

In Fig. 3.3a and 3.3b the values of Plausibility and Belief of the single values and, respectively, the upper and lower bounds of cumulative distributions are plotted.

A viable solution to this problem has been proposed by Yager [24], but under further restrictive conditions. Certainly, a solution is preferable that maintains consonance and allows the development of a theory of approximate reasoning completely within fuzzy set theory; this is even more valuable if one remembers that any non consonant structure can be enclosed into an approximate consonant structure.

From a historical point of view, the problem of the combination of uncertain information conveyed by fuzzy sets has been developed by Zadeh since the origins of fuzzy set theory. This has been performed in an autonomous way with respect to Bayesian procedures or procedures developed in the ambit of evidence theory, as pointed in § 1.3.

In the ambit of fuzzy logic, the fundamental problem relative to the combination of information pertaining to two distinct bodies of evidence has been frequently interpreted as the application of the union and complementation operations as extended to fuzzy sets, thus extending eq. (3.1) and (3.2) in the case A and B are two fuzzy sets.

But, since these operations are not uniquely defined within Fuzzy Set Theory (even when they are chosen in a dual way, i.e. respecting De Morgan's laws; see Klir, [17, p. 83), important difficulties arise about the interpretation of the results obtained. For this reason, sometimes fuzzy logic has been charged with incoherence, explicitly by scholars educated within the probabilistic paradigm.

It seems more natural to derive the composition rule for consonant structures as an "extension of the extension principle", i.e. as an extension of the rule of composition (see eq. 3.8) of information obtained from consonant structures (information A) with deterministic information (information B) conveyed by a point-valued function .

When information B is expressed by means of a consonant structure, the natural extension of eq. (3.10) is given by the following relation (*fuzzy composition rule*):

$$\mu C (x, y) = \min (\mu_A (x, y), \mu_B (x, y)) \tag{3.13}$$

Example 3.6

Let us suppose that information about a concrete mix is given by means of a fuzzy set whose values $\mu_A (c)$ are listed in Tab. 10. The relation between concrete resistance r and mix composition c is given by the fuzzy relation whose values $\mu_A (r, c)$ can be found in Tab. 11.

	$\mu (c_i)$
c_1	0.1
c_2	1
c_3	0.4
c_4	0

Tab. 10

	$\mu_A (r, c_1)$	$\mu_A (r, c_2)$	$\mu_A (r, c_3)$	$\mu_A (r, c_4)$
r_1	1	0.5	0	0
r_2	0.4	1	0.5	0.1
r_3	0.1	0.4	1	0.6
r_4	0	0	0.4	1

Tab. 11

By combining the two pieces of information, A and B, we obtain the consonant structure (fuzzy set) defined by the values $\mu_C (r) = \max \min (\mu_A (c), \mu_B (r, c))$, as shown in Tab.12.

	$\mu (r_i)$
r_1	max(min(1, 0.1), min (0.5,1).min(0,0.4) , min (0,0) = 0.5
r_2	1
r_3	0.4
r_4	0.4

Tab. 12

In the case of two fuzzy sets that induce the possibility distributions $\mu_A (x)$ e $\mu_B (x)$, gained in an independent way, on the same space X, the fuzzy composition rule is simply expressed as in (3.10) by the relation:

$$\mu C (x) = \min (\mu_A (x), \mu_B (x)) \tag{3.14}$$

which coincides with the standard definition of intersection $C = A \cap B$ of two fuzzy sets A and B (relation 1.21a) and therefore it is the extension of eq. (3.1). Moreover, with the standard definitions of complementation (1.21c) the extension of (3.2) is as follows:

$$\mu C (x) = 1 - \min(1 - \mu_A(x), 1 - \mu_B(x)) = \max(\mu_A(x), \mu_B(x)) \qquad (3.15)$$

This coincidence would be lost if different definitions of union, intersection and complementation (even if dual with respect to one another) are used in order to extend (3.1) and (3.2). Thus, the privileged role of the standard definitions initially given by Zadeh is strengthened in the ambit of a formulation obtained as a natural extension of Bayesian theory.

Example 3.7

With reference to the example of Fig.3.2 and previously solved by means of Dempster's rule, relation (3.14) gives the result depicted in Fig. 3.4.

This combination has the following consonant structure:

$$m(C_1 = \{r_1, r_2, r_3, r_4\}) = 0.2 ;$$
$$m(C_2 = \{r_3, r_4\}) = 0.2 ;$$
$$m(C_2 = \{r_3\}) = 0.6$$

whose probabilistic content is shown in Fig.3.5 and is therefore different from that of Fig. 3.3.

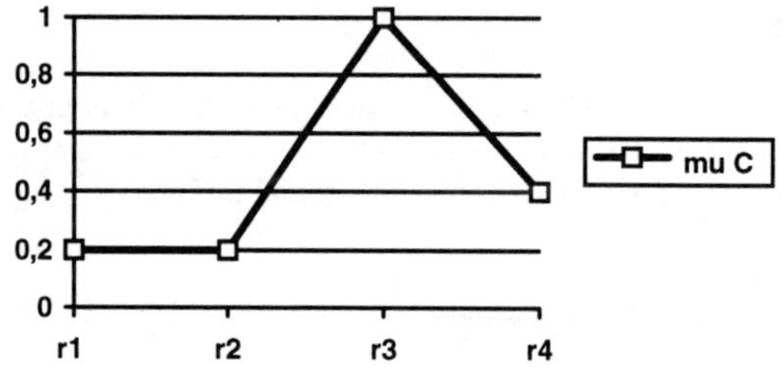

Fig. 3.4 - Possibility distribution of the combination of consonant random sets plotted in Fig. 2, obtained by means of the fuzzy composition rule

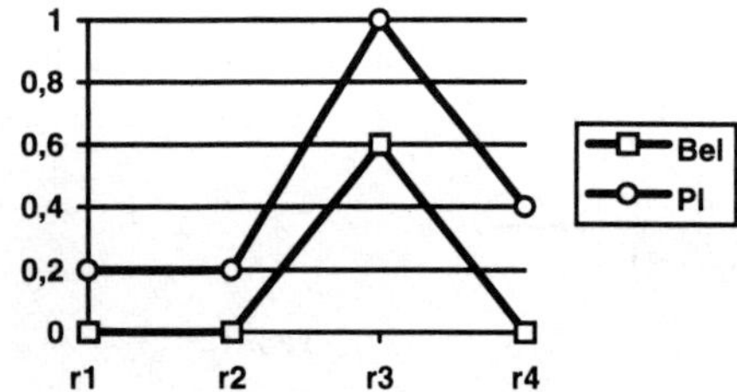

Fig. 3.5a - Plausibility and Belief of the singletons $\{r_i\}$

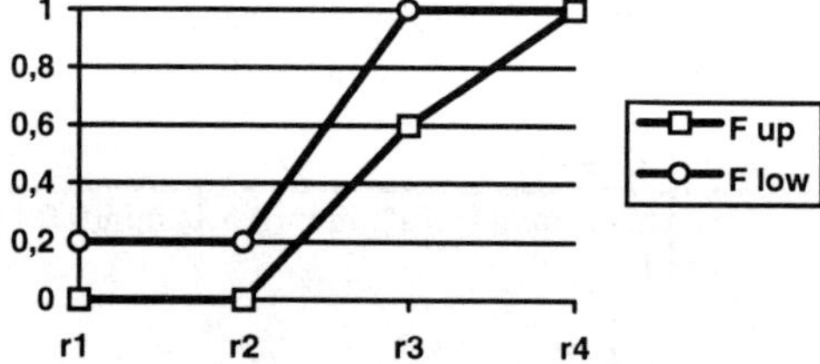

Fig.3.5b - Upper and lower cumulative distribution functions

It is worth noting that both the result obtained by means of the Dempster's rule and the result yielded by the fuzzy composition have to be considered reasonable representations of the evidence combined on the basis of the available information, but not actual frequentistic forecasts of objectively observable occurrences, even when such are the basic data to be combined.

4. HOW TO APPLY RANDOM AND FUZZY SETS IN STRUCTURAL ENGINEERING?

4.1 INTRODUCTION

Both the procedure of fuzzy calculus based on the extension principle (§ 2.4) and fuzzy logic (§ 3.6) have been applied by numerous researchers, and they are potentially of great interest when:
- the uncertainty about the parameters to be considered is very great, involving conflict and non-specificity of the available data;
- the considered system is very complex, and important factors must be taken into account, which can be described only in a qualitative manner, through the value (small, great, very great, etc..) assumed by linguistic variables.
- many constraints or objectives , generally contrasting each other, must be combined or balanced: for example, in the design of a structure, safety related objectives and economy related objectives.

Particularly frequent and interesting appear the applications to problems in geothecnical engineering (particularly the soil structure interactions [25]), earthquake engineering (as an example the design of aseismic structures, their protection with active or passive isolation systems, the evaluation of vulnerability and seismic risk of existent buildings [26, 27] or lifelines), transport engineering (the optimisation of transport networks and the modelling of human behaviours), environmental engineering (problems of pattern recognition and classifications).

The most promising applications seem to be the implementation of experts systems and the design of automatic procedures for on line control of machines and robots, whereas up to now the real applications to the classical problems of civil and structural engineering have been rare.

In the following two §§ 4, 5 some brief and absolutely incomplete information about the above mentioned applications will be given. Particularly in the present § 4 we are going to show applications of the procedures of fuzzy calculus; they seem interesting to analyse systems of which it is possible to give, with acceptable precision, a deterministic model, while some parameters appear to be extremely uncertain. For example in the field of the structural engineering, a substantially deterministic model of the structure is frequently possible, particularly when the behaviour remains in the elastic range, while a great uncertainty is related to the maximum loads during its expected service time.

4.2 STRUCTURAL RESPONSE TO UNCERTAIN LOADING

The evaluation of the   response to loading is a fundamental problem to be solved in structural analysis, particularly to determine the worst combination of the loads in safety analysis.

Generally the maximum intensity of every independent action is assigned by the Codes through nominal values, established by means of simplified probabilistic models, taking into account the expected life of the structure. However the available statistical data are frequently incomplete; more than one precise probability distribution of the maximum intensity or a wide class of possible probability distributions, could better describe the actual uncertainty, i.e. a random set derived through two alternative procedures:

   a) evaluating upper and lower bounds of probability distributions derived from different subsets of the same population of structures, and searching for a consonant or not consonant structure of focal elements and basic probabilistic assignment coherent with the observed bounds (Fig. 4.1b).

   b) combining statistical information of intervals of recorded intensities (Fig. 4.1a)

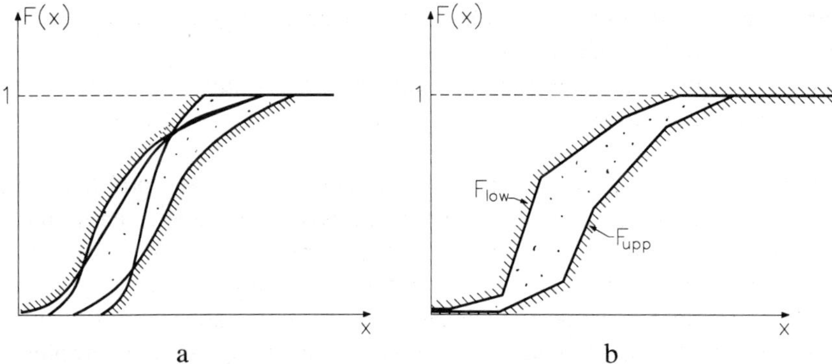

Fig. 4.1 Statistical derivation of a random set measure: a) from envelopes of probability distributions; b) from statistics of set-valued data.

In any case the random set so obtained can be included, on the safe side, in a consonant structure, i.e. can be described by a fuzzy set. It is reasonable to suppose stochastically independent (and therefore not-interactive) the fuzzy sets measuring the intensities of different actions or loading conditions.

We can observe that in many cases the constitutive laws of the engineering materials are strictly monotone increasing (or at least monotone increasing) up to the stress or the strain corresponding to the limit-state; so, when local buckling effects or softening behaviours are excluded, action effects (bending moments, axial loads, shear, displacements...) monotonically increase or decrease with the intensity of each loading condition . So the hypotheses indicated in § 2.4.3, are respected and the action effects due to any selected combinations of actions (e.g. the fundamental loading combinations indicated by the Eurocodes for the ultimate or serviceability limit states) can be measured by fuzzy sets derived through the simple above described procedure based on the vertex method. The simplicity can be appreciated contrasting the Monte Carlo sampling procedures generally required to solve the same problem in the probabilistic ambit.

Here the following steps can be used to find the solution, for a combination of n independent actions:

a- evaluate the action effect corresponding to the 2^n vertices of the n-dimensional box given by the Cartesian product of the α-cuts of the fuzzy sets restricting the load intensities, for any particular selected value of α (e.g. $\alpha = 0$);

b- evaluate the indexes R and L of the critical vertices corresponding to the maximum and minimum action effect;

c- calculate the action effects for a list of discrete values of α and the loading corresponding to the critical vertices with the previously determined indexes;

d- evaluate the membership function of the fuzzy set restricting the action effect through the decomposition theorem (relation 1.23).

Example 4.1

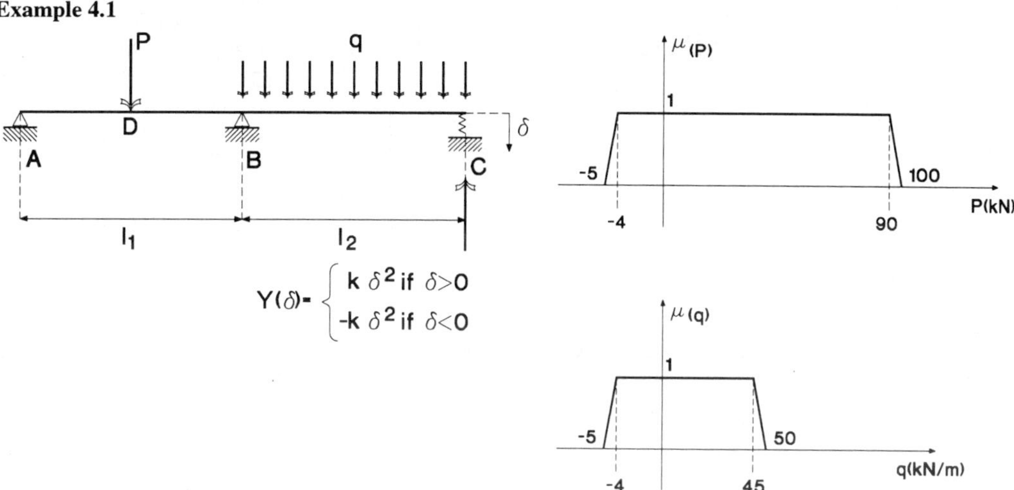

Fig. 4.2 - A two-span linearly elastic beam restrained by two rigid support and a non linear elastic spring.

Fig. 4.3 Fuzzy measures of the concentrated load and of the uniformly distributed load q.

Let us consider the non-linear elastic system shown in Fig. 4.2, a two-span linearly elastic beam, with equal Young modulus E, restrained by two rigid support and a non linear elastic spring whose stiffness monotonically increases with the displacements. The load intensities P and q are restricted by the fuzzy sets shown in Fig. 4.2b, and we want to determine the displacement of the supporting spring.

The displacement δ_C is related to the external forces P and q and to the spring reaction Y by means of a linear deterministic relation, whose coefficients depend on the elastic and geometrical properties of the beams ($l_1 = 5.2$ m , $l_2 = 4.3$ m ; $E = 27$ GPa ; $J_1 = 0.7^3 \cdot 0.4 / 12$ m^4 ; $J_2 = 0.4^3 \cdot 0.4 / 12$ m^4)

$$\delta_C = - D_1\, P + D_2\, q - D_3\, Y = \qquad f(q, P) - D_3\, Y \quad =$$

$$= \quad - 0.02354 \text{ mm/kN} \cdot P + 0.9651 \text{ mm/(kN/m)} \cdot q - 0.5639 \text{ mm/kN} \cdot Y \quad \text{mm}$$

$$D_1 = \frac{l_1^2 \cdot l_2}{16 \cdot E \cdot J_1} ; \quad D_2 = \frac{l_2^4}{8 \cdot E \cdot J_2} + \frac{l_1 \cdot l_2^3}{6 \cdot E \cdot J_1} ; \quad D_3 = \frac{l_2^3}{3 \cdot E \cdot J_2} + \frac{l_1 \cdot l_2^2}{3 \cdot E \cdot J_1} ;$$

while the condition of compatibility between the deformations of the beam and of the springgives the solution:

$$Y = \begin{cases} (c \cdot f + 1 - \sqrt{1 + 2 \cdot c \cdot f}\;) / (c \cdot D_3) & \text{if } f \geq 0 \\ (c \cdot f - 1 + \sqrt{1 - 2 \cdot c \cdot f}\;) / (c \cdot D_3) & \text{if } f < 0 \end{cases} \qquad c = 2 k D_3$$

In Table 1a the values of f corresponding to the $2^2 = 4$ vertices for $\alpha = 0$ (with $k = 31.4$ kN / mm^2) are listed : so L= 3 and R=2 are the critical vertices, because Y is monotone with f. The fuzzy set of Y reconstructed by its α-cuts is displayed in Table 1b.

index j	q	P	f(q, P)	Y	critical indexes
1	- 5	- 5	-4.708		
2	50	- 5	48.375	82.900	R
3	- 5	100	- 7.180	- 11.651	L
4	50	100	45.903		

Table 1a. Interval analysis of the reaction Y for $\alpha = 0$.

Table 1b. α-cuts of the reaction Y.

α-cut	Y_L	Y_R
0	- 11.651	82.900
0.1	- 11.448	82.055
0.2	- 11.245	81.210
0.3	- 11.041	80.365
0.4	- 10.838	79.520
0.5	- 10.635	78.675
0.6	- 10.432	77.830
0.7	- 10.229	76.986
0.8	- 10.027	76.141
0.9	- 9.824	75.296
1	-9.621	74.452

4.3 DECISION MAKING

Frequently in design situations the engineer is required to make decisions in an uncertain environment. For example, the engineer could have at his disposal a deterministic model of the system and moreover some incomplete information about the variables; in this case the procedures exposed in § 3 allow to describe or restrict the uncertainty on the response: nevertheless the engineer should finally give a not ambiguous sentence, a crisp, not fuzzy, statement about the reliability of the designed or retrofitted structure.

Otherwise in the design of a structure the engineer, taking into account the uncertainties about the data, must make a clear, non ambiguous choice about the design variables, assigning precise values to the geometrical dimensions of the structural elements and to the resistance of the materials to be employed. In this choice he should take into account one or more conflicting objectives to be satisfied (multi-objective optimisation).

Otherwise the engineer want to design a device for the automatic active control of a dynamic system , when the relation between the variables of the system to be controlled and its response cannot be defined in a deterministic manner.

In this ambit of problems the methodologies to treat the uncertainty based on the classical set theory often seem unsuitable, while the applications of probabilistic methods prove to be very difficult both for the lack of complete statistical data, and for computational   heaviness when applied to systems of great complexity. Moreover the choice between contrasting objectives requires in any case an empirical criterion, based on the judgement of the operator.

4.3.1 Reliability analyses

We will now consider the problem of how to judge the safety of the structure, whose uncertain response to loading has been determined as shown in § 2, separately considering the different models of the uncertainty.

4.3.1.1 Set Theory

When defining the class of the safe structures with respect to a limit state as the set of structures with positive or zero value of the safety margin z on Z ($z \in Z^S$) and indicating with B the subset of Z containing the possible response to the uncertain loads, the following three cases can arise:

$$a\text{-}\ B \cap Z^S = B \quad : \text{then the structure is safe;}$$
$$b\text{-}\ B \cap Z^S = \varnothing : \text{then the structure is unsafe;}$$
$$c\text{-}\ \varnothing \subset B \cap Z^S \subset B : \text{then the structure is } \textit{not-known} \text{ regarding safety.}$$

There are two basic problems in this solution:
 i) in case c no definite conclusion can be drawn by the decision maker, unless not-known should be considered, on the safe side, equivalent to unsafe; nevertheless not-known structures can be discriminated through safety and unsafety indexes

$$I_s = |B \cap Z^S| \ / \ |B| \quad ; I_u = |B \cap (Z^{Sc} = Z^U)| \ / \ |B| \tag{4.1}$$

where | . | indicates an additive measure of subsets, e.g. cardinality of finite subsets. According to the Law of excluded middle their sum equals 1:

$$|B \cap Z^S| / |B| + |B \cap Z^{Sc}| / |B| \ = \ |B \cap (Z^S \cup Z^{Sc})| \ / |B| \ = \ |B \cap Z| / |B| = \ |B| / |B| = 1$$

ii) in case a (or b) all safe (or unsafe) structures have to be considered equivalent respect to safety, and therefore the results cannot be used to compare safety with other objectives in the design, e.g. cost of the structure.

4.3.1.2 Probability Theory

Generally speaking a nominal admissible probability of failure $P_{fail,\ adm}$ of limit states is established or given by a Code, and the structure is considered to be safe if

$$F(z = 0) \leq P_{fail,\ adm} \tag{4.2}$$

where F(z) is the cumulative distribution function of the exit z, determined combining a probabilistic model of the variables and a deterministic model of the system and its limit state.

Any design can be compared with regard to safety: relations (4.1) can be interpreted as conditional probability of safety (reliability) and unsafety (P_{fail}) given B if | . | indicates the additive measure of Probability (see relation (3.3)). However structures with the same value of F(0) should be considered equivalent with respect to safety, although with different values of F for $z \neq 0$, e.g. with different mean value of z, E[z] (Fig. 4.4).

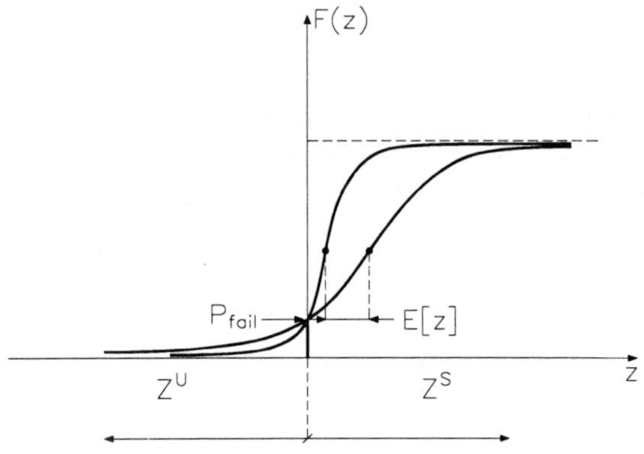

Fig. 4.4 - Comparison between two systems with equivalent nominal reliability but different cumulative probability distribution of the safety margin z of a limit state.

Of course the application of probability theory is frequently hindered by lack of data about the variables and computational complexity of the procedures, as pointed out above.

4.3.1.3 Random set Theory

Generally the result of the analysis, as described in § 2, is a random set on Z, or more simply a fuzzy set B, with membership function $\mu(z)$, including the actual random set. In this case, on the safe side, the decision maker can simply give a judgement on safety based on the admissible probability of failure $P_{fail,\ adm}$ of the limit states by means of the condition:

$$\max_{z \in Z^U} \mu(z) = Pl\,(Z^U = \{z \in Z \mid z \leq 0\}\,) \ \leq \ P_{fail,\ adm} \tag{4.3}$$

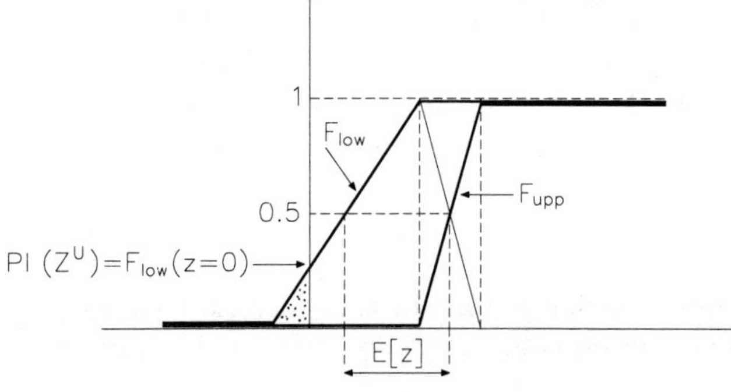

Fig. 4.5 - Reliability analysis through the fuzzy set of the safety margin of a limit state.

In fact, according to the results given in § 1.2 (Fig. 4.5):

$$F_{upp} (z=0) = Bel (Z^U) \leq F(z=0) = P_{Fail} \leq Pl (Z^U) = F_{low} (z=0) \qquad (4.4)$$

Other parameters can be derived by the fuzzy set to guide the choice of the decision maker, e.g. the interval containing $E[z]$ (Fig. 4.5) or the values of z with the maximum plausibility (the α-cut for $\alpha = 1$). Moreover the safety and unsafety indexes (4.1) can be used, where: (i) B is the obtained fuzzy set on Z, (ii) Z^U and Z^S are complementary classical subsets of Z defined by their characteristics function $\chi^U(z)$ and $\chi^S(z)$, (iii) intersection between fuzzy sets and classical sets is defined by the $\min(\mu(z) , \chi(z))$ and (iiii) | . | is the cardinality of the fuzzy sets (relation 1.19).

Example 4.2

Reconsidering the above exposed analysis of the response of an elastic non-linear system subjected to uncertain loading, we are now considering the limit state corresponding to the displacement of the spring equal to 1.6 (mm). That is, the safety margin z is given by the relation $z = 1.6 - \delta$, monotonically decreasing with f.
The critical vertices are calculated in Table 2 for $\alpha = 0$. In Table 3 the fuzzy set measuring z is displayed through the extremes of its α-cuts.

index j	q	P	f(q, P)	z	critical indexes
1	- 5	- 5	-4.708		
2	50	- 5	48.375	-0.025	L
3	- 5	100	- 7.180	2.209	R
4	50	100	45.903		

Table 2. Interval analysis of the safety margin z for $\alpha = 0$

Tab. 3 - α-cuts of the safety margin z of the spring displacement.

α-cut	z_L	z_R
0	-0.025	2.209
0.1	-0.017	2.204
0.2	-0.008	2.198
0.3	0.000	2.193
0.4	0.009	2.188
0.5	0.017	2.182
0.6	0.026	2.176
0.7	0.034	2.171
0.8	0.043	2.165
0.9	0.051	2.159
1	0.060	2.154

The Probability of failure is bounded in the interval [0, 0.3] ,while the Expectation $E[z]$ is bounded in the interval [0.017, 2.182].

4.3.2 Mono-objective optimisation of uncertain structures

Let us consider the following optimisation problem for the design variables $\mathbf{x} = (x_1, x_2, ...$ $x_s) \in \mathbf{X} = X_1 \times X_2 \times .... X_s$ of a structure.

$$\text{minimise } f(\mathbf{x}, \mathbf{u})$$
$$x \in \mathbf{X} \qquad (4.5)$$
$$\text{subjected to: } g_i (\mathbf{x}, \mathbf{u}) \leq 0 \quad i = 1,2...n$$

where $f(\mathbf{x},\mathbf{u})$ is an objective deterministic function, $g_i (\mathbf{x}, \mathbf{u})$ are n deterministic function describing the constraints and $\mathbf{u}$ is a vector of k uncertain parameters.

4.3.2.1 Set theory: worst case optimisation

Let us assume that $\mathbf{u}$ is known to belong to a bounded set C_u . The indeterminacy of set representation of the uncertainty discussed in the § 3.1.1 can be eliminated if we assume, on the safe side, that if it is not possible to demonstrate that the constraint is respected or the objective function is minimum the constraint has to be considered not respected or the minimum not attained. So, as proposed in [28], we can try to solve the problem (4.5) for the worst case of $\mathbf{u} \in C_u$ (anti-optimisation), i.e. (4.5) is rewritten as :

$$\text{minimise} \quad \max \quad f(\mathbf{x}, \mathbf{u}) \mid \mathbf{u} \in C_u \tag{4.6a}$$
$$\mathbf{x} \in \mathbf{X}$$
$$\text{subjected to :} \qquad \max \quad g_i(\mathbf{x}, \mathbf{u}) \leq 0 \mid \mathbf{u} \in C_u \quad i = 1,2...n \tag{4.6b}$$

As we have discussed in § 2.4, if each parameter u_j, $j=1,...,k$, is bounded in the interval $[u_{L,j}, , u_{R,j},]$ then C_u is a k-dimensional box; if, moreover, f and g_i depend linearly on $\mathbf{u}$ [28] or are monotone functions with respect to each parameter u_j (at least in the space $\mathbf{X_u}$ of the *feasible set* satisfying (4.6b)), the extreme cases can be found at the vertices $\mathbf{u}_t$ ($t = 1$, 2, 3... 2^k) of the box and (4.6) can be replaced by (4.7) , where a variable $y \in Y$ is introduced to treat in a symmetrical manner the constraints and the objective function:

$$\text{minimise} \quad y \tag{4.7a}$$
$$\mathbf{x} \in \mathbf{X} ; y \in Y$$
$$\text{subjected to:} \quad g_i(\mathbf{x}, \mathbf{u}_t) \quad \leq 0 \quad ; \quad i = 1,2...n ; t = 1, 2, 3... 2^k \tag{4.7b}$$
$$f(\mathbf{x}, \mathbf{u}_t) - y \leq 0 \quad ; \qquad t = 1, 2, 3... 2^k \tag{4.7c}$$

As $\mathbf{u} \in R^k$, problem (4.7) have $2^k \cdot (n+1)$ constraints : a big number when k and/or n are large.

However the results given in § 2.4.3 suggest a simple procedure to reduce the constraints to (n+1), i.e. the same number of the deterministic case, that is the case when the values of the parameters u_j are exactly known. In fact, taking into account that the functions f and g_i are monotone functions with respect to each parameter u_j , increasing or decreasing with u_j in any point $\mathbf{x} \in \mathbf{X_u}$, it is enough to calculate the values of the functions f and g_i in any point $\mathbf{x_0}$ for all the vertices of C_u ; then it is possible to evaluate the critical indexes t_{gi} and t_f of the vertices giving the maximum of each function and the corresponding values of $\mathbf{u}$: $\mathbf{u}_{gi}$ and $\mathbf{u}_f$.

In Fig. 4.6 the geometrical meaning of this property is indicated in the space of two design variables (x_1, x_2): the subset $\mathbf{X_u}$ of the feasible points respecting all the constraints depend on $\mathbf{u}$, and for any value of $\mathbf{u} \in C_u$, includes the subset $\mathbf{X}^*$ whose boundaries are given by the relations

$$g(\mathbf{x}, \mathbf{u}_{gi}) = 0 \quad ; \quad i = 1,2,... n \tag{4.8a}$$

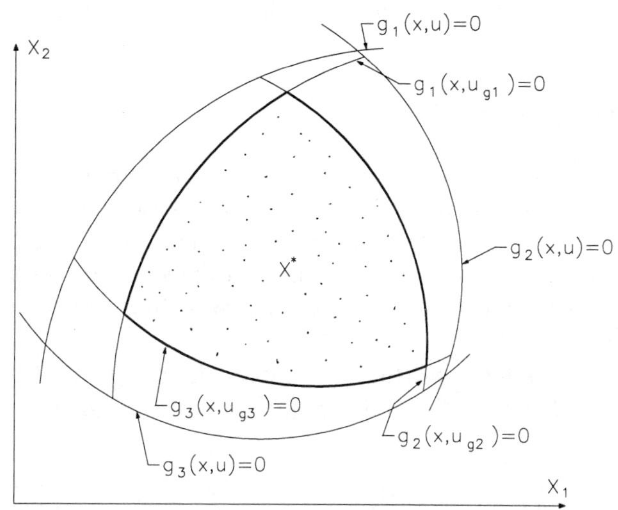

Fig. 4.6 Feasible set in the space of the design variables, monotonically dependent from an uncertain parameter u

The same reasoning could be repeated in the space $\mathbf{X} \times Y$, for the subset defined by (4.8a) and moreover by the relation

$$f(\mathbf{x}, \mathbf{u}_f) - y = 0 \tag{4.8b}$$

Then (4.7) can be replaced by the following problem:

$$\begin{array}{c} \underset{\mathbf{x} \in \mathbf{X}, y \in Y}{\text{minimise}} \quad y \\ \text{subjected to:} \quad g_i(\mathbf{x}, \mathbf{u}_{gi}) \leq 0 \quad ; \quad i = 1,2,\dots n \\ f(\mathbf{x}, \mathbf{u}_f) - y \leq 0 \end{array} \tag{4.9}$$

If on the contrary the functions are not monotone (increasing or decreasing with $\mathbf{u}$) in the overall space $\mathbf{X}$, but separately in some subsets of a partition of the space $\mathbf{X}$, it is possible, as shown in [29], to determine an iterative procedure starting from any point $\mathbf{x}_0$, solving (4.9) with the function determined in the critical vertices in the selected subset, and updating the vertices if the obtained optimum design $\mathbf{x}^*$ belongs to another subset

4.3.2.2 Reliability constrained optimisation

When the uncertainty is modelled in a probabilistic context, the parameters $\mathbf{u}$ are random variables measured by a joint probability distribution $p(\mathbf{u})$. Given a list of n+1 admissible probabilities $P_{gi,\,adm}$ and $P_{f,\,adm}$ that a constraint should not be respected or the function f should not be a minimum, the problem can be solved with the well established procedure of the reliability based optimisation [30].

Lack of data on $p(\mathbf{u})$ and/or computational complexity could however give obstacles that not easily can be overtaken.

4.3.2.3 Optimisation with not-interactive fuzzy parameters

Similarly to the approach presented in § 4.3.1, the application of the random sets theory leads to generalise both the classical set solution and the probability based solution. Moreover the computational weight is drastically reduced at least when the data about the uncertain parameters $\mathbf{u}$ have a consonant structure or are included, on the safe side, in a consonant structure . Under this hypothesis the data are described by a fuzzy relation defined by a membership function $\mu(\mathbf{u})$.

Particularly simple is the case of not-interactive fuzzy variables. In fact to find a solution it is enough: (i) to establish a value of α, (ii) to determine the corresponding α-cuts $^{\alpha}U_j$ and the k-dimensional box

$$C_u = {}^{\alpha}U_1 \times {}^{\alpha}U_2 \times ..... \, {}^{\alpha}U_k \qquad\qquad (4.10)$$

and then (iii) to solve the problem (4.6).   More easily, when the constraints and the objective functions are monotone on $\mathbf{X_u}$, the solution can be obtained searching for the critical vertices in each constraint and then solving (4.9), as above discussed.

Recalling the probabilistic content of the fuzzy sets (as consonant random sets) indicated in § 1.4, it is easy to demonstrate that the probabilities P_{gi} that a constraint should not be respected are bounded in the intervals :

$$0 \leq P_{gi} \leq \alpha \qquad\qquad (4.11)$$

In an analogous manner the probability P_f that the function f should not to be a minimum is bounded in the same interval.

The analysis can be repeated for a selected list of values of α , to show the influence of the required reliability on the optimum design. To this purpose the property indicated in § 2.4.3 is particularly interesting: the indexes of the critical vertices are invariant when the values of α are changed. Moreover different values of α could be prescribed for different constraints or for the objective function.

Example 4.3

The optimisation of the weight of the two-span linearly elastic beam depicted in Fig.4.2 is presented as an example of mono-objective optimisation. In order to make it more manageable, the spring is assumed to have a linear behaviour (with constant stiffness k). In each span, the high h_i of the beam is fixed, while its width x_i (design variable) is to be chosen to meet the design requirements (constraints) and to satisfy the objective function. Constraints g_1 - g_4 are geometrical as they limit the maximum and the minimum width of each span to $x_{i,max}$ and $x_{i,min}$ respectively); they do not depend on fuzzy parameters. Constraints g_5 and g_6 are relative to the behaviour of the system as they bound the maximum and minimum displacement of point C (positive downward) to δ_{max} and δ_{min} respectively; they depend on both fuzzy parameters (loads) P and q (positive downward). The trapezoidal fuzzy sets restricting loads P and q are defined by the left and right extremes of their α-cuts as follows (Fig. 4.3):

$$^{\alpha}q_L = -5 + \alpha \quad (kN/m) \; ; \qquad ^{\alpha}q_R = 50 - 5 \cdot \alpha \quad (kN/m) \; ;$$

$$^{\alpha}P_L = -5 + \alpha \quad (kN) \; ; \qquad ^{\alpha}P_R = 100 - 10 \cdot \alpha \quad (kN)$$

The remaining numerical values are:

$l_1 = 5.2$ m; $\quad l_2 = 4.3$ m; $\quad h_1 = 0.7$ m; $\qquad h_2 = 0.4$ m; $\qquad E = 27$ GPa; $\qquad k = 14$ kN/mm ;

$\delta_{max} = 5$ mm ; $\quad \delta_{min} = -0.7$ mm ; $\qquad x_{1,max} = x_{2,max} = 1.5$ m ; $\; x_{1,min} = x_{2,min} = 0.3$ m .

For a selected value of α, the optimisation problem is:

minimise $\qquad W = x_1 \cdot h_1 \cdot l_1 + x_2 \cdot h_2 \cdot l_2$

subject to:
$$\begin{cases} g_1(x_1, x_2) = x_1 - x_{1,max} \leq 0 \\ g_2(x_1, x_2) = x_{1,min} - x_1 \leq 0 \\ g_3(x_1, x_2) = x_2 - x_{2,max} \leq 0 \\ g_4(x_1, x_2) = x_{2,min} - x_2 \leq 0 \\ g_5(x_1, x_2, q, P) = \delta(x_1, x_2, q, P) - \delta_{max} = \dfrac{D_2(x_1, x_2) \cdot q - D_1(x_1, x_2) \cdot P}{1 + D_3(x_1, x_2) \cdot k} - \delta_{max} \leq 0 \\ g_6(x_1, x_2, q, P) = \delta_{min} - \delta(x_1, x_2, q, P) = \delta_{min} - \dfrac{D_2(x_1, x_2) \cdot q - D_1(x_1, x_2) \cdot P}{1 + D_3(x_1, x_2) \cdot k} \leq 0 \\ \qquad\qquad q \in \left[^{\alpha}q_L, ^{\alpha}q_R \right]; P \in \left[^{\alpha}P_L, ^{\alpha}P_R \right] \end{cases}$$ (4.12)

where D_1, D_2 and D_3 are defined in Example 4.2 and depend on the design variables through the moments of inertia of the beams:

$$J_1 = \frac{x_1 \cdot h_1^3}{12} \; ; \qquad\qquad J_2 = \frac{x_2 \cdot h_2^3}{12}$$

In order to find the critical vertices, it is quite easy in this case to calculate the partial derivatives of g_5 and g_6 with respect to the loads. We get:

$$\frac{\partial g_5(x_1, x_2, q, P)}{\partial q} = \frac{D_2(x_1, x_2)}{1 + D_3(x_1, x_2) \cdot k} > 0 \; \forall(x_1, x_2, q, P) \text{ and therefore the critical value of q for } g_5 \text{ is } q_R;$$

$$\frac{\partial g_5(x_1, x_2, q, P)}{\partial P} = -\frac{D_1(x_1, x_2)}{1 + D_3(x_1, x_2) \cdot k} < 0 \; \forall(x_1, x_2, q, P) \; ; \text{ therefore the critical value of P for } g_5 \text{ is } P_L;$$

$$\frac{\partial g_6(x_1, x_2, q, P)}{\partial q} = -\frac{D_2(x_1, x_2)}{1 + D_3(x_1, x_2) \cdot k} < 0 \; \forall(x_1, x_2, q, P) \; ; \text{ therefore the critical value of q for } g_6 \text{ is } q_L;$$

$$\frac{\partial g_6(x_1, x_2, q, P)}{\partial P} = \frac{D_1(x_1, x_2)}{1 + D_3(x_1, x_2) \cdot k} < 0 \; \forall(x_1, x_2, q, P) \; ; \text{ therefore the critical value of P for } g_6 \text{ is } P_R;$$

So, problem (4.12) becomes the simpler one:

$$\text{minimise} \quad W = x_1 \cdot h_1 \cdot l_1 + x_2 \cdot h_2 \cdot l_2 \tag{4.13a}$$

$$\text{subjected to:} \begin{cases} g_1(x_1,x_2) = x_1 - x_{1,max} \le 0 \\ g_2(x_1,x_2) = x_{1,min} - x_1 \le 0 \\ g_3(x_1,x_2) = x_2 - x_{2,max} \le 0 \\ g_4(x_1,x_2) = x_{2,min} - x_2 \le 0 \\ g_5(x_1,x_2,q_R,P_L) = \delta(x_1,x_2,q_R,P_L) - \delta_{max} \le 0 \\ g_6(x_1,x_2,q_L,P_R) = \delta_{min} - \delta(x_1,x_2,q_L,P_R) \le 0 \end{cases} \qquad (4.13b)$$

Fig. 4.7 displays the feasible domains for different values of α, i.e. of the upper bound on the probability of failure according to eq. 4.11. Also shown are the solutions S to problem (4.12) for each value of α; their numerical values (together with those of the total weight of the beam) are collected in Table 4.3. It can be seen that constraint g_5 is always active, while g_6 becomes active for $\alpha \le 0.9$. Moreover, the increase of the optimum widths (and consequently of the weight) is significantly greater than the increase of the load intervals (200% against 20%) when α goes down from 1 to 0. Obviously, as α diminishes, the feasible domain shrinks and this is because it is ever more difficult to obtain a minimum weight solution. This clearly corroborates the idea of the importance to include uncertainty in optimisation and more generally in design of engineering facilities.

α	x_1 (m)	x_2 (m)	Volume W (m^3)
1	0.33089	0.30846	1.7350
0.9	0.35824	0.37198	1.9438
0.8	0.40069	0.40312	2.1519
0.7	0.44513	0.43397	2.3667
0.6	0.49163	0.46454	2.5885
0.5	0.54029	0.49485	2.8178
0.4	0.59120	0.52490	3.0548
0.3	0.64445	0.55472	3.2999
0.2	0.70014	0.58431	3.5535
0.1	0.75840	0.61369	3.8161
0	0.81932	0.64286	4.0881

Tab. 4.3- Results of the mono-objective optimisation problem.

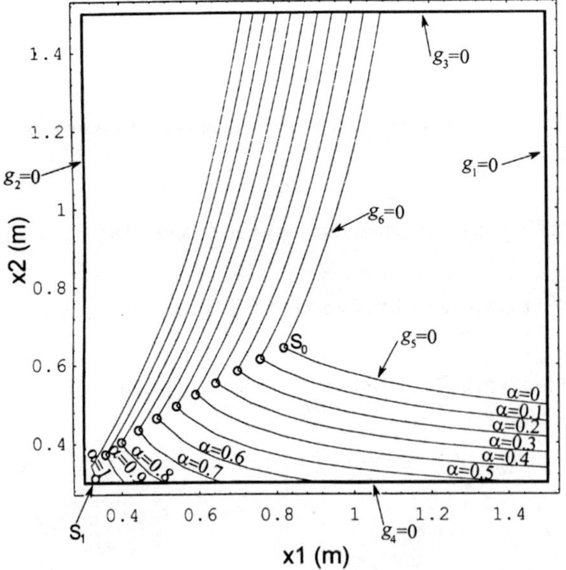

Fig. 4.7- Feasible domains of the optimisation problem for different reliability bounds.

5. HOW TO APPLY FUZZY LOGIC TO APPROXIMATE REASONING?

5.1 INTRODUCTION

Although the theory of random set and the connected interpretation of a fuzzy set as a consonant random set can be very useful in the applications (as an extension of the classical probability theory to the case of incomplete or set-valued data), the most important applications of fuzzy set theory have been based , until its foundation in the sixties, on the powerful extension of the classical logic and of the classical set theory, to give a numerical description of vague and qualitative information, and moreover to combine in a very simple and expressive manner uncertain information   independently given on the same space.

The main problem is to make decisions, on the basis of the forecast of future events, to design a new system or to operate the control of an existing system, in situations of uncertainty, i.e. when two or more alternatives are possible, and a list of objectives or constraints (safety, economic cost, serviceability) is defined.

The basic instruments are firstly the rule of fuzzy composition discussed in   § 3.6, to extend the classical rules of inference, and secondly the definition of an optimal choice when many, generally contrasting, objectives or constraints should be taken into account. The first instrument can be used to develop fuzzy rule based expert systems and fuzzy on line controller of dynamic systems (may be the most popular application of the theory). The second instrument suggests powerful , simple and robust procedure in the fields of pattern recognition, clustering and multi-objective optimisation.

Many thousands of papers and books and many hundreds of alternative procedures have been written or proposed in the last 30 years [see e.g. references in [17]); so I can only give some introductory ideas, trying to clarify the most relevant conceptual aspects, without discussing or classifying the different algorithms and operators proposed to extend the classical logic and set theory.

It is here important to underline that in these applications, as mentioned in § 3.6, the obtained results cannot be interpreted in the sense of expected frequencies of objective phenomena, even in the case when the data to be combined derived from statistics, although incomplete or set-valued, of objective phenomena.

In comparison with fuzzy logic, very rarely has the Dempster's rule (§ 3.5) been used in real applications, in spite of its more extended validity to consonant and non consonant data;   perhaps the reason depends on the computational difficulties arising from the necessity of operating on each focal element, while fuzzy logic allows a point-valued representation.

5.2 INFERENCE FROM CONDITIONAL FUZZY PROPOSITIONS

The implementation of expert systems and on line computer-aided system controller require the development of quick procedures of automatic decision reproducing in any manner the capacity of the human brain to recognise, in a largely uncertain environment, the most relevant information, to compare objects, to evaluate rules of general (but not absolutely universal) validity to be used in the approximate reasoning. When applied to very complex systems the required techniques should be simple and computationally robust, to evaluate directly the main structures and regularity of the data, without passing through and accurate analysis of any particular bit of information.

From this point of view the theory of fuzzy sets seems a particularly powerful instrument, because it enables one to demonstrate, in a very condensed manner, the informative content subtended by a population of individually distinct objects or measures.

As an example a fuzzy relation can summarise a vague or qualitative monotonically increasing dependency between two variables (x, y), expressed in linguistic manner by the propositions:

IF x is SMALL (a fuzzy set A_1 on X) then y is   MEDIUM   (a fuzzy set B_1 on Y)

....

IF x is LARGE (a fuzzy set A_j on X) then y is VERY LARGE (a fuzzy set B_j on Y)

....

and numerically:

$$\mu_R(x,y) = \max_{j} \min (\mu_{A_j}(x) , \mu_{B_j}(y)) \tag{5.1}$$

This rule can be justified observing that the min operator combines, according to the fuzzy composition rule (3.13) the cylindrical extensions of $\mu_{A_j}(x)$ and $\mu_{B_j}(y)$ respectively from X to X x Y and from Y to Y x X, while the max operator gives the standard union of the obtained R_j fuzzy relations on X x Y = Y x X (according to (3.14)).

The rule of fuzzy composition (symbol ° in the following) discussed in § 3.6 builds up the basis for a model of approximate reasoning, extending the rules that in classical logic are given to infer the "truth value" of a dependent proposition. This is performed by combining one or more propositions of universal validity (a deterministic relation) and the evidence of a particular property.

1. The Extended *Modus Ponens*:

Premise 1: $R \subseteq XxY$

Premise 2: $A^* \subseteq X$ $$\tag{5.2}$$

Then: $\mu_{B^*} = A^* \circ R \ (y) = \sup_{x \in X} \min (\mu_{A^*}(x) , \mu_R(x, y))$

2. The extended *Modus Tollens*:

Premise 1: $R \subseteq X \times Y$

$$\text{Premise 2: } B^* \subseteq Y \tag{5.3}$$

$$\text{Then: } \mu_{A^*} = R \circ B^* (x) = \sup_{y \in Y} \min (\mu_{B^*}(y) , \mu_R(x, y))$$

3. The extended *Hypothetical syllogism*:

$$\text{Premise 1: } R_1 \subseteq X \times Y$$
$$\text{Premise 2: } R_2 \subseteq Y \times Z \tag{5.4}$$

$$\text{Then: } \mu_R = R_1 \circ R_2 (x, z) = \sup_{y \in Y} \min (\mu_{R_1} (x, y) , \mu_{R_2} (y, z))$$

Example 5.1

Let $X = Y = \{1, 2, 3, 4\}$, and assume that on X and Y the linguistic judgement SMALL and LARGE correspond to the fuzzy subsets (μ / x means that μ is membership of x to a set):

$$\text{SMALL} = S = \{1 / 1, \ 0.9 / 2, \ 0.1 / 3, \ 0 / 4\}$$
$$\text{LARGE} = L = S^c = \{0 / 1, \ 0.1 / 2, \ 0.9 / 3, \ 1 / 4\}$$

In X x Y a relation is defined as follows:

$$R_1 \cup R_2 = (\text{IF x IS S THEN y is L}) \cup (\text{IF x IS L THEN y is S})$$

Numerically:

x	μ_{A_1}	μ_{R_1}	y=1	2	3	4
		μ_{B_1}	0	0.1	0.9	1
1	1		0	0.1	0.9	1
2	0.9		0	0.1	0.9	0.9
3	0.1		0	0.1	0.1	0.1
4	0		0	0	0	0

Tab.1 - R_1

x	μ_{A_2}	μ_{R_2}	y=1	2	3	4
		μ_{B_2}	1	0.9	0.1	0
1	0		0	0	0	0
2	0.1		0.1	0.1	0.1	0
3	0.9		0.9	0.9	0.1	0
4	1		1	0.9	0.1	0

Tab. 2 - R_2

x	μ_R	y=1	2	3	4
1		0	0.1	0.9	1
2		0.1	0.1	0.9	0.9
3		0.9	0.9	0.1	0.1
4		1	0.9	0.1	0

Tab. 3 - $R = R_1 \cup R_2$

x	μ_{A^*}	$\min(\mu_R , \mu_{A^*})$	y=1	2	3	4
1	0		0	0	0	0
2	1		0.1	0.1	0.9	0.9
3	0.5		0.5	0.5	0.1	0.1
4	0		0	0	0	0
	$\mu_{B^*} = \textbf{sup}$		**0.5**	**0.5**	**0.9**	**0.9**

Tab. 4 - Modus Ponens

Let's suppose to obtain the fuzzy measure on X: $A^* = \{0 / 1, \ 1 / 2, \ 0.5 / 3, \ 0 / 4\}$. Then the rule of the Modus Ponens works as shown in Table 4. The result is the sub-normal fuzzy subset of Y : $B^* = \{0.5 / 1, \ 0.5 / 2, \ 0.9 / 3, \ 0.9 / 4\}$.

5.3 PATTERN RECOGNITION AND CLUSTERING

The classical pattern recognition is generally based on the subdivision of a space $\mathbf{X}$, where some variables x_i assume values, in a standard partition: i.e. a group of disjoint subsets B_j (j=1,2..c) whose union is $\mathbf{X}$; if the actual pattern $\mathbf{x}^*$ is observed , it can be classified according to the standard subsets or patterns. Formally its membership to the standard patterns is given by

$$S (\mathbf{x}^*, B_j) = \chi_j (\mathbf{x}^*) \tag{5.5}$$

indicating with $\chi_j (\mathbf{x})$ the characteristic function on X of the subset B_j (standard pattern j)

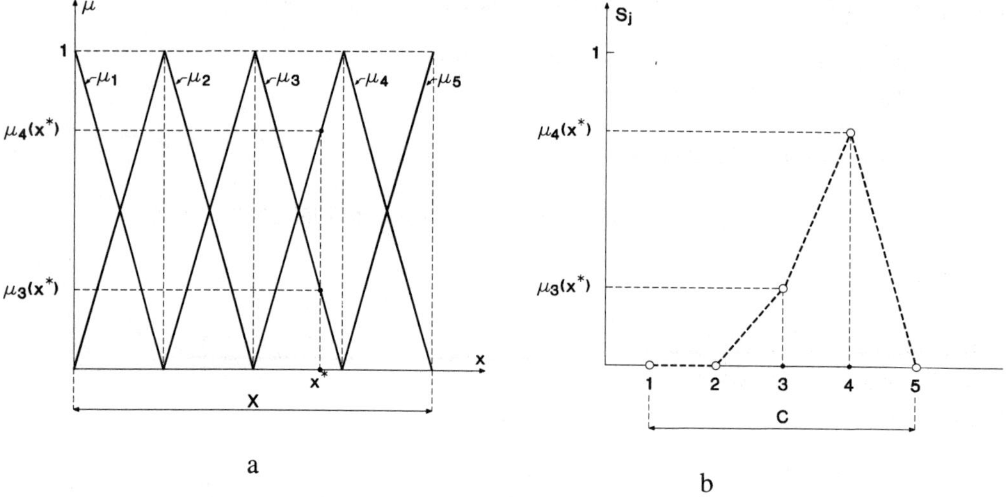

Fig. 5.1 Pattern recognition of a crisp measure x^* : a) fuzzy pseudo-partition; b) fuzzy classifier

A group of c fuzzy relations R_j (j = 1,2..c) on the same space X can be assumed to define a list of c standard patterns; the patterns are frequently given as a fuzzy pseudo-partition of X ([17], p. 359), if the following condition is respected:

$$\forall \mathbf{x} \in \mathbf{X} : \quad \Sigma_j \; \mu_j (\mathbf{x}) = 1 \tag{5.6}$$

If the actual pattern $\mathbf{x}^*$ is crisp, i.e. observed without uncertainty, it can be classified according to the standards by means of a *fuzzy classifier*, a fuzzy subset of the set $C = \{1, 2, ...c\}$ whose membership function is given by the extension of (5.5):

$$S (\mathbf{x}^*, R_j) = \mu_j (\mathbf{x}^*) \tag{5.7}$$

If a crisp classification is required, the pattern J with maximum of S can be selected ([17], pp.367-369).

On the contrary if the actual observed pattern is uncertain, i.e. measured by a random or fuzzy relation B on X, a criterion of comparison of actual and standard pattern is required. The comparison of fuzzy set is based on the idea of weak inclusion, or *degree of sub-sethood* of fuzzy sets, derived by extending the relation (4.1) to two fuzzy sets A and B on the same space X:

$$S (A(x), B(x)) = | A \cap B | / | A | \tag{5.8}$$

In another context the operator S can be considered a *filtering* of A through the *filter* B [31]. As an example the fuzzy set B on the space of a limit state function z could define the class of safe structures, extending the deterministic definition assumed in the relation (2.2a) to a more reasonable transition from safety to unsafety, as shown in Fig. 5.2.

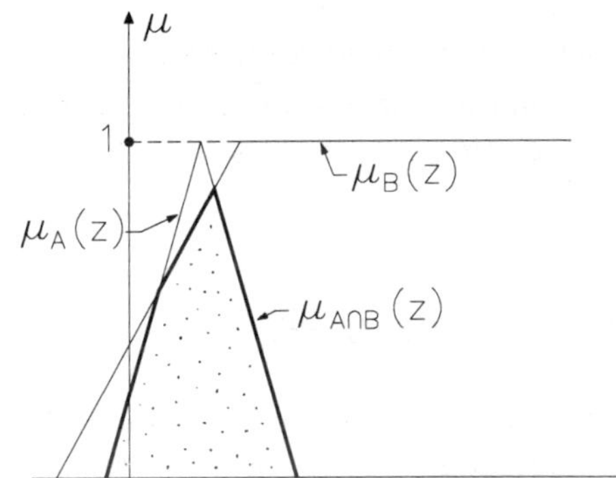

Fig. 5.2 - Filtering the fuzzy measure A of the safety margin z through the fuzzy filter B (defining safety).

It is possible to observe that, assuming for union and intersection the standard max and min operators (dual with respect to the complementation rule given by (1.21c)), the Law of the excluded middle is not respected : if B is a not trivial (i.e. not classic) fuzzy subset of the universal set X (Fig. 5.3)

$$A \cup A^c \subset X \tag{5.9}$$

and therefore, for any couple of not trivial fuzzy sets:

$$S(A, B) + S(A^c, B) < 1 \tag{5.10}$$

That is : the judgement on safety and unsafety cannot be derived one from the other, as in probability theory. In some application this asymmetry is not reasonable or is computationally heavy; so it can be eliminated assuming the following symmetric operator (normalised from 0 to 1 ; Bignoli A. J. [31]) in the comparison of fuzzy sets:

$$K(A, B) = (1 + S(A,B) - S(A^c, B))/2 \qquad (5.11)$$

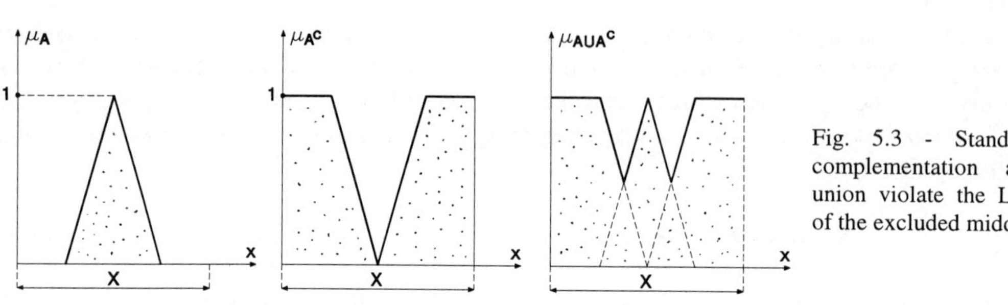

Fig. 5.3 - Standard complementation and union violate the Law of the excluded middle

The same operators (5.10) or (5.11) can be used to compare a fuzzy set or relation A (the actual observed pattern) with a list of fuzzy sets or relations (the standard patterns B_1 , B_2 ,... B_c) of the universal set X, by assuming the values of S or K as membership of the fuzzy classifier.

Fuzzy clustering can be considered the dual problem to the pattern recognition. Here a list of crisp or fuzzy data (x_1 , x_2 , ... x_q) is given and a fuzzy partition is required, according to a criterion of optimality of the number c of clusters, or , if c is selected, of the membership functions of the clusters ([17], pp. 358-365).

5.4 FUZZY MODEL OF MULTI-OBJECTIVE DECISION MAKING

In the space of the variables **X** any objective or any constraint gives a restriction to the values to be chosen by the decision maker. If the restriction is given by two classical subsets A and B (Fig. 5.4), their intersection gives the range of the values **x** satisfying both the objectives and constraints, i.e. it gives the subset D of the decisions through the characteristic function:

$$\chi_D = \chi_{A \cap B}(x) = \min(\chi_A(x), \chi_B(x)) \qquad (5.12)$$

Fig. 5.4 - Classical Decision set.

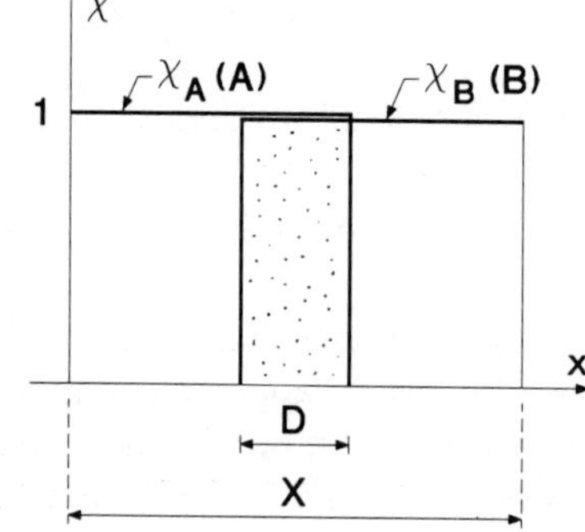

All the decisions in D are completely equivalent for the decision maker, so he or she appears as the donkey described by Giovanni Buridano, that died of hunger having two lots of hay exactly at the same distance on the left and on the right.

Alternatively the restriction could be according to a probabilistic model, i.e. through the conditional probabilities to satisfy the objectives (or constraint) A and B with respect to any choice $x \in X$.

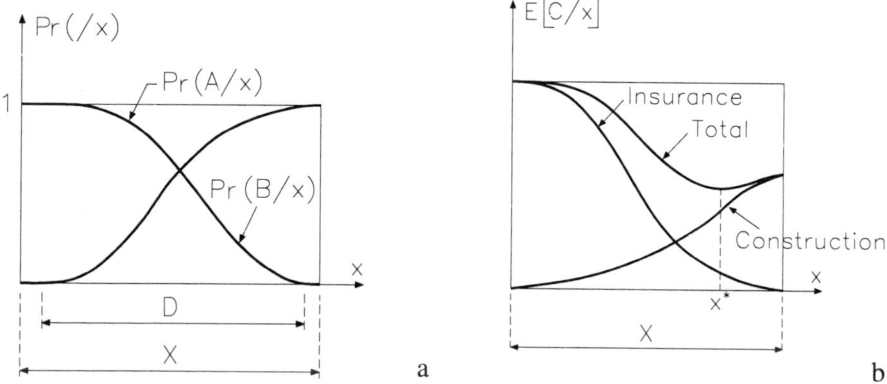

Fig. 5.5 - a) Probabilistic Decision Set ; b) crisp decision combining the objectives.

Even in this case we can derive a range D of the choices with positive probability to satisfy both A and B (5.5a), but not a univocal criterion for a crisp choice; to this purpose the multi-objective problem should be reduced to a mono-objective problem, selecting a priority or combining the objectives.

For example A is the cost of construction of a structure to be minimised and B the safety with respect to a limit-state to be maximised; we could make the choice of minimising the expectation of the total cost in the life of the structure: cost of construction plus cost of the insurance of the expected damages due to exceeding the limit-state (Fig. 5.5b).

If the restrictions are given by two fuzzy sets or relations, the decision subset is the fuzzy set or relation D with membership function given by the extension of relation (5.11)

$$\mu(x) = \min(\mu_A(x), \mu_B(x)) \qquad (5.15)$$

i.e. D is the intersection of the fuzzy sets A and B, when the standard operator min is selected for intersection. Moreover it is quite reasonable for the decision maker to select the crisp decision x^* such that (Fig. 5.6a):

$$\mu(x^*) = \max_{x \in X} \min(\mu_A(x), \mu_B(x)) \qquad (5.16)$$

According to the De Morgan's Laws, with the standard definition of union and complementation, such decision minimise also the membership of the union of the complementary fuzzy sets (Fig. 5.6b):

$$\mu(x^*) = \min_{x \in X} \ \max(1 - \mu_A(x), 1 - \mu_B(x)) \qquad (5.17)$$

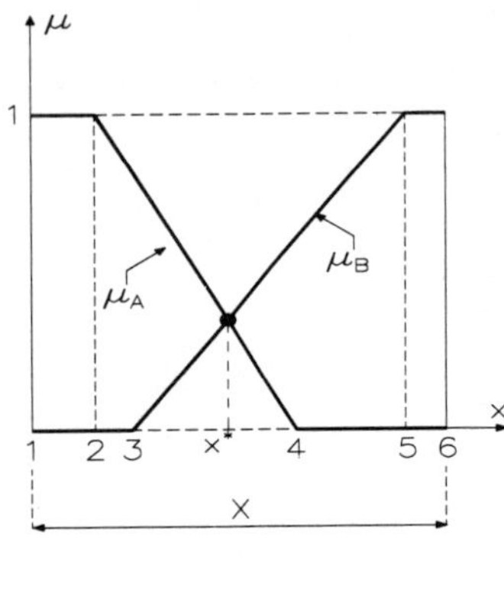

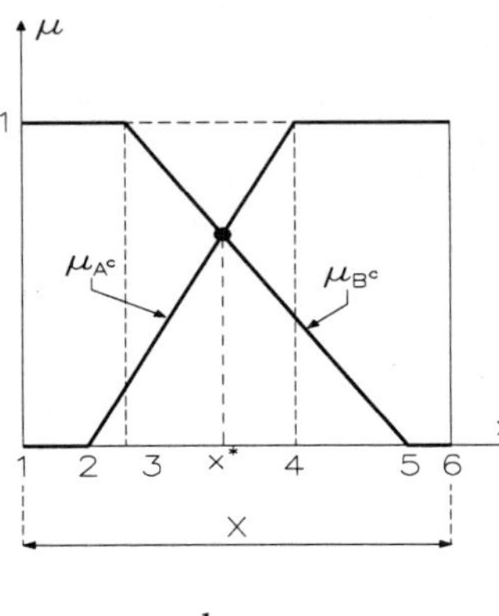

a   b

Fig. 5.6   Optimal decision in the decision space: a) maximising the advantage ; b) minimising the disadvantage.

A geometric interpretation of the solution can be given in the *space of the membership functions* (Fig. 5.7a) or of their complements to 1 (Fig. 5.7b), i.e. the multi-dimensional Cartesian product of their intervals of variation [0, 1] (the unit square with two objectives or constraint as in Fig. 3).In these spaces the point (1, 1,..1) and respectively the point (0, 0, ..0) are the *ideal points*, and the point corresponding to x* should be the point closest to the ideal points.

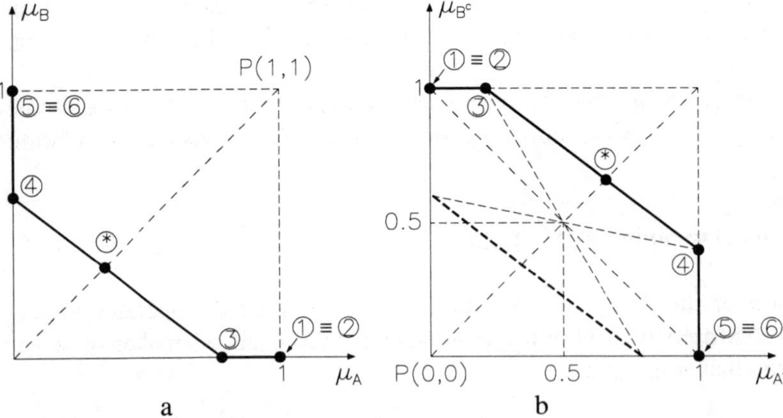

a   b

Fig. 5.7 Optimal decision in the membership space: a) maximising the advantage ; b)   minimising the disadvantage.

Of course a metric should be chosen in these spaces to measure the distance, and there is no reason in preferring an Euclidean metric to any other metric. A large class of possible metrics is the l_p class, defined by the relation:

$$d(P_1, P_2) = \left(\Sigma_j \, |\mu_{j1} - \mu_{j2}|^p \right)^{1/p} \tag{5.18}$$

The Euclidean metric belongs to this class with $p = 2$. The relations (5.16) and (5.17) can be derived by measuring the distances in the memberships spaces through a metric l_∞, i.e. if the distance between points P_1 $(\mu_{11}, \mu_{21},.. \mu_{j1}...)$ and P_2 $(\mu_{12}, \mu_{22},.. \mu_{j2}...)$ is defined by:

$$d(P_1, P_2) = \max_j |\mu_{j1} - \mu_{j2}| \tag{5.19}$$

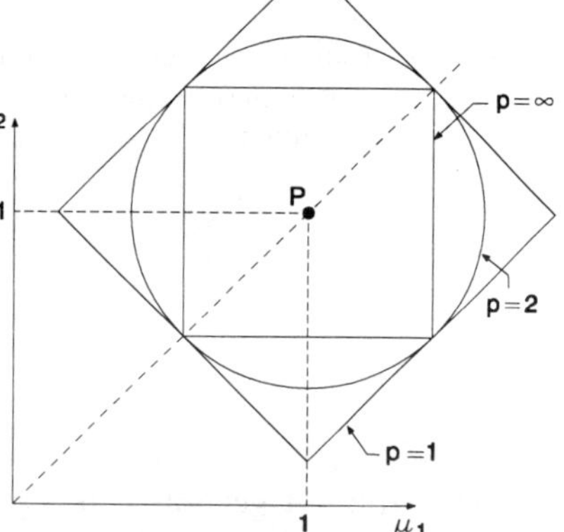

Fig. 5.8 . Points at equal distance from the ideal point P(1, 1) according to different l_p metrics.

Using this metric, the points at equal distance from the ideal point are on the perimeter of a square (with two objectives; a multidimensional square with more objectives) centred on the ideal point, with each side orthogonal to one axis (Fig. 5.8). As a comparison, in the same Figure the points at equal distance from the ideal point are displayed for $p = 1$ and $p=2$ (Euclidean metric).

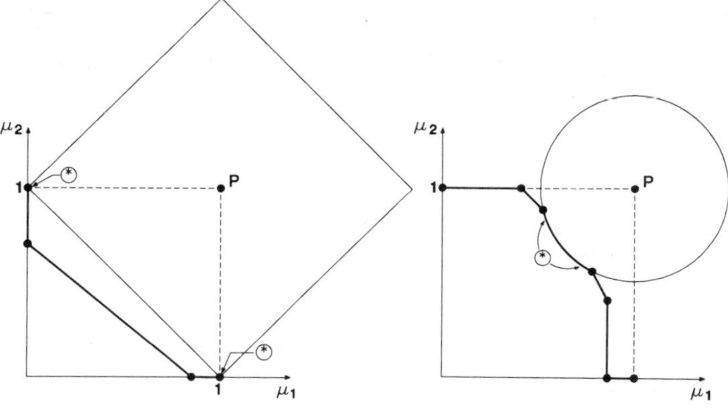

Fig. 5.9a - Not unique decision with p=1   Fig. 5.9b - Not unique decision with p=2

It is possible to demonstrate (21) that with the choice $p = \infty$, if the functions μ_j are strictly monotone continuous functions (increasing or decreasing) in X, the optimal decision is unique. If we use different metrics the optimal decision could be not unique (Fig. 5.9a, b), also in the case of strictly monotone membership functions

The above indicated solution can be extended to the case of many variables $\mathbf{x} = (x_1, x_2, ...x_n)$ and more than two objective or constraint, when each objective or constraint is measured by a fuzzy relation R_j with joint membership function $\mu_j(\mathbf{x})$, $j = 1,2,...m$) (see Fig. 5.10 for n=2 and m=2, and Fig. 5.11 for n= 1, m=3).

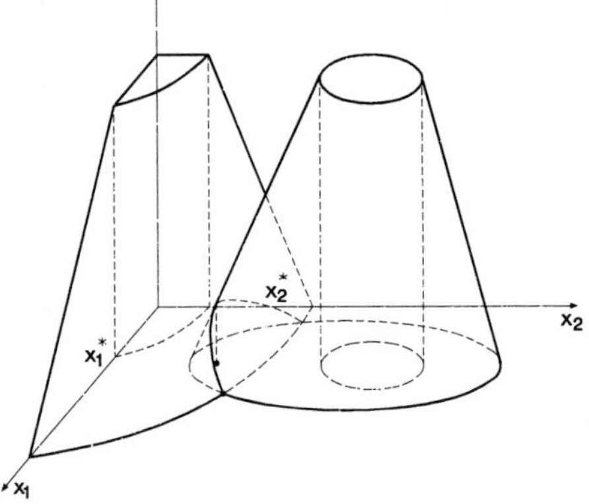

Fig. 5.10 - Optimal decision in the space X_1xX_2 as a compromise between one objective and one constraint.

When $m-n > 1$ generally the optimal solution is determined by some *active* objectives or constraints, while the others are satisfied with a membership greater than the membership obtained for the active ones (Fig. 5.11). As a general procedure, we must search for the maximum value of membership, α, such that all the objectives or constraints are satisfied at least with membership α. That is:

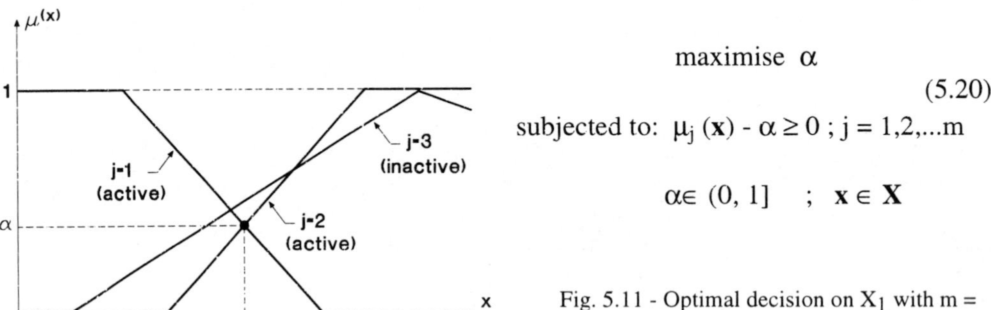

$$\text{maximise } \alpha$$

$$\tag{5.20}$$

$$\text{subjected to: } \mu_j(\mathbf{x}) - \alpha \geq 0 \; ; \; j = 1,2,...m$$

$$\alpha \in (0, 1] \quad ; \quad \mathbf{x} \in \mathbf{X}$$

Fig. 5.11 - Optimal decision on X_1 with m = 3

When the membership function $\mu_j(\mathbf{x})$, $j = 1,2,...m$ are linear function of the variables $(x_1, x_2, ...x_n)$ for $0 < \alpha < 1$ (as shown in Fig. 1.11), relations (5.20) define a standard linear programming problem with respect to the variables x_1, x_2,...x_n, α, to be solved with the Simplex algorithm.

5.5 MULTI-OBJECTIVE OPTIMISATION

5.5.1 Statement of the problem

By combining the results given in § 4.3.2 with the above described definition of optimal decision, we can now indicate a possible procedure to solve the multi-objective constrained optimisation problem with uncertain parameters:

$$\begin{array}{ll} \text{minimise} & f_j(\mathbf{x}, \mathbf{u}), \qquad (j = 1,2,....m) \\ x \in \mathbf{X} \\ \text{subjected to:} & g_i(\mathbf{x}, \mathbf{u}) \le 0 \quad i = 1,2...n \end{array} \qquad (5.21)$$

where $f_j(\mathbf{x,u})$ are m objective deterministic functions; $g_i(\mathbf{x}, \mathbf{u})$ are n deterministic function describing the constraints; $\mathbf{u}$ is a vector of k uncertain parameters.

5.5.2 Deterministic solution

Let us assume that $\mathbf{u}$ is known or prescribed, i.e. we suppose we are designing a deterministic system. The optimum choice x^* of the design variables can be reconnected to the solution of a problem of decision making taking into account both objectives and constraints, according to the procedures discussed in § 4.3.2 and the following steps:
1. Evaluate the *feasible set*:

$$\mathbf{X_g} = \{\, \mathbf{x} \in \mathbf{X} \mid g_i(\mathbf{x}) \le 0 \, , i = 1,2...n \} \qquad (5.22)$$

i.e. the subset of the values of design variables satisfying all the constraints (Fig. 5.12 for $\mathbf{X} = X_1 \times X_2$ and n = 3).

2. The objective functions f_j with values in F_j induce a point-valued (here supposed regular, i.e. one to one) transformation:

$$\mathbf{H} : \mathbf{X} \to \mathbf{F} = F_1 \times F_2 \,....\, F_m \qquad (5.23)$$

Therefore $\mathbf{H}$ gives an *attainable criterion set*, i.e. the image $\mathbf{F_g}$ of $\mathbf{X_g}$ trough $\mathbf{H}$ in the *criterion space* $\mathbf{F}$ (Fig. 5.13 for $\mathbf{F} = F_1 \times F_2$), and moreover their *projections* $F_{g,j}$:

$$F_{g,j} = \{\, f_j(\mathbf{x}) \mid \mathbf{x} \in \mathbf{X} \} \qquad (5.24)$$

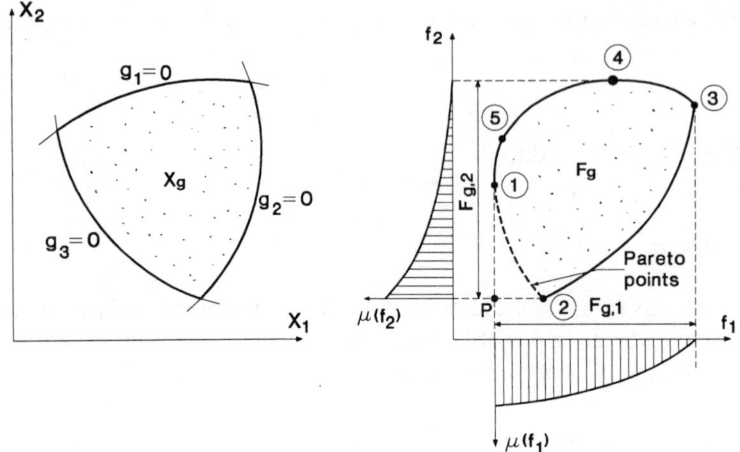

Fig. 5.12 - Feasible set in the design space.

Fig. 5.13 - Attainable criterion set $\mathbf{F_g}$ in the criterion space; Pareto points and fuzzy subsets of the projections of $\mathbf{F_g}$ modelling the relative satisfaction of any objective.

The objective functions are generally globally conflicting, i.e. their minimum is obtained at different points $\mathbf{x}$ and in the criterion space the ideal point $P(f_{1,min}, f_{2,min}, \ldots f_{m,min})$ does not belong to $\mathbf{F_g}$ (Fig. 5.13). Therefore we will be satisfied enough if we obtain a particular subset of $\mathbf{X_g}$, the *Pareto optima*, so defined: any point $\mathbf{x}^* \in \mathbf{X_g}$ such that for no other point $\mathbf{x} \in \mathbf{X_g} : f_j(\mathbf{x}) \leq f_j(\mathbf{x}^*)$, $j = 1, 2, \ldots m$, with $f_j(\mathbf{x}) < f_j(\mathbf{x}^*)$ for at least one j. In Fig. 5.13 the *Pareto points* are displayed on the boundaries of the attainable criterion set $\mathbf{F_g}$: they correspond to the Pareto optima $\mathbf{x}^*$ through the transformation $\mathbf{H}$.

The Pareto points are totally equivalent for the decision maker: so the several methods proposed to solve the multi-objective optimisation problem can be seen as an aim to obtain a further ordering among the Pareto optima. Following Sakawa [32] the ordering can be obtained by assigning for each objective a fuzzy subset of the set (the interval) $F_{g,j}$ through a membership function

$$\mu_j(f_j) : F_{g,j} \rightarrow [0, 1] \tag{5.25}$$

i.e. a mathematical model of the degree of satisfaction of the attainable value of the j-th objective function (Fig. 5.13). It is reasonable to assume for μ_j a decreasing or better strictly decreasing function, modelling a more or less severe judgement about the values of f_j over the minimum. To model the membership function, the maximum and minimum of each f_j on the feasible set $\mathbf{X_g}$ should be determined, by solving the corresponding anti-optimisation problems.

3. After the m membership functions have been fixed, we can define another (regular, if the membership functions are strictly decreasing) space transformation $\mathbf{L}$ from the criterion space to the *membership space*, i.e. the m-dimensional unit cube $\mathbf{M}$:

$$\mathbf{L}: \mathbf{F} \rightarrow \mathbf{M} = [0, 1] \times [0, 1] \quad \ldots [0, 1] \tag{5.26}$$

and therefore the *attainable membership set*, the image $\mathbf{M_g}$ of $\mathbf{F_g}$ through $\mathbf{L}$ (Fig. 5.14).
The transformation $\mathbf{L}$ stretches the set $\mathbf{F_g}$ according to the judgements and priorities of the decision maker, and moreover normalises the set $\mathbf{F_g}$, so that the final decision is regardless of the units, frequently incommensurable, adopted for the objective functions. This indifference is not guaranteed by all those methods acting on criterion space directly.
In this space we can define a criterion of optimality with respect to the ideal point, that is now the point P(1, 1, ...1) extending the definition of Pareto optimum. We can therefore characterise the *M-Pareto optima* as follows: any point $\mathbf{x}^* \in \mathbf{X_g}$ such that for no other point $\mathbf{x} \in \mathbf{X_g} : \mu_j (f_j (\mathbf{x})) \geq \mu_j (f_j (\mathbf{x}^*)) , j = 1, 2, ...m$, with $\mu_j (f_j (\mathbf{x})) > \mu_j (f_j (\mathbf{x}^*))$ for at least one j.

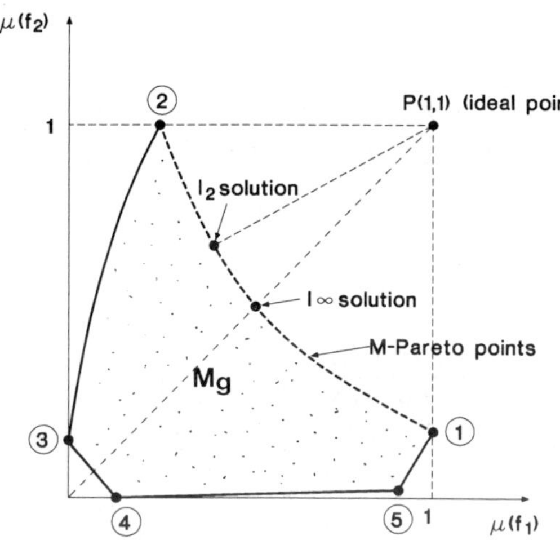

Fig. 5.14 - Membership set $\mathbf{M_g}$ in the membership space; M-Pareto points and optimal solution according to the l_∞ metric assumed for the distances from the ideal point P(1,1).

In Fig. 5.14 the *M-Pareto points* are displayed on the boundaries of the attainable membership set $\mathbf{M_g}$. They correspond to the M-Pareto optima $\mathbf{x}^*$ through the combined transformation $\mathbf{H} \bullet \mathbf{L}$. It is possible to demonstrate [21] that when $\mathbf{H}$ and $\mathbf{L}$ are regular transformations any M-Pareto optimum is a Pareto optimum.

4. The solution can now be obtained through the decision making procedure. In Fig. 5.14 the l_∞ metric is used to determine the solution. As shown in § 4 the M-Pareto point nearest to the ideal point, when l_∞ metric is employed to measure the distances, is unique: if moreover $\mathbf{H}$ and $\mathbf{L}$ are regular transformations, the optimum solution $\mathbf{x}^*$ in the design space is unique.

5.5.3 Solution with fuzzy non interactive parameters

The procedures shown in § 4.3.2.3, can now be combined with the above discussed solution for a deterministic system, to solve the problem (5.21), at least when the uncertain parameters $\mathbf{u}$ are restricted by means of not-interactive fuzzy sets. For a selected value of α

the feasible set $\mathbf{X_u}^\alpha$ and the attainable criterion set $\mathbf{F_g}^\alpha$ can be obtained for the worst case combination (the critical vertices) of the parameters. Finally the M-Pareto best solution can be derived from the membership set $\mathbf{M_g}$ and the ideal point in membership space. As shown in § 4.3.2.3, the selected value of α furnishes an upper bound to the probability that any constraint should not be respected or any objective function should not be a minimum.

Example 5.2

Let us reconsider the numerical example 4.3 and suppose that we are interested in minimising not only the total weight of the beam, but also the downward displacement δ_D of point D. This could be the case of an external beam of a building under which a very stiff and fragile cladding is to be fixed. For a prescribed level α, the mono-objective problem (4.13) becomes the following multi-objective problem:

$$\text{minimise} \begin{cases} f_1(x_1,x_2) = W = x_1 \cdot h_1 \cdot l_1 + x_2 \cdot h_2 \cdot l_2 \\[2mm] f_2(x_1,x_2,q,P) = \delta_D(x_1,x_2,q,P) = \dfrac{l_1^2 \cdot l_2^2}{16 \cdot E \cdot J_1} \left(\dfrac{P \cdot l_1}{3 \cdot l_2^2} - \dfrac{q}{2} + \dfrac{D_2(x_1,x_2) \cdot q - D_1(x_1,x_2) \cdot P}{1/k + D_3(x_1,x_2)} \right) \end{cases}$$

$$\text{subjected to:} \begin{cases} g_1(x_1,x_2) = x_1 - x_{1,\max} \leq 0 \\[1mm] g_2(x_1,x_2) = x_{1,\min} - x_1 \leq 0 \\[1mm] g_3(x_1,x_2) = x_2 - x_{2,\max} \leq 0 \\[1mm] g_4(x_1,x_2) = x_{2,\min} - x_2 \leq 0 \\[1mm] g_5(x_1,x_2,q,P) = \delta(x_1,x_2,q,P) - \delta_{\max} = \dfrac{D_2(x_1,x_2) \cdot q - D_1(x_1,x_2) \cdot P}{1 + D_3(x_1,x_2) \cdot k} - \delta_{\max} \leq 0 \\[3mm] g_6(x_1,x_2,q,P) = \delta_{\min} - \delta(x_1,x_2,q,P) = \delta_{\min} - \dfrac{D_2(x_1,x_2) \cdot q - D_1(x_1,x_2) \cdot P}{1 + D_3(x_1,x_2) \cdot k} \leq 0 \\[3mm] \qquad\qquad\qquad\qquad q \in \left[{}^\alpha q_L, {}^\alpha q_R \right] ; P \in \left[{}^\alpha P_L, {}^\alpha P_R \right] \end{cases} \tag{5.27}$$

The second objective function depends on fuzzy parameters and therefore deserves some attention. Proceeding as in example 4.3, it is easy to verify that δ_D is strictly monotone with respect to both fuzzy parameters (loads) and that the critical vertex of δ_D is (P_R, q_L) ; this result can be achieved simply by using influence line theory. So, problem (5.27) can be rewritten as a crisp one in the following way:

$$\text{minimise} \begin{cases} f_1(x_1,x_2) = W = x_1 \cdot h_1 \cdot l_1 + x_2 \cdot h_2 \cdot l_2 \\[1mm] f_2(x_1,x_2,q_L,P_R) = d_D(x_1,x_2,q_L,P_R) \end{cases}$$

$$\text{subjected to:} \begin{cases} g_1(x_1,x_2) = x_1 - x_{1,\max} \leq 0 \\[1mm] g_2(x_1,x_2) = x_{1,\min} - x_1 \leq 0 \\[1mm] g_3(x_1,x_2) = x_2 - x_{2,\max} \leq 0 \\[1mm] g_4(x_1,x_2) = x_{2,\min} - x_2 \leq 0 \\[1mm] g_5(x_1,x_2,q_R,P_L) = d(x_1,x_2,q_R,P_L) - d_{\max} \leq 0 \\[1mm] g_6(x_1,x_2,q_L,P_R) = d_{\min} - d(x_1,x_2,q_L,P_R) \leq 0 \end{cases} \tag{5.28}$$

Let the numerical values be the same as in example 4.3, so that the feasible domains are those of Fig. 4.7 or 5.15. For a specific level α, we will perform the following steps:

i) calculate the maximum $f_{i,max}$ (reached at point $(x_1,x_2)_{i,max}$) and the minimum $f_{i,min}$ (reached at point $(x_1,x_2)_{i,min}$) of the i-th objective function;

ii) adopt a linear membership function:

$$\mu_i = \frac{f_{i,max} - f_i}{f_{i,max} - f_{i,min}} \tag{5.29}$$

iii) find the feasible design $(x_1,x_2)_{opt}$ closest to point $(1,1)$, in the two dimensional membership space, in accordance with l_∞ metric:

$$\text{minimise max}\ (\ 1 - \mu_1\ ,\ 1 - \mu_2\ ) \tag{5.30}$$

subjected to the constraints given in (5.28).
As the objective functions and the membership functions are one-to-one mappings, transformations **H** and **L** (eq. 5.23 and 5.26) are regular and the solution to (5.30) is unique [21]. Tables 5.5 and 5.6 summarise the results obtained.

α	$(x_1,x_2)_{1,max}$ (m;m) Q in Fig.5.15	$(x_1,x_2)_{1,min}$ (m;m) S in Fig.4.7 and 5.15	$(x_1,x_2)_{2,max}$ (m;m)	$(x_1,x_2)_{2,min}$ (m;m) Q in Fig.5.15	$(x_1,x_2)_{opt}$ (m;m) P in fig. 5.15	$(\mu_1, \mu_2)_{opt}$
1	(1.5;1.5)	(0.3309;0.3084)	(0.3177;0.3405)	(1.5;1.5)	(0.7888;0.3000)	(0.7380;0.7380)
0.9	(1.5;1.5)	(0.3582;0.3720)	(0.3582;0.3720)	(1.5;1.5)	(0.8396;0.3000)	(0.7329;0.7329)
0.8	(1.5;1.5)	(0.4007;0.4031)	(0.4007;0.4031)	(1.5;1.5)	(0.8885;0.3000)	(0.7286;0.7286)
0.7	(1.5;1.5)	(0.4451;0.4334)	(0.4451;0.4334)	(1.5;1.5)	(0.9366;0.3000)	(0.7253;0.7253)
0.6	(1.5;1.5)	(0.4916;0.4645)	(0.4916;0.4645)	(1.5;1.5)	(0.9840;0.3000)	(0.7232;0.7232)
0.5	(1.5;1.5)	(0.5403;0.4948)	(0.5403;0.4948)	(1.5;1.5)	(1.0231;0.3320)	(0.7173;0.7173)
0.4	(1.5;1.5)	(0.5911;0.5249)	(0.5911;0.5249)	(1.5;1.5)	(1.0599;0.3723)	(0.7106;0.7106)
0.3	(1.5;1.5)	(0.6444;0.5547)	(0.6444;0.5547)	(1.5;1.5)	(1.0966;0.4125)	(0.7046;0.7046)
0.2	(1.5;1.5)	(0.7001;0.5843)	(0.7001;0.5843)	(1.5;1.5)	1.1333;0.4526)	(0.6990;0.6990)
0.1	(1.5;1.5)	(0.7584;0.6137)	(0.7584;0.6137)	(1.5;1.5)	(1.1699;0.4927)	(0.6950;0.6950)
0	(1.5;1.5)	(0.8193;0.6429)	(0.8193;0.6429)	(1.5;1.5)	(1.2065;0.5328)	(0.6913;0.6913)

Table 5.5- Results of the multi-objective optimisation problem.

α	$f_{1,max}$ (m^3)	$f_{1,min}$ (m^3)	$f_{2,max}$ (mm)	$f_{2,min}$ (mm)	$f_{1,opt}$ (m^3)	$f_{2,opt}$ (mm)	$\text{max}\ (1-\mu_1,\ 1-\mu_2) =$ $= 1-\mu_1 = 1-\mu_2$
1	8.0400	1.7350	1.0397	0.2366	3.3874	0.4469	0.2620
0.9	8.0400	1.9438	0.9383	0.2396	3.5720	0.4261	0.2671
0.8	8.0400	2.1519	0.8534	0.2425	3.7503	0.4083	0.2714
0.7	8.0400	2.3667	0.7813	0.2455	3.9254	0.3926	0.2747
0.6	8.0400	2.5885	0.7192	0.2485	4.0978	0.3787	0.2768
0.5	8.0400	2.8178	0.6652	0.2515	4.2952	0.3684	0.2827
0.4	8.0400	3.0548	0.6178	0.2544	4.4985	0.3596	0.2894
0.3	8.0400	3.2999	0.5758	0.2574	4.7012	0.3514	0.2954
0.2	8.0400	3.5535	0.5383	0.2604	4.9036	0.3439	0.3010
0.1	8.0400	3.8161	0.5046	0.2634	5.1058	0.3369	0.3050
0	8.0400	4.0880	0.4742	0.2663	5.3082	0.3304	0.3087

Table 5.6- Results of the multi-objective optimisation problem.

As for the above results, it is worth noting that:

1) the objective functions are highly contrasted as one achieves its absolute minimum where the other is maximum and viceversa; this is true for each α but $\alpha=1$ for $f_{1,min}$ and $f_{2,max}$, in which case points $(x_1,x_2)_{1,min}$ $(x_1,x_2)_{2,max}$ are however very closed (second row of columns 3 and 4 in Table 5.5);

2) as α diminishes, the optimum solution is such that the optimum weight increases (column 6 of Table 5.6), but the optimum displacement diminishes (column 7 of Table 5.6); this is due to the fact that the first objective function attains its minimum on the boundary of the feasible domain controlled by constraints 5 and 6, while the second objective function reaches here its maximum. As the upper bound of the probability of failure dwindles, the feasible domain shrinks; constraints 5 and 6 on the displacement of point C force the system to augment the width of the beams and consequently to decrease the positive displacement of point D too, notwithstanding the simultaneous increase of the loads;

3) for $\alpha > 0.5$, optimum widths are constrained by g_4; starting from $\alpha = 0.5$, they are constrained by g_5;

4) although the optimum displacement diminishes as α diminishes, the overall relative satisfaction yielded by the optimum solution (measured by the distance from point (1,1)) decreases, i.e. the higher reliability is, the more difficult it proves to get a satisfactory solution. This is because the membership functions take into account the extrema of the objective functions relative to each level of reliability;

5) the increase of the optimum widths is not as dramatic as it is when attempting to minimise total weight alone, but it is remarkable in any case (50%);

6) the relative satisfaction guaranteed by the optimum solution to the i-th objective function (measured by μ_i) is the same for both objective functions; this is so since we set the ideal point on (1,1) and used l_∞ metric. Moreover, μ_i can be considered "good" for every value of α, as it is around 0.7.

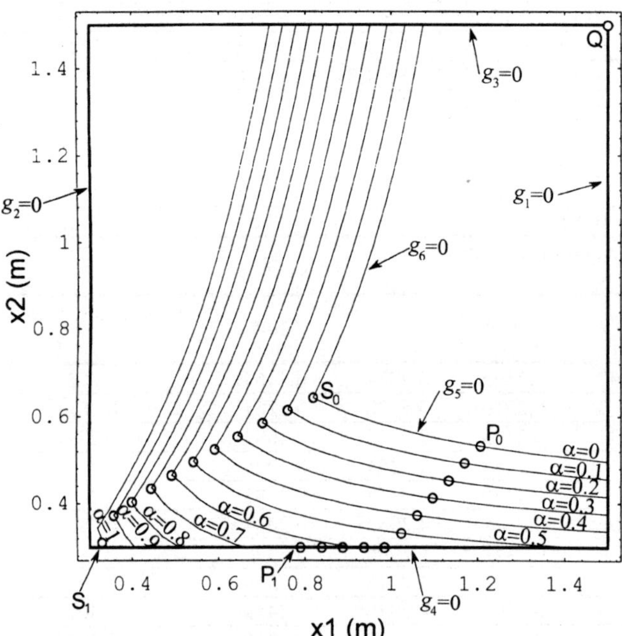

Fig. 5.15 Feasible domains of the multi-optimisation problem and optimal solutions.

ACKNOWLEDGEMENTS

I wish to express my gratitude to Arturo J. Bignoli, Buenos Aires, Argentina and to Isaac Elishakoff, Florida Atalantic University at Boca Raton, Florida, US. Both of them in the last years strongly stimulated my interest on the theoretical foundation of fuzzy sets, fuzzy logic and set-based approaches to uncertainty. Many of the ideas here presented derive directly from our informal discussions. Moreover, I like to acknowledge the important contributions of Fulvio Tonon, Treviso, Italy, firstly preparing his thesis of degree in Civil Engineering on multi-objective optimisation of engineering structures and later as a co-author of papers on applications of random/fuzzy sets theory to engineering problems. Many of the theorems and procedures described in this paper was firstly demonstrated by Eng. Tonon. Moreover, he actively co-operated in the planning and discussions of the paper and preparation of the numerical examples. Finally, I wish to thank Ugo Toffano for his skilful drawings.

REFERENCES

1.   Klir, G.J.: The Many Faces of Uncertainty, in: Uncertainty Modelling and Analysis: Theory and Applications (Eds. Ayyub, B.M. and Gupta, M.M.), North-Holland - Elsevier, Amsterdam, 1994, 3-19

2. Moore, R. E.: Interval analysis, Prentice-Hall, Englewood Cliffs, NJ, 1966

3. Ben-Haim, Y. and Elishakoff, I.: Convex Models of Uncertainty in Applied mechanics, Elsevier Science Pub., Amsterdam 1990

4. Dubois, D. and Prade, H.: Fuzzy Sets, Probability and Measurement, European Journal of Operational Research, 40 (1989), 135-154

5. Wang, Z.. and Klir, G.J.: Fuzzy measure theory, Plenum Press, New York, 1992.

6. Shafer, G.: Belief functions and Possibility measures, in: Analysis of fuzzy information. Vol. 1: Mathematics and Logic (Ed. Bezdek, J. C.), CRC Press, Boca Raton, Florida 1987

7. Robbins, H.E.: On the Measure of Random Set, I & II, Annals of Mathematical Statistics, 15 (1944), 70-74,; 16 (1945), 342-347,.

8. Matheron, G.: Random Sets and Integral Geometry, Wiley, New York, 1975.

9. Peng, X. T., Wang, P. and Kandel, A.: Knowledge Acquisition by Random Sets, International Journal of Intelligent Systems, Vol. 11(1996), 113-147

10. Dempster, A. P.: Upper and lower probabilities induced by a multivalued mapping, Ann. Math. Stat., 38 (1967), 325-339

11. Shafer, G.: A mathematical theory of evidence, Princeton University Press, Princeton, NJ , 1976

12. Dubois, D. and Prade H.: Possibility theory: an approach to computerized processing of uncertainty, Plenum Press, New York, 1988

13. Tonon, F., Bernardini, A. and Elishakoff I.: Concept of Random Sets as Applied to the Design of structures and Analysis of Expert Opinions for Aircraft Crash, Jour. of Educational Engineering (submitted)

14. Zadeh, L. A.: Fuzzy sets, Information and Control, 8 (1965), 338-353

15. Zadeh, L. A.: Fuzzy sets as a basis for a theory of possibility, Fuzzy Sets and Systems, 1 (1978), 3-28

16. Dubois, D. and Prade, H.: Fuzzy sets and statistical data, Eur. Journ. of Operational Research , 25 (1986), 345-356

17. Klir, G. J. and Yuan, B.: Fuzzy sets and Fuzzy logic. Theory and Applications, Prentice Hall PTR, Upper Saddle River, NJ , 1995

18. Yager, R. R.: The entailment principle for Dempster-Shafer granules, Int. J. Intelligent Systems, 1 (1986), 247-262

19. Dubois, D. and Prade, H.: Random sets and fuzzy interval analysis, Fuzzy Sets and Systems, 42 (1991), 87-101

20. Dong, W. and Shah, H. C.: Vertex method for computing functions of fuzzy variables, Fuzzy Sets and Systems, 24 (1987), 65-78

21. Tonon, F.: Ottimizzazione multiobiettivo in ambiente incerto: un approccio attraverso la teoria degli insiemi sfuocati con applicazione al progetto del rivestimento delle gallerie in roccia, Università di Padova, Tesi di Laurea in Ingegneria Civile, 1995

22. Shafer, G.: Non-additive probabilities in the work of Bernoulli and Lambert, Archive of History of Exact Sciences, 19 (1978), 309-370

23. Walley, P.: Statistical reasoning with imprecise probabilities, Chapman and Hall, London etc. , 1991.

24. Yager, R. R.: Aggregating fuzzy sets represented by belief structures, Journ. of Intelligent and Fuzzy Systems, 1 (1993), 215-224

25. Tonon , F., Mammino, A. and Bernardini, A.: A random set approach to the uncertainties in rock engineering and tunnel lining design, Proc. ISRM International Symposium on Prediction and performance in Rock mechanic and Rock Engineering EUROCK '96, 2-5 September 1996, Torino, Balkema, Rotterdam, 1996

26. Bernardini, A., Gori, R. and Modena, C. : Application of coupled analytical models and experiential knowledge to seismic vulnerability analyses of masonry buildings, in : Earthquake damage evaluation & vulnerability analysis of building structures, (Ed. Koridze, A.), Omega Scientific, Oxon, UK ,1990

27. Bernardini, A., Gori, R. and Modena C.: A knowledge based methodology for a-priori estimates of earthquake induced economical losses in old urban nuclei, Proc. CERRA-ICASP6, v. 2, 1037-1044, Mexico City, 1991

28. Elishakoff, I., Haftka, R.T. and Fang, J.: Structural Design Under Bounded Uncertainty- Optimization with Anti-Optimization. Computers and Structures, 53 (1994), 6, 1401-1405

29. Bernardini, A. and Tonon, F.: A random set approach to the optimisation of uncertain structures, Computers&Structures (submitted)

30. Frangopol, D. M. and Moses, F.: Reliability-based structural optimisation in: Advances in Design Optimization (Ed. Adeli, H.), Chapman and Hall, London etc. , 1994

31. Bignoli, A. J.: Teoria Elemental de los Conjuntos Borrosos, Academia Nazional de Ingenieria, Buenos Aires, 1991

32. Sakawa, M.: Fuzzy Sets and Interactive Multiobjective Optimization, Plenum Press, New York, 1993

33. Bernardini , A. and Tonon, F.: Multiobjective optimisation of uncertain structures through fuzzy set and random set theory, Microcomputers in Engineering (submitted).

CHAPTER 3

**WHAT IS ELLIPSOIDAL MODELLING AND HOW TO USE IT
FOR CONTROL AND STATE ESTIMATION?**

F.L. Chernousko
Russian Academy of Sciences, Moscow, Russia

ABSTRACT

The paper is devoted to the overview of the method of ellipsoids for modelling and estimating uncertainties in dynamical systems. The method was developed during the last decades and belongs to the guaranteed (minimax or set-membership) approach to description of uncertain factors such as disturbances, measurement errors, indeterminate parameters, etc. By contrast to the well-known probabilistic approach, the guaranteed approach does not require precise knowledge of probability distributions (which are seldom available in practical problems) and yields reliable estimates for the system behavior. Whereas the guaranteed approach operates with the sets of uncertain parameters, the ellipsoidal technique essentially simplifies the estimation procedure by replacing general n-dimensional sets by their inner and outer ellipsoidal estimates. Thus, two-sided optimal estimates are obtained for reachable sets of dynamical systems. The fundamentals of the method of ellipsoids are presented, and the basic properties of the approximating ellipsoids are discussed. Various applications of the method to problems in control and state estimation are given.

1. WHAT IS THE GUARANTEED APPROACH TO UNCERTAINTIES?

1.1 Probabilistic and Guaranteed Approaches

Let us discuss briefly basic assumptions of the probabilistic and guaranteed approaches to uncertainties. Suppose that x is the uncertain n-dimensional vector, i.e. the vector of parameters which are not known exactly.

In the probabilistic approach, we introduce the probability distribution $p(x)$ for the vector x where the function $p(x)$ is assumed to be known or given, and after that we deal with this function.

In the guaranteed, or minimax, or set-membership approach, we assume that the uncertain vector x belongs to a set D: $x \in D$, and this set is assumed to be known. Further, we are to operate with sets.

Let us consider a linear dynamical system governed by a differential equation and initial condition

$$\dot{x} = A(t)x + B(t)\xi(t) + f(t) \qquad (1.1)$$

$$x(0) = x^0 \qquad (1.2)$$

Here, $x = (x_1, \ldots, x_n)$ is the n-dimensional vector of state of the system, $\xi = (\xi_1, \ldots, \xi_m)$ is the m-dimensional vector of disturbances, t is time, $A(t)$ and $B(t)$ are given $n \times n$ and $n \times m$ matrices dependent on time, $f(t)$ is a given n-dimensional vector function, x^0 is a given initial vector of state, and dot denotes time derivative.

In the probabilistic approach, the disturbance $\xi(t)$ is assumed to be a stochastic process (for example, a white noise), and equation (1.1) is treated as a stochastic differential equation. Its solution $x(t)$ is also a stochastic process. There exists a vast literature treating the theory of stochastic processes.

By contrast, in the guaranteed approach we consider the disturbance $\xi(t)$ as an unknown but bounded function; its values are restricted by the constraint $\xi(t) \in G(t)$, where $G(t)$ for each $t \geq 0$ is a given closed set in R^m. For each admissible disturbance $\xi(t)$, there exists a corresponding solution $x(t)$ of the initial value problem (1.1), (1.2). Thus, we have here a family of possible solutions $x(t)$. Different problems associated with the guaranteed approach to the system (1.1) are studied in the theory of control.

The probabilistic approach is more suitable for the cases where there is a priori knowledge about the uncertain disturbances $\xi(t)$. This approach yields probabilistic estimates which are important if there are many realizations of the process. Strictly speaking, in the probabilistic approach all statements concerning the behavior of the system are true either on the average or with a certain probability.

By contrast, the guaranteed approach does not require any knowledge about the disturbances $\xi(t)$ besides their bounds, i.e. the sets $G(t)$. Note that the probability distributions for the unknown disturbances are seldom available is practical problems. Moreover, the guaranteed approach produces results which are applicable to any individual realization of the process. Thus, the guaranteed approach requires less information and is more reliable. However, if there is a lot of statistical data for the uncertainties, the probabilistic approach can give better estimates.

Therefore, each approach has its strong and weak sides. There are also some intermediate versions (see [1, 2]). For example, one can use the probabilistic approach with probability distributions depending on unknown parameters and assume that these parameters lie within given boundaries. Such a technique was successfully used for determination of space trajectories based on measurement data.

We shall focus on the guaranteed approach to dynamical systems subject to uncertainties and/or control.

1.2 Dynamical Systems

Consider a dynamical system governed by a vector differential equation

$$\dot{x} = f(x, u, t) \tag{1.3}$$

Here, $x = (x_1, \ldots, x_n)$ is the n-vector of state, or phase coordinates, of the system; $u = (u_1, \ldots, u_m)$ is the m-vector of control or disturbance; t is time; $f = (f_1, \ldots, f_n)$ is a given vector-valued function. Hereinafter, the vector u may denote either control parameters or disturbances acting on the system. In the first case, the vector u is free and can be chosen whereas in the second case it is an uncertain disturbance. In both cases, we shall often call it control in accordance with the terminology of the control theory.

Suppose that at any time instant, the vector u is subject to the constraint

$$u(t) \in U \tag{1.4}$$

where U is a given closed set in R^m. We shall also consider constraints

$$u(t) \in U(x(t), t) \tag{1.5}$$

more general than (1.4). Here $U(x, t)$ is a closed set depending on state x and time t. Constraints (1.5) are called mixed because they involve both control u and state x. Controls satisfying (1.4) or (1.5) are called admissible. The controlled system (1.3) with constraint (1.5) (or (1.4)) can be described by the following differential inclusion

$$\dot{x} \in X(x, t) \tag{1.6}$$

Here, $X(x, t)$ is the set of all possible values of the right-hand side of (1.3) with controls satisfying the constraint (1.5) (or (1.4)), i.e.

$$X(x, t) = f(x, U(x, t), t) \tag{1.7}$$

Therefore, our controlled system can be described either by equation (1.3) with constraint (1.5) (or (1.4)), or by the differential inclusion (1.6). The second description is more general: each set $X(x, t)$ in (1.6) may correspond to various systems (1.3), (1.5) satisfying (1.7). The first description is useful when we are to choose the control law and/or to analyze individual motions of the system while the second approach is more convenient for describing the whole set of possible motions. Further on we use both approaches.

The dynamical system (1.3) and constraints (1.4) (or (1.5)) as well as the differential inclusion (1.6) are defined for $t \geq s$, where s is the initial time instant. Suppose the following initial condition is given

$$x(s) \in M \tag{1.8}$$

Here, M is a given closed set in R^n containing all possible initial states of the system. In particular, the set M can be a point: $M = \{x^0\}$.

1.3 Reachable Sets

For each point $x(s) \in M$, there exist many trajectories of the system (1.3) which correspond to different controls. We shall sometimes denote the trajectories by the symbol $x(\cdot)$ showing that we are interested in the entire function and not in its specific value at some instant.

Definition 1.1. The set of the end points $x(t)$ at $t \geq s$ of all trajectories $x(\cdot)$ of the system (1.3), (1.5) (or (1.6)) under the initial condition $x(s) \in M$ (see (1.8)) is called the reachable, or attainable set of the system and denoted by $D(t, s, M)$.

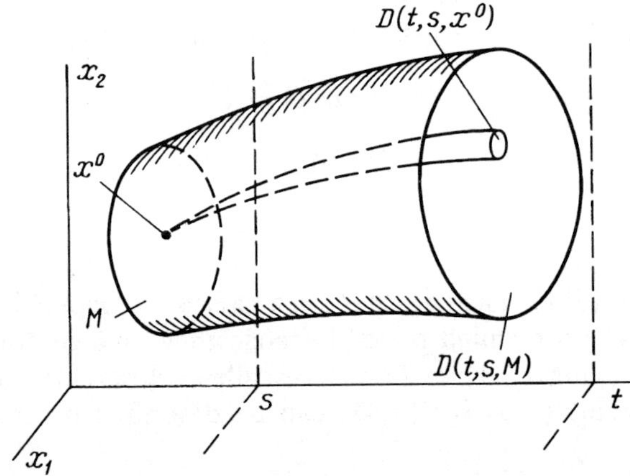

Fig. 1.1, A reachable set

The initial set M and the reachable set $D(t, s, M)$ are shown in Fig. 1.1 for $n = 2$. Here we can also see the reachable set $D(t, s, x^0)$ corresponding to the point initial set $M^0 = \{x^0\}$. The set $D(t, s, M)$ is the union of all sets $D(t, s, x^0)$, where $x^0 \in M$.

For controlled dynamical systems, reachable sets are fundamental characteristics playing the same role as solutions of initial value problems for differential equations.

1.4 Applications of Reachable Sets

Reachable sets play an important role in the control theory and its applications. Many basic problems of this theory can be stated and solved in terms of reachable sets. Below we consider some of these problems.

a) Evaluation of control possibilities. If we know sets $D(t, s, M)$, we can readily evaluate possibilities of control. For example, the following q estion often arises: is it possible to bring the control system to a certain prescribed state x^* at a given instant $T > s$? To answer this question, it is sufficient to check whether the vector x^* belongs to the set $D(T, s, M)$, i.e., the inclusion $x^* \in D(T, s, M)$ holds.

b) Controllability. Consider the set $\Delta(s, M)$ which is the union of reachable sets $D(T, s, M)$ for all $T \geq s$. If the set $\Delta(s, M)$ coincides with R^n, the system (1.3) is called completely controllable. For linear systems, conditions of controllability were first obtained by R.Kalman [3]. Later, controllability became an important concept of the control theory [4–7].

c) Evaluation of disturbances and practical stability. Let the vector u in (1.3) be some disturbance subject to (1.4). Then the reachable set $D(t, s, M)$ is a domain into which our system can be brought by admissible disturbances. Thus, reachable sets characterize deviations of trajectories caused by disturbances, or the accuracy with which we can predict the motion of the system in the presence of disturbances. Sometimes, estimation of possible deviations of the perturbed motion from the ideal trajectory is referred to as a problem of practical stability.

d) Terminal control. Consider the following problem of optimal control for system (1.3) under constraint (1.4) (or (1.5)).

Find an admissible control $u(t)$ and an initial vector $x(s) \in M$ such that the minimal value of the cost functional

$$J = F(x(T)), \quad T > s \tag{1.9}$$

is attained. Here, $F(x)$ is a given scalar function, T is a given time instant. This problem can be formulated as the following minimization problem

$$F^* = \min F(x), \quad x \in D(t, s, M) \tag{1.10}$$

For a given reachable set $D(T, s, M)$, problem (1.10) can be regarded as an ordinary problem of nonlinear programming which is much simpler than the initial optimal control problem. Note that the reachable set $D(T, s, M)$ represents all the necessary information about the behavior of the system. Suppose this set is known; then, if we change the function $F(x)$ in (1.9), we need only to solve another problem of nonlinear programming (1.10) for the same set $D(T, s, M)$.

e) Time-optimal control. Time-optimal control problems are often considered both in theory and applications. Suppose we are to find an admissible control $u(t)$ and an initial vector $x(s) \in M$ such that the system reaches a certain set $N \subset R^n$

in minimal possible time. The terminal time instant T^* for the time-optimal trajectory can be found as the minimal T for which the condition $N \cap D(T, s, M) \neq \emptyset$ holds.

f) Differential games. Let us discuss possible applications of reachable sets to the theory of differential games [8, 9]. Consider two controlled systems

$$X : \dot{x} = f(x, u, t), \quad u \in U, \quad x(s) = x^0 \tag{1.11}$$

$$Y : \dot{y} = g(y, v, t), \quad v \in V, \quad y(s) = y^0$$

Here, x and y are the state vectors of the systems X and Y respectively, f and g are given functions, u and v are controls, U and V are closed sets, and x^0 and y^0 are the initial states of the systems. Let the terminal time $T > s$ be fixed and the terminal cost functional be given by

$$J = \Phi(x(T), y(T)) \tag{1.12}$$

The players X and Y can choose their controls as strategies depending on the states

$$u = u(x, y, t), \quad v = v(x, y, t) \tag{1.13}$$

The first player X seeks to minimize while the second player Y seeks to maximize the functional (1.12). Suppose our differential game has a solution in some class of strategies (1.13), i.e., there exists a pair of optimal strategies u^*, v^* of the form (1.13). The value J^* of the functional (1.12) corresponding to these strategies cannot be decreased by the player X or increased by the player Y.

Let $D_x = D_x(T, s, x^0)$ and $D_y = D_y(T, s, y^0)$ be the reachable sets of the players X and Y, respectively. Then it is clear that the optimal value J^* of the functional (1.12) is bounded by the following inequalities

$$\max_{y \in D_y} \min_{x \in D_x} \Phi(x, y) \leq J^* \leq \min_{x \in D_x} \max_{y \in D_y} \Phi(x, y) \tag{1.14}$$

Maximin in the left-hand side of (1.14) corresponds to the case, where the player Y chooses his open-loop control $v(t)$ beforehand and makes it known to the player X who can choose his optimal control $u(t)$ afterwards. Minimax in the right-hand side of (1.14) corresponds to the opposite situation, namely, the player X first chooses his control $u(t)$, and then Y chooses $v(t)$ (knowing $u(t)$ already). Instead of the choice of controls, the players can choose points in their respective reachable sets. The inequalities (1.14) mean that the optimal case lies between the two cases which are considered above and correspond to the discrimination of one of two players. Thus, by means of reachable sets we have obtained two-sided bounds on the optimal value J^* of the cost functional. If the function $\Phi(x, y)$ has a saddle point, the lower and upper bounds in (1.14) coincide and are equal to the optimal value J^*.

g) Games of pursuit and evasion. Now consider an important class of differential games called games of pursuit and evasion. Let the systems (1.11) be linear

$$X : \dot{x} = A(t)x + u, \quad u \in U(t), \quad x(s) = x^0$$

$$Y : \dot{y} = B(t)y + v, \quad v \in V(t), \quad y(s) = y^0$$

Here, $A(t)$ and $B(t)$ are given matrix-valued functions, $U(t)$ and $V(t)$ are bounded closed convex sets depending on t. To be definite, we assume that the vectors $x, y, u,$ and v are of the same dimension n. The player X seeks to attain the meeting condition $x(T) = y(T)$ at minimal possible T. The player Y desires to avoid meeting, or, if the avoidance is impossible, to maximize T.

According to the rule of extremal aiming [5, 9] which expresses the solution of the game of pursuit in terms of reachable sets, under some general regularity conditions, the optimal time of pursuit T^* can be found as the minimal T for which the reachable set of the pursuer X contains the reachable set of the evader Y, i.e.

$$T^* = \min T, \quad D_x(T, s, x^0) \supset D_y(T, s, y^0)$$

h) Guaranteed estimation (filtering) in dynamical systems. In the theory of guaranteed estimation (filtering), external disturbances and observation errors are assumed to be unknown but bounded functions of time [5, 10–15]. It will be shown in Section 6 that the problem of the guaranteed state estimation can be formulated and solved in terms of reachable sets.

Hence, efficient methods for constructing and approximating reachable sets may be a useful tool in the control theory and its applications.

1.5 Convex Sets

We shall recall here briefly some definitions and concepts used below and related to the theory of sets and, especially, to convex sets. More details can be found, e.g., in [16, 17]. We consider points (vectors) $x = (x_1, \ldots, x_n)$ in R^n. The scalar product of vectors x and y is denoted by $(x, y) = x_1 y_1 + \ldots + x_n y_n$. The norm of a vector is denoted by

$$| \, x \, | = (x, x)^{1/2} = (x_1^2 + \ldots + x_n^2)^{1/2}$$

The set C of all points belonging to at least one of two sets, A or B, is called the union of sets and denoted by $C = A \cup B$. Analogously, the union $C = A_1 \cup \ldots \cup A_k$ of any number of sets $A_1, \ldots, A_k$ is introduced. The set C consisting of points belonging to both sets A and B simultaneously, is called the intersection of sets and denoted by $C = A \cap B$. In the same way, the intersection $C = A_1 \cap \ldots \cap A_k$ is defined. The set of all points $x = x_1 + x_2$ such that $x_1 \in A, x_2 \in B$ is called the sum (or the Minkowsky sum) of the sets A and B denoted by $C = A + B$.

Suppose $f(x)$ is a vector-valued function of x, where $x \in R^n$ and $f(x) \in R^m$, and let A be a set in R^n. Then the set $B \subset R^m$ of all vectors $y = f(x), x \in A,$

i.e., the set $B = \{y : y = f(x), x \in A\}$ is called a function of a set and denoted by $B = f(A)$.

A set A is called closed, if it contains all its limit points. A set is called bounded, if it belongs to some finite ball. A bounded closed set is called a compact set.

The ball in R^n with the center z and radius $r \geq 0$ is denoted by $S_r(z)$. In other words, $S_r(z)$ is a set of all $x \in R^n$ such that $|x - z| \leq r$.

A set A is called convex, if the inclusions $x_1 \in A$ and $x_2 \in A$ entail the inclusion $\lambda x_1 + (1 - \lambda)x_2 \in A$ for any $\lambda \in [0, 1]$. In other words, a convex set contains the segment connecting any two points of this set. The empty set $\emptyset$, the point set, a segment, a ball, etc., are examples of convex sets. The intersection $A \cap B$ and the sum $A + B$ of two (or more) convex sets are also convex sets. The intersection of all convex sets containing a set A is called the convex hull of A and denoted by $C = \mathrm{co}A$. The convex hull $\mathrm{co}A$ is the smallest convex set containing A. A set A is convex whenever it coincides with its convex hull: $A = \mathrm{co}A$. The convex hull $\mathrm{co}A$ of a bounded closed set A is a bounded closed convex set. If a set A consists of a finite number of points $x_1, \ldots, x_n$, then its convex hull is a convex polyhedron.

Let A be a non-empty set in R^n. The scalar function $\rho(z, A)$ defined by

$$\rho(z, A) = \sup_{x \in A}(x, z), \quad z \in R^n$$

is called the support function of the set A. For a bounded closed set A, this equation can be rewritten as follows

$$\rho(z, A) = \max_{x \in A}(x, z) \tag{1.15}$$

The support function $\rho(z, A)$ of a non-empty set A is a positive homogeneous and convex function of z. Therefore

$$\rho(\alpha z_1 + \beta z_2, A) \leq \alpha \rho(z_1, A) + \beta \rho(z_2, A) \tag{1.16}$$

for any $z_1 \in R^n$, $z_2 \in R^n$, $\alpha > 0$, and $\beta > 0$.

Let A and B be non-empty closed convex sets in R^n. If $x \in A$, then $(x, z) \leq \rho(z, A)$ for any $z \in R^n$. Inversely, if the latter inequality holds for any $z \in R^n$, then $x \in A$. The inclusion $A \subset B$ is equivalent to the inequalities $\rho(z, A) \leq \rho(z, B)$ which hold for all $z \in R^n$. Suppose $C = A + B$; then $\rho(z, C) = \rho(z, A) + \rho(z, B)$.

The hyperplane defined by the equation $(x, z) = \rho(z, A)$ for some $z \in R^n$ is called the support hyperplane of the set A; the vector z is called the support vector. Let x^* be a point on the boundary of the closed convex set A. There exists at least one support hyperplane containing the point x^*. We have $(x, z) \leq (x^*, z)$ for any $x \in A$, i.e., the set A lies to one side of the support hyperplane.

1.6 Example: Support Function of an Ellipsoid

To illustrate the concepts introduced above, let us find the support function of an ellipsoid. Consider the ellipsoid E in R^n defined by the inequality

$$E : (Q^{-1}(x - a), (x - a)) \leq 1 \tag{1.17}$$

Here, $a \in R^n$ is the center of the ellipsoid, and Q is a symmetric positive definite $n \times n$-matrix. Since E is a bounded closed set, we have, in accordance with (1.15)

$$\rho(z, E) = \max_{x \in E}(x, z) \tag{1.18}$$

It is obvious that the maximum (1.18) is attained on the boundary of the ellipsoid (1.17). Thus, we come to the following extremal problem with a constraint

$$(x, z) \to \max, \qquad (Q^{-1}(x - a), (x - a)) \leq 1 \tag{1.19}$$

The Lagrange function for the problem (1.19) is

$$L = (x, z) + \lambda(Q^{-1}(x - a), (x - a))$$

where λ is a Lagrange multiplier. We equate the gradient of the Lagrange function to zero and obtain: $z + 2\lambda Q^{-1}(x - a) = 0$. From this equation we find

$$x = a - (2\lambda)^{-1}Qz \tag{1.20}$$

Substituting (1.20) into (1.19) we obtain

$$(Qz, z) = 4\lambda^2, \qquad \lambda = \pm(1/2)(Qz, z)^{1/2}$$

Inserting this expression for λ into equation (1.20) we get

$$x = a \pm (Qz, z)^{-1/2}Qz \tag{1.21}$$

Here, the signs "$-$" and "$+$" correspond to the minimal and maximal values of the scalar product (x, z), respectively. Since we seek the maximum in (1.19), we choose "$+$". Substituting (1.21) into (1.18), we obtain finally

$$\rho(z, E) = \max_{x \in E}(x, z) = (Qz, z)^{1/2} + (a, z) \tag{1.22}$$

1.7 Properties of Reachable Sets

Theorem 1.1. Reachable sets introduced by Definition 1.1 have the following property

$$D(t, s, M) = D(t, \tau, D(\tau, s, M)) \tag{1.23}$$

Here, τ is any instant, $\tau \in [s, t]$.

The proof of Theorem 1.1 can be found in [14]. The property (1.23) is called evolutionary because it describes the evolution of reachable sets in time.

Consider the system (1.3) with the constraint (1.4) and the initial condition $x(s) = x^0$.

Theorem 1.2. Suppose the function f in (1.3) is continuously differentiable, the set U is a compact set, all measurable functions $u(t) \in U$, $t \in [s, T]$, are admissible controls, the set $X(x, t) = f(x, U, t)$ is a convex compact set for any x and $t \in [s, T]$, and for all trajectories we have $\mid x(t) \mid < b, t \in [s, T], b = \mathrm{const} > 0$. Then the reachable set $D(t, s, x^0)$ is a compact set depending continuously on $t \in [s, T]$.

Stronger statements can be obtained for linear systems. Consider a linear controlled system

$$\dot{x} = C(t)x + K(t)u + g(t), \quad u(t) \in U, \quad x(s) = x^0 \qquad (1.24)$$

As before, $x \in R^n$ is the state vector, $u \in R^m$ is the control, x^0 is the given initial vector, and s is the initial time instant. Suppose the coefficients of the system (1.24) satisfy the following conditions.

1. The matrices $C(t), K(t)$ and the vector $g(t)$ are measurable functions of time, and their norms are Lebesgue integrable on any compact set belonging to the t-axis.

2. The control $u(t)$ is a bounded measurable function, $u(t) \in U, t \in [s, T]$.

Theorem 1.3. Suppose the linear controlled system (1.24) satisfies the conditions 1, 2 for $t \in [s, T]$, and U is a compact set. Then the reachable set $D(t, s, x^0)$ is a convex compact set depending continuously on $t \in [t, T]$. The reachable set of system (1.24) under the constraint $u(t) \in \mathrm{co}U$ coincides with the reachable set $D(t, s, x^0)$ of the initial system for all $t \in [s, T]$.

The proof of Theorems 1.2 and 1.3 is given in [6].

The statement of Theorem 1.3 on the convexity of reachable sets holds, if we replace the initial condition in (1.24) by the condition $x(t) \in M$ where M is a closed convex set. This assertion is proved in [13, 14].

1.8 Optimal Control and Reachable Sets

Reachable sets can be constructed by means of the optimal control methods. Consider the following auxiliary optimal control problem: find the minimal value of the linear terminal functional $J = (c, x(T))$ for the control system (1.3) under the constraint (1.4) and initial condition $x(s) = x^0$. Here, c is a given n-vector, T is a fixed time instant, $T > s$. Pontryagin's maximum principle reduces this problem to the nonlinear boundary-value problem for the system of order $2n$ which includes the original system (1.3) and the adjoint system. Knowing the solution of this problem, in particular the optimal trajectory $x^*(t)$ and the minimal value J^* of the functional J, we can obtain the support function of the reachable set $D(T, s, x^0)$ at the point

$x^*(T)$ as follows

$$\rho(-c, D(T, s, x^0)) = -J^* = -(c, x^*(T)) \qquad (1.25)$$

Taking a number of different values for the vector c, we can, in principle, obtain an approximation of the reachable set $D(T, s, x^0)$.

In the case of the linear system (1.24), this approach has some advantages. First, for these systems the two-point boundary-value problem of the maximum principle is reduced to two initial value problems. Second, the reachable set for the linear system is convex, and thus is fully characterized by its support functions for different c. Let $x_i^*(T)$ be the end points of optimal trajectories corresponding to the values c_i of the vector c, $i = 1, \ldots, N$. Then we have the following two-sided estimates on the reachable set: $\Gamma^- \subset D(T, s, x^0) \subset \Gamma^+$ where Γ^- and Γ^+ are inner and outer approximating polyhedra. The inner convex polyhedron $\Gamma^- = \mathrm{co}(x_1^*(T), \ldots, x_N^*(T))$ has $x_i^*(T)$ as its vertices, whereas the boundary of the outer convex polyhedron Γ^+ is formed by the support hyperplanes (1.25) with c and $x^*(T)$ replaced by c_i and $x_i^*(T)$, respectively.

If the number N is large enough, we can obtain sufficiently good approximation of the reachable set at given $t = T$. However, for large n this procedure requires a great amount of computation because N must grow fast with n. Moreover, we are to repeat the whole procedure anew for each value of T. Also, it must be noted that polyhedral approximations of reachable sets are not convenient for applications, because the boundaries of polyhedra are not smooth.

The procedure described above becomes much more complicated and is not always feasible for nonlinear systems because of the following serious difficulties. The nonlinear boundary-value problem of the maximum principle is not reduced here to two initial value problems and can be very sensitive and time consuming. This problem can have more than one solution, and we are not sure that the obtained solution is optimal one because the maximum principle is only a necessary (not sufficient) optimality condition. The reachable set may be non-convex and is not specified by its support functions. Thus, methods of optimal control do not seem to be efficient for constructing or approximating reachable sets for different time instants, if n is large.

1.9 Example of a Reachable Set

We present here a simple example in which the reachable set is obtained analytically by the procedure described above. The system of the second order is described by the equations and initial conditions

$$\dot{x}_1 = x_2, \quad \dot{x}_2 = u, \quad |u| \le 1 \qquad (1.26)$$

$$x_1(0) = x_2(0) = 0$$

The reachable set $D(T, 0, 0)$ for the system (1.26) is specified by the inequalities

$$\frac{(x_2 + T)^2}{4} - \frac{T^2}{2} \le x_1 \le \frac{T^2}{2} - \frac{(x_2 - T)^2}{4}, \quad |x_2| \le T \qquad (1.27)$$

This set is bounded by two arcs of parabolas and has two angular points.

2. WHAT ARE THE FUNDAMENTALS OF THE ELLIPSOIDAL MODELLING?

2.1 Approximation of Reachable Sets

The approach described in Section 1.8 makes it possible to obtain two-sided polyhedral approximations of reachable sets. The number N of faces of the outer approximating polyhedron (equal to the number of vertices of the inner approximating polyhedron) is determined by the number N of different values ascribed to the vector c. Hence, N is also the number of optimal control problems which are to be solved in the frames of this approach. If the dimension n is sufficiently large, the number N must be very large for a satisfactory approximation, namely, $N \sim \exp(kn)$ where k is a positive constant. Thus, this approach needs a lot of computation for large n, and these computations must be repeated for each time instant T. Besides, it is not convenient to use polyhedral approximations of reachable sets in applications.

The alternative approach implies approximation of reachable sets by a class of domains having a certain shape and depending on a limited number of parameters. In this case, we cannot attain convergence of approximating domains to the reachable set which is possible for polyhedra as $N \to \infty$. On the other hand, there are certain advantages. For suitable classes of approximating domains, we can essentially simplify both the approximation procedure and various applications.

To realize this approach, some class of approximating domains must be chosen and the basic operations necessary for transforming these domains should be developed. These operations form the algebra for the chosen class of domains and include affine transformations, addition, and intersection of domains. Operations of affine transformations and addition are used directly in the approximation procedure for reachable sets. To show this, let us consider a linear system (1.24) and rewrite it in a finite-difference form

$$x(t + h) = x_1 + x_2 + o(h), \quad x_1 = h[K(t)u(t) + g(t)] \qquad (2.1)$$

$$x_2 = [I + hC(t)]x(t)$$

Here, h is a small time step, I is the unit matrix. Suppose we know that at some instant t the vectors of state $x(t)$ and control $u(t)$ belong to certain sets. To obtain the set containing the vector $x(t + h)$, we are to find the affine transformations of sets and the sum of two sets containing vectors x_1 and x_2. Intersection of sets is needed for the guaranteed state estimation based on observational data.

In what follows, reachable sets are systematically approximated by means of ellipsoids. We shall consider some properties of ellipsoids which are essential for ellipsoidal approximations of sets and discuss advantages of ellipsoids compared to other classes of sets.

2.2 Class of Ellipsoids

Consider closed ellipsoidal sets in n-dimensional Euclidean space. An ellipsoid can be defined by the following inequality:

$$(Q^{-1}(x-a),(x-a)) \leq 1 \tag{2.2}$$

Here, a is the n-dimensional vector of the center of the ellipsoid, and Q is a symmetric positive definite $n \times n$-matrix. For the sake of brevity, we shall denote the ellipsoid (2.2) by $E(a,Q)$. Thus,

$$E(a,Q) = \{x : (Q^{-1}(x-a),(x-a)) \leq 1\} \tag{2.3}$$

Ellipsoids (2.2) or (2.3) depend on $n + n(n+1)/2$ parameters. Of these, $n(n+1)/2$ parameters are elements of the symmetric matrix Q, whereas n parameters are components of the vector a. By rotation of the coordinate system, we can make its axes parallel to the principal axes of the ellipsoid (2.3). Then the matrix Q becomes diagonal

$$Q = \operatorname{diag}(Q_{11}, \ldots, Q_{nn}) \tag{2.4}$$

The inequality (2.2) in the case (2.4) is reduced to

$$\sum_{i=1}^{n} Q_{ii}^{-1}(x_i - a_i)^2 = \sum_{i=1}^{n} \frac{(x_i - a_i)^2}{c_i^2} \leq 1 \tag{2.5}$$

where c_i are the semi-axes of the ellipsoid. Thus, for (2.4)

$$Q_{ii} = c_i^2, \qquad i = 1, \ldots, n \tag{2.6}$$

If $Q = r^2 I$, where $r > 0$ is a scalar, the ellipsoid (2.3) is a ball in n-dimensional space. Its center is placed at the point a and its radius is equal to r. Using notation of Section 1.5, we have

$$E(a, r^2 I) = S_r(a) \tag{2.7}$$

Sometimes we shall also consider degenerate ellipsoids for which one of the two matrices (Q or Q^{-1}) is non-negative definite rather than positive definite. If Q is not positive definite, some of the semi-axes of the ellipsoid are zero ($c_i = 0$ in (2.6)), and its volume is also equal to zero. A disk in three-dimensional space can serve as an example for this case. The second case (Q^{-1} is not positive definite) implies that some of the semi-axes of the ellipsoid are infinite ($c_i = \infty$ in (2.6)), and the volume of the ellipsoid is infinite as well. For example, a cylinder is such a degenerate ellipsoid. If the center of an ellipsoid and the length of one of its axes go to infinity simultaneously, this ellipsoid degenerates into a half-space. According to (2.2), the case $Q \to 0$ corresponds to the point set: $E(a,Q) = \{x = a\}$. Formally speaking, the inequality (2.2) and other formulas obtained below are not valid for degenerate ellipsoids. However, these ellipsoids will be quite useful, and careful consideration shows that many results obtained for non-degenerate ellipsoids are also valid for their degenerate limits.

2.3 Affine Transformation of Ellipsoids

An affine transformation in R^n is defined by

$$y = Ax + b \tag{2.8}$$

where A is a non-singular $n \times n$-matrix and b is an n-vector. The transformation (2.8) applied to any ellipsoid (2.3) gives an ellipsoid. Consider an arbitrary vector $x \in E(a, Q)$. Let us solve (2.8) with respect to x and insert $x = A^{-1}(y - b)$ into (2.3). Thus, we obtain

$$(Q^{-1}(A^{-1}(y - b) - a), (A^{-1}(y - b) - a))$$

$$= ((AQA^T)^{-1}(y - (Aa + b)), (y - (Aa + b))) \leq 1$$

where superscript T denotes the transposed matrix. Hence, y belongs to the ellipsoid with the center $Aa + b$ and the matrix AQA^T, i.e.

$$y = Ax + b \in E(Aa + b, AQA^T), \quad x \in E(a, Q) \tag{2.9}$$

Using functions of sets (see Section 1.5), we can rewrite (2.9) as follows

$$AE(a, Q) + b = E(Aa + b, AQA^T) \tag{2.10}$$

Formulas (2.9) and (2.10) define the affine transformation of ellipsoid (2.3) corresponding to the affine transformation (2.8) of vectors. These formulas are also valid for various kinds of degenerate ellipsoids, as well as for the degenerate transformation (2.8) with $\det A = 0$.

Transformation (2.8) can be chosen so as to make the principal axes of ellipsoid (2.3) parallel to the coordinate axes. To achieve that, we impose the following condition $AQA^T = D$ on the matrix A, where D is a diagonal matrix. It is well-known that such transformation can be performed by an orthogonal matrix, i.e. by rotation of the coordinate system.

Ellipsoid (2.3) can be converted into the unit ball $E(0, I) = S_1(0)$ by the affine transformation (2.8). According to (2.9), this transformation can be obtained if $AQA^T = I, Aa + b = 0$. These conditions are satisfied if $A = Q^{1/2}, \quad b = -Q^{1/2}a$. Here, $Q^{1/2}$ is the square root of the symmetric positive definite matrix Q. This square root exists and is a symmetric positive definite matrix itself [18].

In what follows we shall need the formula for the volume of an n-dimensional ellipsoid (2.3)

$$V_E = \frac{\pi^{n/2}(\det Q)^{1/2}}{\Gamma(\frac{n}{2} + 1)} \tag{2.11}$$

Here, $\Gamma(z)$ is the Euler gamma-function.

2.4 Approximation of Convex Sets by Ellipsoids

We begin with some general results presented in [19, 20].

Theorem 2.1. For an arbitrary convex set D in R^n, there exists an ellipsoid $E(a,Q) \subset D$ such that $E(a,n^2Q)$ contains D, i.e.

$$E(a,Q) \subset D \subset E(a,n^2Q) \tag{2.12}$$

The proof of Theorem 2.1 is given in [21].

Theorem 2.2. (F. John [22]). Let D be a convex set in R^n which has the property of central symmetry with respect to some point a. There exists an ellipsoid $E(a,Q) \subset D$ such that $E(a,nQ)$ contains D, i.e.

$$E(a,Q) \subset D \subset E(a,nQ) \tag{2.13}$$

The ellipsoids $E(a,n^2Q)$ and $E(a,nQ)$ in (2.12), (2.13) are homothetic to the ellipsoid $E(a,Q)$, and their axes are, respectively, n and $n^{1/2}$ times longer, as follows from (2.6). Theorems 2.1 and 2.2 provide some estimates for the approximation accuracy attainable in the class of ellipsoids. The property of central symmetry makes it possible to substantially improve this accuracy and to decrease gaps between convex sets and approximating ellipsoids.

There exist convex sets for which coefficients (n^2 and n) in (2.12) and (2.13) cannot be made smaller. These sets are n-dimensional simplexes for Theorem 2.1, and n-dimensional parallelepipeds for Theorem 2.2. The proof of this statement can be found in [14].

To approximate sets, we will use inner ellipsoids E^- of the maximal volume and outer ellipsoids E^+ of the minimal volume. The following theorems [23] are useful.

Theorem 2.3. For any bounded set D in R^n, there exists a unique ellipsoid E^+ of the minimal volume containing D, $E^+ \supset D$.

Theorem 2.4. For any closed convex set D in R^n, there exists a unique ellipsoid E^- of the maximal volume contained in D, $E^- \subset D$.

Problems of finding the outer ellipsoid of the minimal volume and the inner ellipsoid of the maximal volume can be stated as the following extremal problems.

Find such vectors a and symmetric positive definite matrices Q which satisfy the conditions

$$D \subset E(a,Q) = E^+, \qquad \det Q \to \min \tag{2.14}$$

$$D \supset E(a,Q) = E^-, \qquad \det Q \to \max \tag{2.15}$$

Theorem 2.3 states that the problem (2.14) has the unique solution for any bounded set, and Theorem 2.4 states that the problem (2.15) has the unique solution for any closed convex set D. The following Corollary 2.1 stems from Theorems 2.3 and 2.4. Its proof is given in [14].

Corollary 2.1. Let the bounded set D be symmetric with respect to some coordinate hyperplane: if $x \in D$ then also $x' \in D$, where x' differs from x by a sign of some component x_i, i.e. $x'_j = x_j, x'_i = -x_i, j \neq i, 1 \leq j \leq n$. Then both the outer ellipsoid E^+ of the minimal volume and (if D is a closed convex set) the inner ellipsoid E^- of the maximal volume for the set D, have x_i as one of their axes of symmetry.

2.5 Operations of Addition for Ellipsoids

We shall give here two-sided (inner and outer) ellipsoidal approximations for sets which are the Minkowsky sums of two arbitrary ellipsoids. These approximations are optimal in the sense of volume: the inner ellipsoid has the largest, whereas the outer ellipsoid has the smallest possible volume. Thus, we shall obtain inner and outer optimal operations of addition for ellipsoids which will be employed for approximation of reachable sets. Consider two ellipsoids in n-dimensional Euclidean space R^n. According to (2.3), these ellipsoids are defined by

$$E(a_1, Q_1) : (Q_1^{-1}(x - a_1), (x - a_1)) \leq 1 \tag{2.16}$$

$$E(a_2, Q_2) : (Q_2^{-1}(x - a_2), (x - a_2)) \leq 1$$

Here, a_i are the n-dimensional vectors of the centers of ellipsoids, Q_i are symmetric positive definite $n \times n$-matrices, $i = 1, 2$.

The Minkowsky sum S of two ellipsoids (2.16) is (see Section 1.5) the set of all points x such that $x = x_1 + x_2$, where x_1 and x_2 belong to ellipsoids (2.16), i.e.

$$x = x_1 + x_2 \in S = E(a_1, Q_1) + E(a_2, Q_2) \tag{2.17}$$

$$x_1 \in E(a_1, Q_1), \qquad x_2 \in E(a_2, Q_2)$$

The set S is bounded, closed and convex, but, in general, it is not an ellipsoid. Consider two examples for a two-dimensional case, $n = 2$.

1. Let the both ellipses (2.16) degenerate into segments: one axis of each ellipse has zero length. We assume also that the centers of these segments coincide with the origin of coordinates $x = 0$. Then the sum (2.17) of our segments is a parallelogram with sides parallel to the segments.

2. Let one of the ellipses be a segment $[-1, 1]$ of the axis x_1, whereas the other ellipse be a circle of the radius r, i.e.

$$E(a_1, Q_1) = \{x : |x_1| \leq 1, x_2 = 0\} \tag{2.18}$$

$$E(a_2, Q_2) = \{x : x_1^2 + x_2^2 \leq r^2\}$$

The sum (2.17) of ellipses (2.18) is a domain S consisting of a rectangle $|x_1| \leq 1, |x_2| \leq r$ and two half-circles of the radius r with the centers placed at the points $x_1 = \pm 1, x_2 = 0$.

Two following problems relate to the outer and inner addition of ellipsoids.

Problem 2.1. Find the ellipsoid $E(a^+, Q^+)$ of the minimal volume containing the sum S of two ellipsoids (2.17), i.e. an ellipsoid such that $S \subset E(a^+, Q^+)$, $\det Q^+ \to \min$.

Problem 2.2. Find the ellipsoid $E(a^-, Q^-)$ of the maximal volume contained in the sum S of two ellipsoids (2.17), i.e. an ellipsoid such that $E(a^-, Q^-) \subset S$, $\det Q^- \to \max$.

According to Theorems 2.3 and 2.4, Problems 2.1 and 2.2 have unique solutions.

These solutions in the general case were first obtained in [24] and are presented below in Theorems 2.5 and 2.6 for respective Problems 2.1 and 2.2.

Theorem 2.5. The parameters of the ellipsoid $E(a^+, Q^+)$ of the minimal volume containing the sum S of the ellipsoids (2.17), one of which may be degenerate (the matrix Q_1 is non-negative definite, and the matrix Q_2 is positive definite), are given by

$$a^+ = a_1 + a_2, \quad Q^+ = (p^{-1} + 1)Q_1 + (p + 1)Q_2 \qquad (2.19)$$

Here, $p > 0$ is the unique positive root of the algebraic equation

$$\sum_{j=1}^{n} \frac{1}{p + \lambda_j} = \frac{n}{p(p+1)} \qquad (2.20)$$

where $\lambda_j \geq 0, j = 1, \ldots, n$, are the roots of the characteristic equation

$$\det(Q_1 - \lambda Q_2) = 0 \qquad (2.21)$$

The roots λ_j are counted according to their multiplicity.

Theorem 2.6. The parameters of the ellipsoid $E(a^-, Q^-)$ of the maximal volume contained in the sum S of the ellipsoids (2.17), one of which may be degenerate (the matrix Q_1 is non-negative definite, and the matrix Q_2 is positive definite), are given by

$$a^- = a_1 + a_2$$

$$Q^- = Q_1 + Q_2 + 2Q_2^{1/2}(Q_2^{-1/2}Q_1Q_2^{-1/2})^{1/2}Q_2^{1/2} \qquad (2.22)$$

$$= Q_1 + Q_2 + 2Q_1^{1/2}(Q_1^{-1/2}Q_2Q_1^{-1/2})^{1/2}Q_1^{1/2}$$

Here, both expressions for Q^- are equivalent if both matrices Q_1 and Q_2 are positive definite. If the matrix Q_1 is not positive definite, the first expression (2.22) for Q^- should be used.

The proofs of Theorems 2.5 and 2.6 can be found in [13, 14, 24].

Thus, for any two ellipsoids $E(a_1, Q_1)$ and $E(a_2, Q_2)$, we have determined ellipsoids $E(a^+, Q^+)$ and $E(a^-, Q^-)$ such that the following inclusions are true

$$E(a^-, Q^-) \subset E(a_1, Q_1) + E(a_2, Q_2) \subset E(a^+, Q^+) \qquad (2.23)$$

Here, the ellipsoid $E(a^-, Q^-)$ has the minimal possible volume, and the ellipsoid $E(a^+, Q^+)$ has the maximal possible volume. Therefore, we have obtained the optimal (in the sense of volume) two-sided (inner and outer) ellipsoidal approximations for the Minkowsky sum of two ellipsoids. In other words, the approximating ellipsoids

$E(a^-, Q^-)$ and $E(a^+, Q^+)$ are the results of the optimal (in the sense of volume) operations of the inner and outer addition in the class of ellipsoids. Note that according to (2.19) and (2.22), the centers of the both resulting ellipsoids coincide: $a^+ = a^- = a_1 + a_2$.

The addition operations for ellipsoids possess the commutativity property: if the ellipsoids $E(a_1, Q_1)$ and $E(a_2, Q_2)$ change places, the resulting ellipsoids $E(a^-, Q^-)$ and $E(a^+, Q^+)$ undergo no change [14]. However, these operations do not possess the property of associativity. If more than two ellipsoids are added up, the resultant ellipsoids $E(a^+, Q^+)$ and $E(a^-, Q^-)$ depend on the order of the added ellipsoids.

The addition operations introduced by Theorems 2.5 and 2.6 possess the following invariance property. Let the original ellipsoids $E(a_1, Q_1)$ and $E(a_2, Q_2)$ be subjected to a linear transformation with a non-singular matrix C. Let us find the inner and outer approximations for the sums of the transformed ellipsoids, using Theorems 2.5 and 2.6, and then subject these approximations to the inverse linear transformation with the matrix C^{-1}. The resultant ellipsoids will coincide with the results of the direct addition of the original ellipsoids. The proof of this property is given in [14]. This invariance property is very important. It means that the results of our addition operations do not depend on the choice of the coordinate system (one may perform arbitrary affine transformations) and, in particular, on the choice of units for the state variables.

2.6 Example

Let the ellipsoids (2.16) be homothetic, i.e.

$$E(a_i, Q_i) : (Q_i^{-1}(x - a_i), (x - a_i)) \leq 1 \tag{2.24}$$

$$i = 1, 2, \qquad Q_1 = \nu Q_2, \qquad \nu > 0$$

where ν is a scalar. The characteristic equation (2.21) in the case (2.24) is reduced to $\det[(\nu - \lambda)Q_2] = 0$ and has the multiple root $\lambda_j = \nu, j = 1, \ldots, n$. In this case, equation (2.20) takes the form $p + \nu = p(p + 1)$ and has a positive root $p = \nu^{1/2}$. According to (2.19) and (2.24), we obtain

$$a^+ = a_1 + a_2, \qquad Q^+ = (\nu^{1/2} + 1)^2 Q_2 \tag{2.25}$$

From (2.22) and (2.24) we get

$$a^- = a_1 + a_2$$

$$Q^- = \nu Q_2 + Q_2 + 2\nu^{1/2} Q_2 = (\nu^{1/2} + 1)^2 Q_2 \tag{2.26}$$

In the case of (2.24) and only in this case, the sum S of two ellipsoids (2.17) is an ellipsoid itself. That is why both the inner and outer ellipsoids coincide here with each other and with the set S, compare (2.25) and (2.26). Of course, this is always true for the one-dimensional case $(n = 1)$, where all ellipsoids are, in fact, segments, and also in the case, where ellipsoids (2.24) are balls.

2.7 Discussion

There are also other approaches to the outer approximation of the sum S of two ellipsoids (2.17). In [10], the outer ellipsoid is taken as follows

$$S \subset E(a, Q), \quad a = a_1 + a_2. \quad Q = \gamma_1 Q_1 + \gamma_2 Q_2 \qquad (2.27)$$

where positive constants γ_1 and γ_2 are such that the inclusion $S \subset E(a, Q)$ holds. For example, one can take $\gamma_1 = \gamma_2$ or $\gamma_1 = \gamma^{-1}$, $\gamma_2 = (1-\gamma)^{-1}$ where $\gamma \in (0, 1)$. Comparing formulas (2.19) and (2.27), we see that in the optimal approach given by Theorem 2.5, the coefficients γ_1 and γ_2 depend on the matrices Q_1 and Q_2, so that Q is a non-linear function of Q_1 and Q_2. Numerical results presented in [14] show a considerable advantage of the optimal operation given by Theorem 2.5 over the approach described by (2.27): optimal ellipsoids have much less volume than ellipsoids (2.27). Ellipsoidal approximations which are not necessarily optimal are considered in [15]. Note that the Minkowsky sum of two ellipsoids can be approximated by an ellipsoid which is optimal in some sense other than volume. The case of the general optimality criterion will be considered below.

Let us sum up those properties of the class of ellipsoids which led us to choose this class for the approximation of reachable sets.

1. Ellipsoids can give a satisfactory two-sided approximation of arbitrary convex sets. Theorems 2.1 and 2.2 evaluate the maximal possible error of such approximation. Theorems 2.3 and 2.4 guarantee the existence and uniqueness for the approximating ellipsoids of the extremal volumes.

2. The class of ellipsoids is invariant with respect to affine transformations.

3. An ellipsoid in R^n is defined by a comparatively small number of parameters equal to $n(n + 1)/2 + n$.

4. Ellipsoids have smooth boundaries and are defined by simple analytical formulas. This property makes them suitable for various operations needed in applications.

5. The class of ellipsoids is in many aspects similar to the class of quadratic Lyapunov functions (in the theory of stability) and to the class of Gaussian random distributions (in the theory of probability). These similarities manifest themselves in the invariance with respect to affine transformations, in the same number of parameters, and in the close connection of all these objects with positive definite quadratic forms. Note that the surfaces of constancy for the quadratic Lyapunov functions and Gaussian distributions are ellipsoids.

6. We have obtained constructive inner and outer operations of addition in the class of ellipsoids. These operations given by Theorems 2.5 and 2.6 are optimal in the sense of volume. Using these operations, we shall find the two-sided ellipsoidal approximations of reachable sets. Further on, we shall also consider approximations optimal with respect to a general criterion.

Note the importance of Property 2 which means the invariance of ellipsoids. Suppose the units of some state variables are changed. This change of scales is equivalent to some linear transformation. It is desirable that our approximations of sets should be independent of this change. However, in the absence of Property 2, our approximations will not be invariant to scaling.

Consider some other classes of approximating sets.

The class of balls is the simplest but rather poor. It can only provide very rough approximation and lacks Property 2. Moreover, operations with balls require, quite unexpectedly, more computations than operations with ellipsoids.

The class of parallelepipeds and the class of simplexes posses the invariance property, but are less convenient than ellipsoids, see Property 4. Besides, parallelepipeds and simplexes are defined by $n + 1$ vectors, i.e., by $n(n + 1)$ parameters, which is almost twice as much as the number needed for ellipsoids, see Property 3.

The class of rectangular parallelepipeds is defined by the same number of parameters as ellipsoids. This class, however, is not invariant with respect to linear transformations.

All these considerations led us to choose the class of ellipsoids for approximation of reachable sets.

3. HOW DOES THE ELLIPSOIDAL MODELLING WORK?

3.1 Evolutionary Properties of Approximating Ellipsoids

In Section 3, we shall obtain our main result: we shall derive ordinary differential equations governing the evolution of ellipsoids which present two-sided (inner and outer) optimal approximations of reachable sets. First, we shall introduce important evolutionary properties of these approximations.

According to Theorem 1.1, reachable sets $D(t, s, M)$ of the controlled systems has the evolutionary property (1.23) which implies that the reachable set for the instant t can be obtained from the same set for the instant τ by prolongation of all trajectories, starting from the instant τ up to the instant t.

Definition 3.1. Sets $D^-(t)$ are called *subreachable* sets for the controlled system (1.3) and (1.5) (or (1.6)) with the initial condition (1.8), if the following inclusion holds for all $\tau \in [s, t]$

$$D^-(t) \subset D(t, \tau, D^-(\tau)) \tag{3.1}$$

and, besides, if $D^-(S) \subset M$.

Definition 3.2. Sets $D^+(t)$ are called *superreachable* sets for the controlled system (1.3) and (1.5) (or (1.6)) with the initial condition (1.8), if the following inclusion holds for all $\tau \in [s, t]$

$$D^+(t) \supset D(t, \tau, D^+(\tau)) \tag{3.2}$$

and, besides, if $D^+(s) \supset M$.

The properties (3.1) and (3.2) similar to the evolutionary property (1.23) of the reachable set will be called *evolutionary* properties of the approximating sets.

It follows from (3.1), (3.2) that for all time instants

$$D^-(t) \subset D(t, s, D^-(s)) \subset D(t, s, M) \subset D(t, s, D^+(s)) \subset D^+(t)$$

Therefore, subreachable and superreachable sets provide two-sided approximations of the reachable set $D(t, s, M)$ for all $t \geq s$, i.e.

$$D^-(t) \subset D(t, s, M) \subset D^+(t) \tag{3.3}$$

We shall obtain the two-sided approximations (3.3) by means of ellipsoidal sets $D^-(t)$ and $D^+(t)$ with evolutionary properties. These properties impose certain conditions on sets and thus diminish the approximation accuracy. To clarify this remark, consider the following situation. Assume that there is some approximation error at the instant τ, i.e., the set $D^-(\tau)$ is smaller or the set $D^+(\tau)$ is larger than $D(\tau, s, M)$. This error cannot be reduced; moreover, it will accumulate for $t \geq \tau$, according to (3.1) and (3.2). However, sets with the evolutionary properties (3.1) and (3.2) have also some important advantages. These sets can be obtained by means of step-by-step procedures, they can be described by differential equations and are convenient for applications.

3.2 Linear Systems

Consider a linear system described by the following vector differential equation and constraint

$$\dot{x} = C(t)x + K(t)u + f(t), \qquad u \in E(0, G(t)) \tag{3.4}$$

Here, $x \in R^n$ is the state vector; $u \in R^m$ is the vector of control or disturbance; the n-vector $f(t)$, $n \times n$-matrix $C(t)$, $n \times m$-matrix $K(t)$, and $m \times m$-matrix $G(t)$ are piecewise continuous functions of time. The matrix $G(t)$ is symmetric and non-negative definite. The initial conditions are given by

$$x(s) \in E(a_0, Q_0) \tag{3.5}$$

Here, s is the initial time instant, a_0 is a given n-vector, and Q_0 is a given symmetric positive definite $n \times n$-matrix.

The reachable set of the system (3.4) with the initial condition (3.5) is

$$D(t, s, M) = D(t, s, E(a_0, Q_0)) \tag{3.6}$$

We shall consider the following problem.

Find two-sided estimates on the reachable sets (3.6) by means of subreachable and superreachable ellipsoids, see Definitions 3.1 and 3.2. In other words, for the sets (3.6), we are to obtain the estimates (3.3), where $D^-(t)$ and $D^+(t)$ are ellipsoids:

$$E(a^-(t), Q^-(t)) \subset D(t, s, E(a_0, Q_0)) \subset E(a^+(t), Q^+(t)) \tag{3.7}$$

Here, $a^-(t)$ and $a^+(t)$ are n-vectors, $Q^-(t)$ and $Q^+(t)$ are symmetric positive definite $n \times n$-matrices. These vectors and matrices are to be found as functions of time for $t \geq s$. We introduce the new n-vector of control $v(t)$ by

$$v = K(t)u + f(t) \tag{3.8}$$

Since u belongs to the ellipsoid (3.4), the new control vector v from (3.8) belongs, according to the formulas (2.9) and (2.10), to the following ellipsoid

$$v \in E(f(t), B(t)), \qquad B(t) = KGK^T \tag{3.9}$$

Here, $B(t)$ is a symmetric non-negative definite $n \times n$-matrix. We rewrite the system (3.4), taking into account (3.8) and (3.9)

$$\dot{x} = C(t)x + v, \qquad v \in E(f(t), B(t)) \tag{3.10}$$

3.3 Equations for Evolution of Ellipsoids

We take a sufficiently small time increment $h > 0$ and replace the differential equation (3.10) by its finite-difference approximation. Omitting terms much smaller that h, we obtain, similar to (2.1)

$$x(t + h) = x_1 + x_2, \quad x_1 = hv(t), \quad x_2 = [I + hC(t)]x(t) \tag{3.11}$$

By virtue of inclusion (3.10) and equation (2.10), we have for the vector x_1

$$x_1 \in E(hf, h^2 B) \tag{3.12}$$

Assume that inclusions (3.7) hold for some $t \geq s$. Since the vector $x(t)$ belongs to the reachable set (3.6), the vector x_2 from (3.11) belongs to the set D' satisfying the following inclusions

$$E((I + hC)a^-, (I + hC)Q^-(I + hC^T)) \subset D'$$

$$\subset E((I + hC)a^+, (I + hC)Q^+(I + hC^T)) \tag{3.13}$$

Here again, we have used (2.10). The argument t in (3.12) and (3.13) is omitted.

The vector $x(t + h)$ belonging to the reachable set $D(t + h, s, E(a_0, Q_0))$ is represented in (3.11) as a sum of the two vectors x_1 and x_2. The first vector (x_1) belongs to the ellipsoid (3.12), whereas the second one (x_2) belongs to the set D' satisfying the two-sided ellipsoidal estimates (3.13). Therefore, we can obtain the two-sided ellipsoidal estimates for the reachable set $D(t + h, s, E(a_0, Q_0))$ in the following way. For the inner estimate, we should use the inner ellipsoidal approximation for the sum of the two ellipsoids: (3.12) and the inner estimate in (3.13). For the outer estimate, we should use the outer ellipsoidal approximation for the sum of (3.12) and the ellipsoid in the right-hand side of (3.13).

In other words, we are to find the approximating ellipsoids at the instant $t + h$ such that

$$E(a^-(t+h), Q^-(t+h)) \subset D(t+h, s, E(a_0, Q_0)) \subset E(a^+(t+h), Q^+(t+h))$$
$$(3.14)$$

where the following inclusions hold

$$E(a^-(t+h), Q^-(t+h)) \subset E(a_1, Q_1) + E(a_2^-, Q_2^-)$$

$$E(a^+(t+h), Q^+(t+h)) \supset E(a_1, Q_1) + E(a_2^+, Q_2^+) \qquad (3.15)$$

$$a_1 = hf, \quad Q_1 = h^2 B, \quad a_2^- = (I + hC)a^-, \quad a_2^+ = (I + hC)a^+$$

$$Q_2^- = (I + hC)Q^-(I + hC^T), \qquad Q_2^+ = (I + hC)Q^+(I + hC^T)$$

The matrix Q_1 here is proportional to the small quantity h^2. We shall use this fact to simplify the general formulas given in Section 2.5, see Theorems 2.5 and 2.6.

For the outer approximation, the roots λ_j of the characteristic equation (2.21) can be represented as

$$\lambda_j = h^2 \lambda_j^0, \qquad j = 1, \ldots, n \qquad (3.16)$$

where λ_j^0 are new unknowns. Substituting (3.16) and $Q_1 = h^2 B$ into (2.21), we obtain the equation $\det(B - \lambda^0 Q_2^+) = 0$. Its roots coincide with the roots of the following equation

$$\det[(Q_2^+)^{-1} B - \lambda^0 I] = 0 \qquad (3.17)$$

This equation as well as (2.21) has n positive roots λ_j^0, if we take into account their multiplicity. The positive root of equation (2.20) is represented as the expansion

$$p = hq + \ldots \qquad (3.18)$$

where dots denote terms of the higher order of h. Inserting (3.16) and (3.18) into equation (2.20) and expanding both its sides into a power series in h, we obtain

$$q^{-1}\left(n - hq^{-1} \sum_{j=1}^{n} \lambda_j^0 + \ldots\right) = nq^{-1}(1 - hq + \ldots)$$

Making equal the terms proportional to h, we get

$$q = \left(n^{-1} \sum_{j=1}^{n} \lambda_j^0\right)^{1/2} \qquad (3.19)$$

The sum of all roots of the characteristic equation (3.17) is equal [18] to the trace of the matrix $(Q_2^+)^{-1}B$. Hence, we have from (3.19)

$$q = \left\{n^{-1}\mathrm{Tr}[(Q_2^+)^{-1}B]\right\}^{1/2} \tag{3.20}$$

where Tr denotes the trace of a matrix. Substituting (3.18) and (3.19) into (2.19) and taking into account our notation (3.15), we obtain after simplifications

$$a^+(t+h) = a_1 + a_2^+ = a^+ + h(Ca^+ + f), \quad Q^+(t+h) \tag{3.21}$$

$$= [(hq)^{-1}+1]Q_1 + (hq+1)Q_2^+ = Q^+ + h(CQ^+ + Q^+C^T + qQ^+ + q^{-1}B)$$

Here, we again neglect terms of the order h^2 and higher, and omit the argument t in the right-hand sides of the obtained equations. Using notation (3.15), we also simplify equation (3.20) for q

$$q = \left\{n^{-1}\mathrm{Tr}[(Q^+)^{-1}B]\right\}^{1/2} \tag{3.22}$$

For the inner approximation, we substitute formulas (3.15) into (2.22), expand the right-hand sides of the resulting equations into series in h, and obtain with the accuracy up to $O(h)$

$$a^-(t+h) = a^- + h(Ca^- + f) \tag{3.23}$$

$$Q^-(t+h) = Q^- + h[CQ^- + Q^-C^T + 2B^{1/2}(B^{-1/2}Q^-B^{-1/2})^{1/2}B^{1/2}]$$

$$= Q^- + h\left\{CQ^- + Q^-C^T + 2(Q^-)^{1/2}[(Q^-)^{-1/2}B(Q^-)^{-1/2}]^{1/2}(Q^-)^{1/2}\right\}$$

As $h \to 0$, the finite-difference equations (3.21) become the following differential equations

$$\dot{a}^+ = Ca^+ + f \tag{3.24}$$

$$\dot{Q}^+ = CQ^+ + Q^+C^T + qQ^+ + q^{-1}B \tag{3.25}$$

Similarly, we obtain from (3.23) as $h \to 0$

$$\dot{a}^- = Ca^- + f \tag{3.26}$$

$$\dot{Q}^- = CQ^- + Q^-C^T + 2B^{1/2}(B^{-1/2}Q^-B^{-1/2})^{1/2}B^{1/2} \tag{3.27}$$

$$= CQ^- + Q^-C^T + 2(Q^-)^{1/2}[(Q^-)^{-1/2}B(Q^-)^{-1/2}]^{1/2}(Q^-)^{1/2}$$

We assumed that inclusions (3.7) hold at some instant $t \geq s$ and then proved that they hold also for $t + h$, $h > 0$, see (3.14). But at the initial time instant $t = s$ inclusions (3.7) are satisfied, if we take, according to (3.5)

$$a^-(s) = a^+(s) = a_0, \quad Q^-(s) = Q^+(s) = Q^0 \tag{3.28}$$

Thus, inclusions (3.7) hold for all $t \geq s$. The parameters of the approximating ellipsoids satisfy the differential equations (3.24)–(3.27) and the initial conditions (3.28). Since the vectors a^- and a^+ satisfy the same equations (3.24), (3.26) and initial conditions, we have $a^-(t) = a^+(t) = a(t)$, and $a(t)$ satisfies the following initial value problem

$$\dot{a} = C(t)a + f(t), \qquad a(s) = a_0 \tag{3.29}$$

Then inclusions (3.7) can be rewritten as follows

$$E(a(t), Q^-(t)) \subset D(t, s, E(a_0, Q_0)) \subset E(a(t), Q^+(t)) \tag{3.30}$$

3.4 The Main Theorem

To give a rigorous formulation of our results, we shall first introduce some definitions. Consider the linear controlled system (3.4) or the equivalent system (3.10) with the initial conditions (3.5). Let the matrix $C(t)$ be piecewise continuous and bounded for each finite time interval, whereas the matrix $B(t)$ defined in (3.9) and the vector $f(t)$ be continuous functions of t. The family of sets depending on t is called smooth, if the support function to the sets of this family is a smooth function of t. The symmetric matrices Q_0 and $B(t)$ for $t \geq s$ are assumed to be positive definite.

Definition 3.3. A smooth family $D^-(t)$ of subreachable ellipsoids (in the sense of Definition 3.1) for the system (3.10), (3.5) is called the *locally optimal* (in the sense of volume) approximation of the reachable sets of this system, if

$$\frac{d}{dt}\mathrm{vol}D^-(t)\Big|_{t=\tau} \geq \frac{d}{dt}\mathrm{vol}D_1(t)\Big|_{t=\tau} \tag{3.31}$$

for any $\tau \geq s$. Here, $D_1(t)$ is an arbitrary smooth family of ellipsoids defined for $t \geq \tau$ and satisfying the initial condition $D_1(\tau) = D^-(\tau)$ and the evolutionary condition (3.1), and vol in (3.31) denotes the volume of ellipsoids.

Similarly, a smooth family $D^+(t)$ of superreachable ellipsoids (in the sense of Definition 3.2) for the system (3.10), (3.5) is called *locally optimal,* if for any $t \geq s$

$$\frac{d}{dt}\mathrm{vol}D^+(t)\Big|_{t=\tau} \leq \frac{d}{dt}\mathrm{vol}D_2(t)\Big|_{t=\tau} \tag{3.32}$$

Here, $D_2(t)$ is an arbitrary smooth family of ellipsoids defined for $t \geq \tau$ and satisfying (3.2) and the initial condition $D_2(\tau) = D^+(\tau)$.

Now we can formulate our main theorem.

Theorem 3.1. Under the assumptions made at the beginning of Section 3.4, there exist unique locally optimal families of subreachable and superreachable ellipsoids $D^-(t) = E(a(t), Q^-(t))$ and $D^+(t) = E(a(t), Q^+(t))$ such that inclusions (3.30) hold for the system (3.4) or (3.10) with the initial condition (3.5). The parameters a, Q^-, Q^+ of these ellipsoids satisfy the respective differential equations (3.29), (3.27), and (3.25), and the initial conditions (3.29), (3.28). The first form of equation (3.27) is applicable, if the matrix $B(t)$ is positive definite, and the second form is applicable, if $Q^-(t)$ is positive definite.

Thus, to obtain the desired two-sided estimates (3.30) for the system (3.4) or (3.10), (3.5) by means of the locally optimal subreachable and superreachable ellipsoids, we should find the solutions of certain initial value problems for ordinary differential equations. The common center of the inner and outer ellipsoids satisfies the linear initial value problem (3.29). The matrices $Q^-(t)$ and $Q^+(t)$ of the ellipsoids are the solutions of the nonlinear initial value problems for the systems (3.27) and (3.25), respectively, with the initial conditions (3.28). Equations (3.29), (3.27), and (3.25) were first deduced in [24–26].

Note that there are usually some gaps between the reachable sets and the approximating ellipsoids. These gaps are due to the fact that our ellipsoids possess evolutionary properties and are chosen from the classes of the subreachable and superreachable ellipsoids. At each time instant, we use the ellipsoidal approximations of the sum of two ellipsoids; these approximations are accompanied by certain errors. The approximation errors accumulate with time and lead to the gaps mentioned above.

3.5 Nonlinear Systems

Let us extend our results to nonlinear systems. Consider a general nonlinear controlled system (1.3) and (1.5) with the initial condition (1.8). Phase trajectories $x(t)$ of this system coincide with the trajectories of the differential inclusion (1.6). As before, we denote by $D(t, s, M)$ the reachable sets of system (1.3) or (1.6). To obtain the estimates on the sets $D(t, s, M)$, we introduce two auxiliary linear systems (3.10) which will serve for comparison with our nonlinear system.

Suppose the sets $X(x, t)$ and M from (1.6), (1.8) satisfy the following inclusions

$$X(x, t) \supset E(C^-(t)x + f^-(t), B^-(t)), \quad M \supset E(a_0^-, Q_0^-) \qquad (3.33)$$

or the inclusions

$$X(x, t) \subset E(C^+(t)x + f^+(t), B^+(t)), \quad M \subset E(a_0^+, Q_0^+) \qquad (3.34)$$

Here, $C^-(t)$ and $C^+(t)$ are $n \times n$-matrices, $f^-(t)$ and $f^+(t)$ are n-vectors, $B^-(t)$ and $B^+(t)$ are symmetric non-negative definite $n \times n$-matrices, a_0^- and a_0^+ are given n-vectors, and Q_0^- and Q_0^+ are given symmetric positive definite $n \times n$-matrices. The matrix-valued and vector-valued functions $C^\pm(t)$, $B^\pm(t)$, and $f^\pm(t)$ are defined, bounded and piecewise continuous in the interval $t \in [s, T]$. We assume that the first inclusions in (3.33) and (3.34) hold at least in that domain (in (x, t)-space) which contains all phase trajectories of our system (1.3) (or (1.6)) for $t \in [s, T]$. Note that the right-hand sides of the first inclusions in (3.33) and (3.34) correspond to linear systems (3.10).

Theorem 3.2. If inclusions (3.33) hold, then the following inner estimates on the reachable sets $D(t, s, M)$ of system (1.3), (1.5) (or (1.6)) are true

$$D(t, s, M) \supset E(a^-(t), Q^-(t)), \quad t \in [s, T] \tag{3.35}$$

If inclusions (3.34) hold, then the outer estimates on the sets $D(t, s, M)$ hold

$$D(t, s, M) \subset E(a^+(t), Q^+(t)), \quad t \in [s, T] \tag{3.36}$$

Here, $a^-(t)$ and $Q^-(t)$ are the parameters of the inner approximating ellipsoid for the linear system

$$\dot{x} = C^-(t)x + v, \quad v(t) \in E(f^-(t), B^-(t)) \tag{3.37}$$

$$x(s) \in E(a_0^-, Q_0^-)$$

whereas $a^+(t)$ and $Q^+(t)$ are the parameters of the outer approximating ellipsoid for the linear system

$$\dot{x} = C^+(t)x + v, \quad v(t) \in E(f^+(t), B^+(t)) \tag{3.38}$$

$$x(s) \in E(a_0^+, Q_0^+)$$

The vectors $a^-(t)$ and $a^+(t)$ are the solutions of the initial value problems (3.29), whereas $Q^-(t)$ and $Q^+(t)$ are the solutions of the initial value problems (3.27), (3.28) and (3.25), (3.28), respectively. Here, the vectors and matrices C, f, B, a, a_0, and Q_0 in equations (3.25), (3.27)–(3.29) are to be taken with the superscripts "–" for a^- and Q^-, and with the superscripts "+" for a^+ and Q^+.

Proof. Inclusions (3.33) imply that the reachable set $D(t, s, M)$ of the original nonlinear system (1.3) (or (1.6)) contains the reachable set $D^-(t, s, E(a_0^-, Q_0^-))$ of the auxiliary linear system (3.37), i.e.

$$D(t, s, M) \supset D^-(t, s, E(a_0^-, Q_0^-)), \quad t \in [s, T] \tag{3.39}$$

Since the system (3.37) belongs to the type of (3.10), the left inclusion in (3.30) holds for the reachable set of this system

$$D^-(t, s, E(a_0^-, Q_0^-)) \supset E(a^-(t), Q^-(t)), \quad t \in [s, T] \tag{3.40}$$

Inclusions (3.39) and (3.40) entail (3.35). Similarly, inclusion (3.36) is proved.

To apply Theorem 3.2, we need to carry out the following operations. First, we should construct the auxiliary linear systems (3.37) and (3.38). Second, we should obtain the approximating ellipsoids for these systems. The second task is, in principle, solved by Theorem 3.1 and reduced to integrating certain differential equations. The first task is difficult to formalize and not always feasible. Note that, as a rule, it is easier to obtain the outer bounds (3.34) than the inner ones (3.33) because the outer estimates can be based on the boundedness of nonlinear terms. We consider below some particular cases, where the bounds (3.33) and (3.34) can be obtained easier that in the general case.

3.6 Particular Cases and Remarks

1. Suppose the original system (1.3) is linear and coincides with (3.10), whereas the set M is not an ellipsoid. In this case we put $C^- = C^+ = C$, $f^- = f^+ = f$, $B^- = B^+ = B$ in (3.33), (3.34), (3.37), and (3.38). To satisfy the conditions of Theorem 3.2, it is sufficient to obtain two-sided ellipsoidal approximations of the initial set M. To obtain the estimates (3.35) and (3.36), we are to integrate systems (3.29) for a^- and a^+, (3.27) for Q^-, and (3.25) for Q^+. The initial value problems for the vectors a^- and a^+ differ in their initial conditions alone (if $a_0^- \neq a_0^+$).

2. Suppose the original system (1.3) is again linear and coincides with (3.10), but the set U in (1.5) and the initial set M are not ellipsoids. Then, to satisfy inclusions (3.33) and (3.34), we are to satisfy the following conditions, respectively

$$U(x,t) \supset E(f^-(t), B^-(t)), \quad M \supset E(a_0^-, Q_0^-) \qquad (3.41)$$

$$U(x,t) \subset E(f^+(t), B^+(t)), \quad M \subset E(a_0^+, Q_0^+)$$

3. If the set U in (1.5) is independent of x, then conditions (3.41) are independent of x as well. It is much easier to find ellipsoids satisfying (3.41) than to satisfy the general conditions (3.33) and (3.34). The problem becomes especially simple, if the set U in (1.4) or (1.5) is constant. In this case, we are only to find inner and outer ellipsoids for the given constant sets U and M.

4. Consider an important case of the nonlinear system (1.3), where

$$f(x, u, t) = C(t)x + \varphi(x, u, t) \qquad (3.42)$$

Here, $C(t)$ is an $n \times n$-matrix, and $\varphi(x, u, t)$ is a given vector-valued function bounded for all x, u and $t \in [s, T]$. Assumption (3.42) means that our system consists of a linear part independent of the control u and a bounded controlled nonlinearity. To fulfil inclusions (3.33) and (3.34), it is sufficient to require that

$$\varphi(x, U(x,t), t) \supset E(f^-(t), B^-(t)) \qquad (3.43)$$

$$\varphi(x, U(x,t), t) \subset E(f^+(t), B^+(t))$$

and, besides, that the conditions (3.33) and (3.34) for the initial set M hold. Since the function φ is bounded, the outer bound (3.43) exists and is feasible. The inner bound (3.43) may occur more difficult to obtain. Note that systems with bounded nonlinearities such as (3.42) are very common in applications.

As an example, consider the following nonlinear differential equation

$$\ddot{x} + \alpha\dot{x} + kx = v(x, \dot{x}, u, t), \quad u \in U \qquad (3.44)$$

Here, α and k are constant coefficients, and v is a specified function. Let this function be bounded for all admissible controls $u \in U$, e.g., let $|v| \leq 1$. Then, to obtain the outer estimates on the reachable sets of our system (3.44), we can use the following linear system of the second order

$$\ddot{x} + \alpha\dot{x} + kx = u, \quad |u| \leq 1 \tag{3.45}$$

as the outer auxiliary system (3.38). Thus, numerical results for the linear equation (3.35) [13, 14] can be used for outer ellipsoidal approximations of the reachable set or equation (3.44).

Remark 3.1. The ellipsoidal estimates (3.35) and (3.36) satisfy the evolutionary conditions: the inner ellipsoids from (3.35) are subreachable, whereas the outer ellipsoids from (3.36) are superreachable sets in the sense of Definitions 3.1, 3.2.

Remark 3.2. The assertion of Theorem 3.1 that the ellipsoidal estimates are optimal is not true for nonlinear systems because we have not imposed any optimality requirements on estimates (3.33) and (3.34). However, the following statement is true: estimates (3.35) and (3.36) are optimal (unimprovable) in the sense of the rate of change of the phase volume with respect to the classes of nonlinear systems satisfying inclusions (3.33) and (3.34), respectively.

3.7 Approximations Using Several Ellipsoids

Suppose the set $X(x,t)$ from (1.6) and M from (1.8) satisfy the following inclusions

$$X(x,t) \supset E(C_i^-(t)x + f_i^-(t), B_i^-(t)) \tag{3.46}$$

$$M \supset E(a_{0i}^-, Q_{0i}^-), \quad i = 1, \ldots, k$$

or the inclusions

$$X(x,t) \subset E(C_i^+(t)x + f_i^+(t), B_i^+(t)) \tag{3.47}$$

$$M \subset E(a_{0i}^+, Q_{0i}^+), \quad i = 1, \ldots, \ell$$

Here, C_i^-, f_i^-, B_i^-, a_{0i}^-, Q_{0i}^-, $i = 1, \ldots, k$ and C_i^+, f_i^+, B_i^+, a_{0i}^+, Q_{0i}^+, $i = 1, \ldots, \ell$ are vectors and matrices specified like the vectors and matrices $C^\pm$, $f^\pm$, $B^\pm$, $a_0^\pm$, and $Q_0^\pm$ in (3.33) and (3.34).

Theorem 3.3. If inclusions (3.46) hold, then the reachable set $D(t,s,M)$ of system (1.3) (or (1.6)) satisfies the inclusion

$$D(t,s,M) \supset \bigcup_{i=1}^{k} E(a_i^-(t), Q_i^-(t)) \tag{3.48}$$

where $a_i^-(t)$ and $Q_i^-(t)$ are the parameters of the inner approximating ellipsoid for the system

$$\dot{x}_i = C_i^-(t)x + v, \quad v(t) \in E(f_i^-(t), B_i^-(t)) \tag{3.49}$$

$$x(s) \in E(a_{0i}^-, Q_{0i}^-), \quad i = 1, \ldots, k, \quad t \geq s$$

If inclusions (3.47) hold, then the reachable set $D(t, s, M)$ satisfies the inclusion

$$D(t, s, M) \subset \bigcap_{i=1}^{\ell} E(a_i^+(t), Q_i^+(t)) \tag{3.50}$$

where $a_i^+(t)$ and $Q_i^+(t)$ are the parameters of the outer approximating ellipsoid for the system

$$\dot{x} = C_i^+(t)x + v, \quad v(t) \in E(f_i^+(t), B_i^+(t)) \tag{3.51}$$

$$x(s) \in E(a_{0i}^+, Q_{0i}^+), \quad i = 1, \ldots, \ell, \quad t \geq s$$

Proof. If the conditions (3.46) are satisfied, then for each $i = 1, \ldots, k$, by virtue of Theorem 3.2, we have the estimate (3.35), where a^- and Q^- should be replaced by a_i^- and Q_i^-, respectively. Since the set $D(t, s, M)$ contains several sets, it contains the union of these sets as well. This entails inclusion (3.48). Analogously, conditions (3.47) imply estimates (3.36) for $i = 1, \ldots, \ell$. Since the reachable set is contained in several ellipsoids, it is contained in their intersection as well. Thus, inclusion (3.50) is true. Theorem 3.3 is proved.

Remark 3.3. In the approach described in Theorem 3.3, it is possible to approximate only one set (X or M) by means of several ellipsoids, while the other (M or X) will be approximated by one ellipsoid.

Remark 3.4. Estimates (3.48) and (3.50) have the evolutionary property: the sets in the right-hand sides of (3.48) and (3.50) are, respectively, subreachable and superreachable for our original system (1.3).

Remark 3.5. To apply Theorem 3.3, it is necessary to construct several approximating ellipsoids, i.e., to solve several initial value problems for systems (3.29), (3.27), and (3.25).

Remark 3.6. Theorem 3.3 extends the possibilities of approximation and can increase its accuracy, since we use here wider classes of sets than the class of ellipsoids. For inner approximations, we employ the class of sets that are unions of several ellipsoids (3.48). Generally, these sets are not convex and have non-smooth boundaries. The class of sets used for outer approximations in (3.50) consists of intersections of several ellipsoids; these sets are convex but their boundaries are not smooth. Theorem 3.3 makes it possible to obtain satisfactory approximations for reachable sets of complicated shapes.

Consider two simple examples showing some advantages of using several ellipsoids. Let the initial set M in the two-dimensional case be a rectangle; then, to approximate it from the outside, we take the intersection of two ellipses with the major axes perpendicular to each other as shown in Fig. 3.1. Let the initial set M is non-convex and L-shaped. The union of two ellipses shown in Fig. 3.2 can be taken as an inner approximating set in this case. In the both examples, several ellipsoids provide better approximations than can be achieved by means of one ellipsoid.

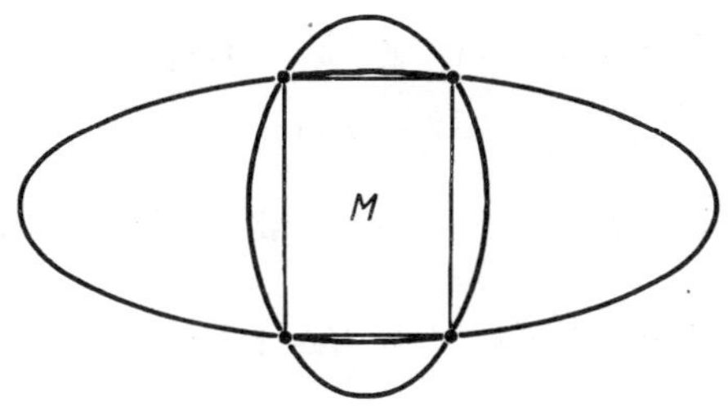 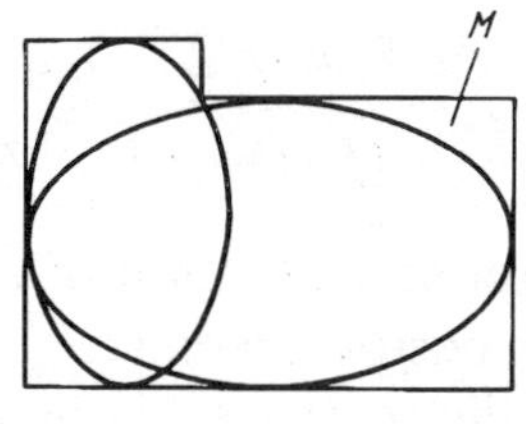

Fig. 3.1, Approximation by the intersection of ellipses.

Fig. 3.2, Approximation by
the union of ellipses.

Numerical examples show [14] that the use of several ellipsoids can considerably improve the approximation of reachable sets.

4. WHAT ARE THE PROPERTIES OF THE APPROXIMATING ELLIPSOIDS?

4.1 Transformation of Equations

It was shown that the parameters a, Q^-, and Q^+ of the ellipsoids approximating, according to (3.30), reachable sets of the system (3.10), satisfy certain differential equations. Equations (3.29) for $a(t)$ are linear, whereas the matrix equations (3.25) for Q^+ and (3.27) for Q^- include nonlinear terms of the square root type which can lead to singularities. We shall simplify equations (3.25) and (3.27) and study their properties.

According to (2.10), the affine transformation of the x-space $y = Ax + b$ implies the following transformation of the parameters of the ellipsoid: $a' = Aa + b$, $Q' = AQA^T$ where primes correspond to the transformed ellipsoid. Keeping this in mind, let us substitute

$$Q^\pm = VZ^\pm V^T \tag{4.1}$$

into equations (3.25) and (3.27). Here, $V(t)$ is a non-singular $n \times n$-matrix, and the matrices $Z^\pm$ are new variables. We shall consider two possibilities for the matrix V which lead to certain simplifications of equations (3.25) and (3.27).

1. We take $V(t)$ equal to the fundamental matrix of the system (3.10), i.e.

$$\dot{V} = C(t)V, \quad V(s) = I, \quad t \geq s \tag{4.2}$$

It can be shown that Z^- and Z^+ satisfy the equations and initial conditions following from (3.25), (3.27), (3.28), and (4.1)

$$\dot{Z}^- = 2B_1^{1/2}(B_1^{-1/2}Z^-B_1^{-1/2})^{1/2}B_1^{1/2}$$

$$= 2(Z^-)^{1/2}[(Z^-)^{-1/2}B_1(Z^-)^{-1/2}]^{1/2}(Z^-)^{1/2} \tag{4.3}$$

$$\dot{Z}^+ = q_1 Z^+ + q_1^{-1} B_1, \quad q_1 = \{n^{-1}\mathrm{Tr}[(Z^+)^{-1}B_1]\}^{1/2}$$

$$B_1(t) = V^{-1}B(V^T)^{-1}, \quad Z^-(s) = Z^+(s) = Q_0$$

The first (or second) form of equation (4.3) for Z^- is applicable, if the matrix B_1 (or Z^-) is positive definite.

2. Let the matrix $B(t)$ be positive definite for $t \geq s$. We take

$$V(t) = [B(t)]^{1/2} \tag{4.4}$$

Substituting (4.1) and (4.4) into (3.25), (3.27), and (3.28), we obtain

$$\dot{Z}^- = C_1 Z^- + Z^- C_1^T + 2(Z^-)^{1/2}$$

$$\dot{Z}^+ = C_1 Z^+ + Z^+ C_1^T + q_1 Z^+ + q_1^{-1} I \tag{4.5}$$

$$C_1(t) = B^{-1/2}(CB^{1/2} - dB^{1/2}/dt), \quad q_1 = \{n^{-1}\mathrm{Tr}[(Z^+)^{-1}]\}^{1/2}$$

$$Z^-(s) = Z^+(s) = Z_0 = B^{-1/2}(s)Q_0 B^{-1/2}(s)$$

These results are summarized in the following theorem proved in [27, 14].

Theorem 4.1. Transformations (4.1), (4.2) and (4.1), (4.4) reduce initial value problems (3.25), (3.27), and (3.28) to the forms (4.3) and (4.5), respectively.

Theorem 4.1 makes it possible to set either $C = 0$ or $B = I$ in (3.25) and (3.27) without loss of generality, i.e., to consider the simplified but equivalent systems (4.3) or (4.5) instead of (3.25) and (3.27). Each system (4.3) and (4.5) depends only on one matrix characterizing the original system: system (4.3) depends on the symmetric non-negative definite matrix B_1, whereas system (4.5) depends on C_1.

4.2 Coincidence of Ellipsoids

We shall give conditions under which the both ellipsoids in estimates (3.30) coincide with each other and, thus, with the reachable set. The centers of the approximating ellipsoids always coincide, see Section 3.3. The initial conditions for the matrices Q^- and Q^+ are also the same, see (3.28). Therefore, in order to have $Q^-(t) = Q^+(t)$, it is necessary and sufficient that the right-hand sides of the systems (3.25) and (3.27) are equal. This condition implies that the right-hand sides of the transformed systems (4.5) coincide for $Z^- = Z^+ = Z$. From this condition, we obtain

$$2Z^{1/2} = qZ + q^{-1}I, \quad \text{or} \quad [(qZ)^{1/2} - (qI)^{-1/2}]^2 = 0$$

Hence, $Z = \lambda^2(t)I$, where $\lambda(t) \geq 0$ is a scalar function. Using (4.1) and (4.4), we obtain $Q = \lambda^2 B$. Consequently, the ellipsoids coincide, if

$$Z(t) = \lambda^2(t)I, \quad Q(t) = \lambda^2(t)B(t), \quad \lambda(t) \geq 0, \quad t \leq s \tag{4.6}$$

The matrix Z from (4.6) must satisfy equations and initial conditions (4.5). Substituting (4.6) into (4.5), we find that the relationships (4.5) are satisfied, if

$$C_1 + C_1^T = \mu(t)I, \quad Q_0 = \lambda_0^2 B(s) \tag{4.7}$$

$$\mu(t) = 2\lambda^{-1}(\dot{\lambda} - 1), \quad \lambda_0 = \lambda(s)$$

Inserting C_1 from (4.5) into (4.7), we obtain

$$CB + BC^T - \dot{B} = \mu(t)B, \quad t \geq s \tag{4.8}$$

The last two equations (4.7) specify the linear initial value problem for $\lambda(t)$. Its solution is

$$\lambda(t) = \int_s^t \exp\left[\int_{t_2}^t \mu(t_1)dt_1\right] dt_2 + \lambda_0 \exp\left[\frac{1}{2}\int_s^t \mu(t_1)dt_1\right] \tag{4.9}$$

As a result, we have the following theorem [27, 14].

Theorem 4.2. The inner and outer ellipsoids (3.30) coincide ($Q^- = Q^+$) for all $t \geq s$, if and only if the conditions (4.8) are satisfied for some scalar function $\mu(t)$, and $Q_0 = \lambda_0^2 B(s)$ for some constant $\lambda_0 \geq 0$. Under these conditions, the matrix $Q(t)$ of the both ellipsoids is given by (4.6) with $\lambda(t)$ defined by (4.9).

4.3 Estimates on the Volumes of Ellipsoids

Our approximating ellipsoids are optimal in the sense of their volume. Consequently, it is natural to estimate the volume of these ellipsoids. Below we present some results concerning these estimates.

Theorem 4.3. The volumes $v^-(t)$ and $v^+(t)$ of the inner and outer approximating ellipsoids $E(a(t), Q^-(t))$ and $E(a(t), Q^+(t))$ satisfy the following equalities

$$d(\ln v^-)/dt = \operatorname{Tr} C + \operatorname{Tr}\left\{[(Q^-)^{-1}B]^{1/2}\right\}$$

$$d(\ln v^+)/dt = \operatorname{Tr} C + \left\{n\operatorname{Tr}[Q^+)^{-1}B]\right\}^{1/2}$$

Theorem 4.4. The following inequalities are true

$$v^-(t) \leq v^+(t), \quad v^+(t)\sigma(t)v_0^{-1} \leq [v^-(t)\sigma(t)v_0^{-1}]^{\sqrt{n}} \tag{4.10}$$

$$\sigma(t) = \exp\left[-\int_s^t \operatorname{Tr} C(\tau)d\tau\right] > 0, \quad t \geq s, \quad v^+(s) = v^-(s) = v_0$$

where v_0 is the volume of the initial ellipsoid $E(a_0, Q_0)$ from (3.5).

Theorem 4.4 enables us to estimate from both sides (from within and from without) one of the volumes v^- or v^+, if we know the other (v^+ or v^-). Therefore, to obtain two-sided estimates on the volume of the reachable set, it is sufficient to integrate only one of the systems (3.25) or (3.27) and to find either $v^-(t)$ or $v^+(t)$. If $C(t) \equiv 0$, the estimates (4.10) become especially simple

$$v^-(t) \le v^+(t), \quad v^+(t)v_0^{-1} \le \left[v^-(t)v_0^{-1}\right]^{\sqrt{n}}$$

Note that, according to Theorem 4.1, the condition $C(t) = 0$ can be adopted without loss of generality. The proof of Theorem 4.3 and 4.4 as well as a lower bound on the volume of the inner ellipsoid and an upper bound on the volume of the outer ellipsoid are given in [28, 14].

4.4 Examples of Approximating Ellipsoids

First we present a case where the exact analytical solution of nonlinear equations (3.25) and (3.27) was obtained [24, 26]. Consider the two-dimensional system

$$\dot{x}_1 = x_2, \quad \dot{x}_2 = u, \quad |u| \le 1, \quad x_1(0) = x_2(0) = 0 \tag{4.11}$$

The initial value problem (3.29) for our system (4.11) becomes

$$\dot{a}_1 = a_2, \quad \dot{a}_2 = 0, \quad a_1(0) = a_2(0) = 0$$

and has the trivial solution $a_1(t) = a_2(t) = 0$.

Equations (3.27) and (3.25) for our system (4.11) take the form

$$dQ_{11}^-/dt = 2Q_{12}^-, \qquad dQ_{12}^-/dt = Q_{22}^- \tag{4.12}$$

$$dQ_{22}^-/dt = 2\left[Q_{11}^- Q_{22}^- - (Q_{12}^-)^2\right]^{1/2} \left(Q_{11}^-\right)^{-1/2}$$

$$dQ_{11}^+/dt = 2Q_{12}^+ + qQ_{11}^+, \quad dQ_{12}^+/dt = Q_{22}^+ + qQ_{12}^+ \tag{4.13}$$

$$dQ_{22}^+/dt = qQ_{22}^+ + q^{-1}, \quad q = (Q_{11}^+/2)^{1/2}[Q_{11}^+ Q_{22}^+ - (Q_{12}^+)^2]^{-1/2}$$

We shall find the solutions of the nonlinear systems (4.12) and (4.13) under the zero initial conditions

$$Q_{ij}^-(0) = Q_{ij}^+(0) = 0, \qquad i,j = 1,2 \tag{4.14}$$

which follow from (4.11) and correspond to the degenerate initial ellipsoid (a point $x = 0$). The desired solution can be obtained in the following form

$$Q_1^\pm = b_{11}^\pm t^4, \quad Q_{12}^\pm = b_{12}^\pm t^3, \quad Q_{22}^\pm = b_{22}^\pm t^2 \tag{4.15}$$

Here, $b_{11}^\pm$, $b_{12}^\pm$, and $b_{22}^\pm$ are coefficients still unknown. Substituting (4.15) into equations (4.12) and (4.13), we find the only non-trivial values of the coefficients

$$b_{11}^- = 1/18, \quad b_{12}^- = 1/9, \quad b_{22}^- = 1/3 \tag{4.16}$$

$$b_{11}^+ = 32/45, \quad b_{12}^+ = 8/9, \quad b_{22}^+ = 4/3, \quad q(t) = 3(2t)^{-1}$$

Formulas (4.15) and (4.16) determine the sought-for matrices $Q^-(t)$ and $Q^+(t)$. The approximating ellipses are conveniently expressed in terms of the variables

$$\xi_1 = x_1 t^{-2}, \qquad \xi_2 = x_2 t^{-1} \tag{4.17}$$

Let us represent our ellipses defined by (4.15) and (4.16), using the variables (4.17)

$$E(a^-, Q^-) : 54\xi_1^2 - 36\xi_1\xi_2 + 9\xi_2^2 \le 1 \tag{4.18}$$

$$E(a^+, Q^+) : \frac{135}{16}\xi_1^2 - \frac{45}{4}\xi_1\xi_2 + \frac{9}{2}\xi_2^2 \le 1$$

The reachable set $D(t, 0, 0)$ of our system (4.11) was obtained in Section 1.9 (see (1.27)). In terms of the variables (4.17), we have

$$D(t, 0, 0) : \frac{(1+\xi_2)^2}{4} - \frac{1}{2} \le \xi_1 \le \frac{1}{2} - \frac{(\xi_2 - 1)^2}{4}, \quad |\xi_2| \le 1 \tag{4.19}$$

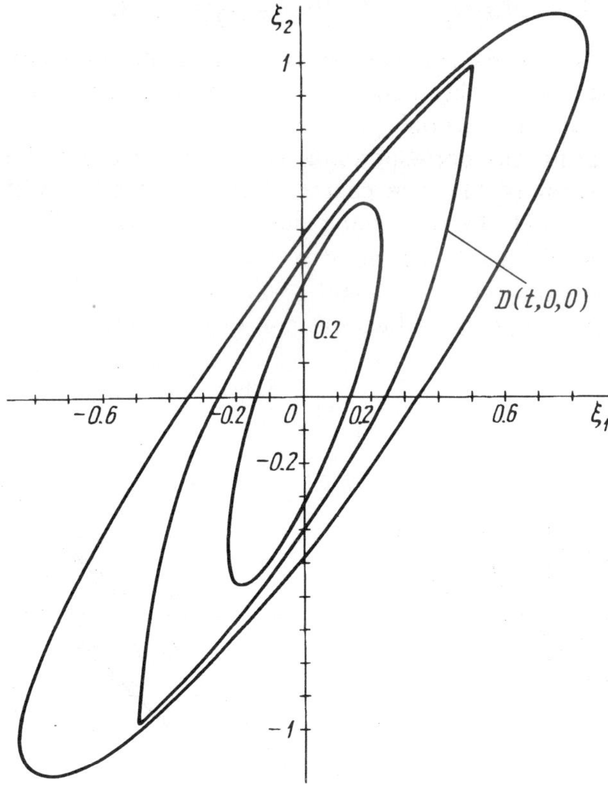

Fig. 4.1, The reachable set of equation (4.11) and the approximating ellipses.

The set (4.19) is bounded by two parabolic arcs. The both approximating ellipses (4.18) and the set (4.19) are shown in Fig. 4.1. Note that the ellipses do not touch the boundary of the reachable set, though the distance between the ellipses and the set $D(t, 0, 0)$ is rather small at some places.

We present now some examples in which the approximating ellipsoids were obtained by numerical integration of equations (3.25) and (3.27). Consider a linear damped oscillator subjected to the bounded control force under the zero initial conditions

$$\ddot{x} + \alpha\dot{x} + kx = u, \quad |u| \le 1, \quad x_1(0) = \dot{x}(0) = 0 \qquad (4.20)$$

Here, x is the scalar coordinate, u is the control, and k and α are constants. Without loss of generality, we suppose that k assumes one of the following three values: $k \in \{-1, 0, 1)\}$; all other cases can be reduced to the above-mentioned by scaling. Thus, any linear controlled equation of the second order with constant coefficients can be reduced to our case (4.20). We rewrite equation (4.20) as follows

$$\dot{x}_1 = x_2, \quad \dot{x}_2 = -kx_1 - \alpha x_2 + u \qquad (4.21)$$

$$|u| \le 1, \quad x_1(0) = x_2(0) = 0$$

For the system (4.21), we constructed the approximating ellipses and the reachable sets. The latter were obtained by means of the procedure described in Section 1.8 and based on the optimal control theory. The approximating ellipses were obtained by integrating the corresponding equations (3.25) and (3.27) numerically. Here, as for system (4.11), the center of the ellipses stays at the initial point $x_1 = x_2 = 0$. For $t = 0$, the right-hand sides of systems (3.25) and (3.27) become uncertain, since the initial set is a point and $Q^-(0) = Q^+(0) = 0$. Hence, we cannot directly use equations (3.25) and (3.27) at $t = 0$. However, for any small $t > 0$, the approximating ellipses become non-degenerate.

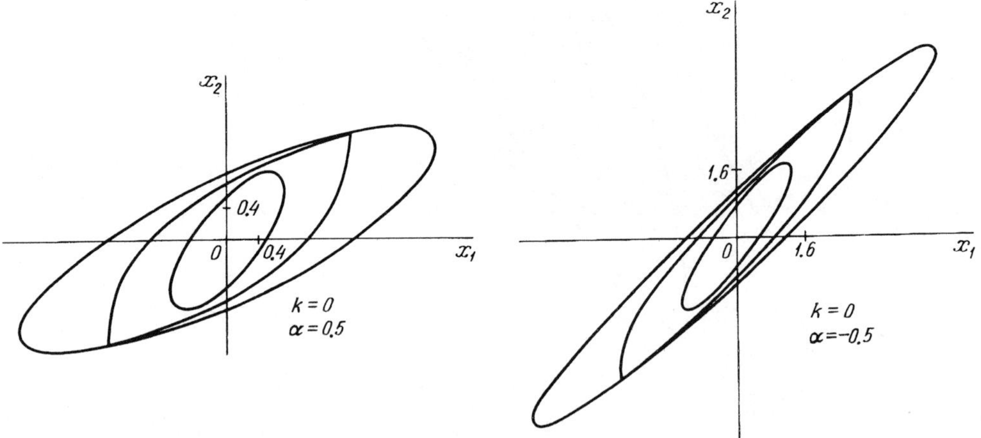

<table>
<tr><td align="center">Fig. 4.2.</td><td align="center">Fig. 4.3.</td></tr>
</table>

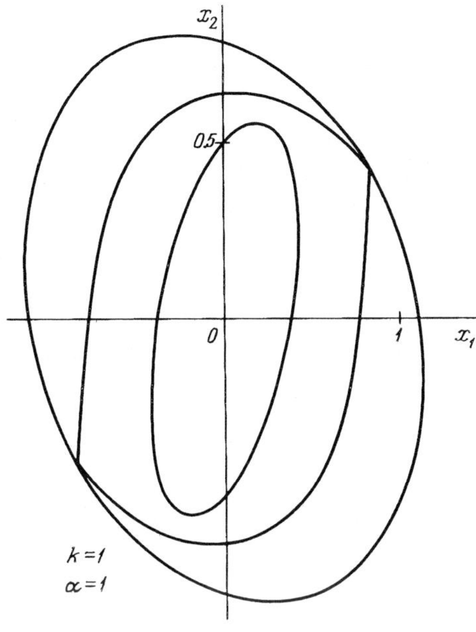

Fig. 4.4.

To start numerical integration, we used the exact solutions (4.15), (4.16) to obtain the initial conditions at small t. After that, we integrated the nonlinear equations (3.25) and (3.27) numerically. Some computational results for $T = 2$ are presented in Figs 4.2–4.4 for different values of the parameters k and α. The reachable set and two approximating ellipses are shown in each figure. More examples can be found in [14].

4.5 Asymptotic Expansions near the Initial Point

We shall consider the important special case, where the initial set (3.5) degenerates into a point $x(s) = a_0$. In this case we have $Q_0 = 0$. Since the right-hand sides of the systems (3.25) and (3.27) contain the inverse matrices $(Q^-)^{-1}$ and $(Q^+)^{-1}$, we cannot straighforwardly calculate the right-hand sides at the initial time instant. Let us analyze the arising singularities for small $t - s$.

Suppose the matrix $B(t)$ is positive definite for $t \geq s$. In accordance with Theorem 4.1, we shall, without loss of generality, consider equations (4.5), instead of (3.25) and (3.27), under the zero initial conditions: $Z^-(s) = Z^+(s) = 0$. Let the following power expansion be true for the matrix $C_1(t)$ from (4.5) in the neighborhood of the initial point

$$C_1(t) = C_{10} + \theta C_{11} + O(\theta^2), \quad \theta = t - s \geq 0 \qquad (4.22)$$

Here, C_{10} and C_{11} are constant matrices. Let us seek the solutions of equations

(4.5) as power series

$$Z^{\pm}(t) = \theta Z_1^{\pm} + \theta^2 Z_2^{\pm} + \theta^3 Z_3^{\pm} + \theta^4 Z_4^{\pm} + O(\theta^5) \qquad (4.23)$$

Here, $Z_1^{\pm}, Z_2^{\pm}, \ldots$ are constant symmetric matrices as yet unknown. We substitute (4.22) and (4.23) into equations (4.5) and expand the both sides of these equations into power series in θ. By equating the coefficients of the obtained expansions in the both sides of our equations, we find the unknown coefficients in (4.23). After rather lengthy but straightforward calculations, we obtain

$$Z_1^- = Z_1^+ = 0, \quad Z_2^- = Z_2^+ = I, \quad Z_3^- = Z_3^+ = D_0$$

$$Z_4^- = (7/12)D_0^2 + (2/3)D_1 \qquad (4.24)$$

$$Z_4^+ = (2/3)(D_0^2 + D_1) + (1/12)n^{-2}(\mathrm{Tr}D_0)^2 I - (1/6)n^{-1}(\mathrm{Tr}D_0)D_0$$

$$D_0 = (1/2)(C_{10} + C_{10}^T), \quad D_1 = (1/2)(C_{11} + C_{11}^T)$$

The obtained results are summed up in the following theorem [27].

Theorem 4.5. Under the condition (4.22), the solutions of equations (4.5) under the zero initial conditions have the asymptotic expansions (4.23) in the neighborhood of the initial point with coefficients given by (4.24).

Note that the expansions (4.23) for Z^- and Z^+ coincide up to the terms $O(\theta^3)$ and differ in $O(\theta^4)$. It follows from (4.23) and (4.24) that

$$Z^+(t) - Z^-(t) = (1/2)\theta^4[D_0 - n^{-1}(\mathrm{Tr}D_0)I]^2 + O(\theta^5)$$

Therefore, $Z^+ - Z^-$ is a non-negative definite matrix for small θ. This assertion is obviously true for all $\theta \geq 0$, since it means that the outer ellipsoid contains the inner one. Using Theorem 4.5 and the transformation defined by (4.1) and (4.4), we can obtain the asymptotic expansions for the solutions of equations (3.25) and (3.27) under the zero initial conditions. The obtained solutions are useful for starting the numerical integration of equations (3.25) and (3.27) (or (4.5)) under these conditions. Asymptotic expansions in a more general case of a degenerate initial ellipsoid as well as asymptotic behavior of approximating ellipsoids at infinity (as $t \to \infty$) can be found in [14].

4.6 General Optimality Criteria for Ellipsoids

Our two-sided ellipsoidal bounds on reachable sets are optimal in the sense of their volume. However, other criteria of optimality, or cost functions, can also be of interest. Here, we consider the generalized optimality criterion as well as some specific cases. Let the ellipsoid $E(a, Q)$ be characterized by the scalar optimality criterion, or the cost function, J which is a given function $L(Q)$ of the matrix Q of the ellipsoid

$$J(E(a, Q)) = L(Q) \qquad (4.25)$$

Suppose the function $L(Q)$ in (4.25) is defined for all symmetric non-negative definite matrices Q, is smooth and monotone. The latter property means that $L(Q_1) \geq L(Q_2)$, if the difference $Q_1 - Q_2$ is a non-negative definite matrix. In other words, if $E(a, Q_1) \supset E(a, Q_2)$, then $J(E(a, Q_1)) \geq J(E(a, Q_2))$. Consider some particular cases of the generalized optimality criterion (4.25).

1) The volume of an ellipsoid is given by (2.11). In this case we have
$$J(E(a, Q)) = L(Q) = (\det Q)^{1/2}.$$

2) In some cases, we are interested not in the volume of some set in the n-dimensional state space, but in the volume of the projection of this set on a subspace. This situation occurs, if some state coordinates are insignificant and need not to be estimated. For example, the volume E_{n-1} of the projection of the ellipsoid on the hyperplane $x_n = 0$, i.e. $J = \operatorname{vol} E_{n-1}$, can be taken as the cost functional. The corresponding formulas for this case are given in [14].

3) The sum of the squared semi-axes of the ellipsoid is given by (see (2.6)) $J = L(Q) = \operatorname{Tr} Q$.

4) The following linear optimality criterion $J = L(Q) = \operatorname{Tr}(CQ)$ where C is a symmetric positive definite $n \times n$-matrix, is a generalization of the previous case. Choosing the appropriate matrix C, we can have approximations more accurate with respect to some space coordinates.

5) The criterion $J = L(Q) = \operatorname{Tr} Q^2$ is equal to the sum of the semi-axes of the ellipsoid in the fourth power.

The generalization of the optimality criterion opens some additional possibilities. Using the generalized criterion (4.25) and its particular cases, we can obtain more flexible approximations and take into account various specific requirements.

We shall present only final results concerning the approximation of reachable sets. All proofs can be found in [29–31], see also [14]. Consider a linear system of differential equations which is a generalization of (3.10)

$$\dot{x} = A(t)x + u, \qquad u = \sum_{i=1}^{r} u^i$$

$$u^i(t) \in E(f^i(t), G_i(t)), \quad i = 1, \ldots, r \tag{4.26}$$

Here, $x \in R^n$ is the state vector, $u \in R^m$ is the vector of control or disturbance, and $A(t)$ is a given $n \times n$-matrix. The control consists of $r \geq 1$ m-dimensional components u^i bounded by ellipsoids. The centers $f^i(t)$ and the symmetric positive definite $n \times n$- matrices $G_i(t)$ of these ellipsoids are assumed to be specified functions of time, $i = 1, \ldots, r$. The initial conditions for our system (4.26) are the same as (3.5). The following definition is a generalization of Definition 3.3.

Definition 4.1. The smooth family

$$E^+(t) = E(a(t), Q(t)) \tag{4.27}$$

of superreachable ellipsoids (in the sense of Definition 3.2) is called *locally optimal,* if for all $\tau \geq s$

$$\frac{d}{dt}L(Q(t))|_{t=\tau} \to \min \tag{4.28}$$

where the minimum is taken over all smooth families of superreachable ellipsoids $E'(t)$ such that $E'(\tau) = E^+(\tau)$.

Definition 4.2. The smooth family (4.27) of superreachable ellipsoids is called *globally optimal* for the given $t = T$, if it solves the minimization problem

$$L(Q(T)) \to \min \tag{4.29}$$

The locally optimal as well as the globally optimal ellipsoids defined above can be determined by means of the following theorem.

Theorem 4.6. The parameters of the both locally and globally optimal super-reachable ellipsoids (4.27) satisfy the following equations and initial conditions

$$\dot{a} = A(t)a + \sum_{i=1}^{r} f^i(t), \quad a(s) = a_0 \tag{4.30}$$

$$\dot{Q} = A(t)Q + QA^T(t) + \sum_{i=1}^{r}(h_iQ + h_i^{-1}G_i) \tag{4.31}$$

$$h_i = \{\mathrm{Tr}[PG_i(t)]/\mathrm{Tr}(PQ)\}^{1/2}, \quad Q(s) = Q_0$$

Here, P is a symmetric $n \times n$-matrix. In the locally optimal case

$$P = \partial L(Q)/\partial Q \tag{4.32}$$

In the globally optimal case, the matrix P satisfies the following differential equation and boundary condition

$$\dot{P} = -PA(t) - A^T(t)P, \quad P(T) = [\partial L(Q)/\partial Q]_{t=T} \tag{4.33}$$

Remark 4.1. The matrices $\partial L/\partial Q$ in (4.32) and (4.33) are given by the following formulas for the respective cases 1, 3–5 listed above

$$1) \quad \partial L/\partial Q = (1/2)[L(Q)]^{-1}Q^{-1}; 3) \quad \partial L/\partial Q = I \tag{4.34}$$

$$4) \quad \partial L/\partial Q = C; \quad 5) \quad \partial L/\partial Q = 2Q$$

Remark 4.2. In the case of the outer locally optimal ellipsoids, Theorem 4.6 is a generalization of Theorem 3.1.

Remark 4.3. To obtain the locally optimal ellipsoids, we are to solve the linear initial value problem (4.30) for the vector $a(t)$ and the nonlinear initial value problem (4.31) for the symmetric positive definite matrix $Q(t)$.

Remark 4.4. For the globally optimal ellipsoids, we are to solve the same initial value problem (4.30) for the vector $a(t)$ and the nonlinear two-point boundary value problem for the matrices $Q(t)$ and $P(t)$ described by equations and boundary conditions (4.31) and (4.33). The boundary-value problem depends on the terminal time T.

Let us consider the case of the linear optimality criterion $L(Q) = \mathrm{Tr}(CQ)$. According to (4.34), we have $\partial L/\partial Q = C$ in this case. Hence, equations (4.31), (4.32) for the locally optimal ellipsoids become

$$\dot{Q} = A(t)Q + QA^T(t) + \sum_{i=1}^{r}(h_i Q + h_i^{-1}G_i) \qquad (4.35)$$

$$h_i = \{\mathrm{Tr}[CG_i(t)/\mathrm{Tr}(CQ)]\}^{1/2}, \quad Q(s) = Q_0$$

For the globally optimal ellipsoids, we obtain from (4.33)

$$P(T) = C \qquad (4.36)$$

Hence, in this case the nonlinear boundary value problem for the matrices Q and P can be reduced to two initial value problems which can be solved consecutively. First, we are to find $P(t)$ from the linear initial value problem (4.33) and (4.36), and after that we should solve the nonlinear initial value problem (4.35) for $Q(t)$. Thus, the case of the linear optimality criterion admits considerable simplifications.

Using Theorem 4.6, we can make outer approximations of reachable sets more flexible and accurate. Especially simple is the case of the linear optimality criterion $L(Q) = CQ$: here we can obtain globally optimal outer approximation of the reachable set by solving only initial value problems. Moreover, we can now consider the case where the disturbance (or control) $u(t)$ is a sum of several independent components bounded by separate ellipsoids (see (4.26)). However, we are to remember that all approximations optimal in the sense of a criterion other than the volume of the approximating ellipsoid are not invariant. In other words, optimal ellipsoids will change, if we make linear transformations of the state x-space. As for inner approximations of reachable sets, the generalization of the optimality criterion does not lead to new non-trivial results [14].

5. HOW TO USE ELLIPSOIDAL MODELLING IN CONTROL?

5.1 Estimation of Control Possibilities

It was shown in Section 1.4 that a number of important problems of control theory can be formulated and solved in terms of reachable sets. Now we shall apply the two-sided ellipsoidal approximations of reachable sets to obtain approximate solutions and estimates in these problems.

Consider the following question: is it possible to bring system (1.3), (1.5) (or (1.6)) to the prescribed terminal state x^* at a given time instant $T > s$, i.e., does the inclusion $x^* \in D(T, s, M)$ hold? Using estimates (3.35) and (3.36), we can assert that if $x^* \in E(a^-(T), Q^-(T))$ then the answer to our question is positive, and if x^* does not belong to $E(a^+(T), Q^+(T))$, the answer is negative. More general problem is to bring system (1.3), (1.5) (or (1.6)) to the given set N in x-space at $T > s$. This problem is solvable, whenever $N \cap D(T, s, M) \neq \emptyset$. Using estimates (3.35) and (3.36), we come to the following solvability conditions for our problem. The sufficient solvability condition is

$$N \cap E(a^-(T), Q^-(T)) \neq \emptyset \tag{5.1}$$

whereas the necessary solvability condition can be written as

$$N \cap E(a^+(T), Q^+(T)) \neq \emptyset \tag{5.2}$$

To verify conditions (5.1) and (5.2), we are to integrate the differential equations for the corresponding ellipsoids $E(a^-(t), Q^-(t))$ and $E(a^+(t), Q^+(t))$.

5.2 Estimation of Disturbances

If $u(t)$ is an uncontrolled external disturbance in (1.3), then the following important problem arises: estimate the set of possible states of the system subjected to the disturbance. In particular, we are often interested whether the set of possible states lies within the given admissible set $P(t)$, i.e., whether the following inclusion

$$D(t, s, M) \subset P(t) \tag{5.3}$$

holds at some given time instants or intervals, or at $t \in \Gamma$, where Γ is a given time set. Using the method of ellipsoids, we come to the following results. If the condition

$$E(a^+(t), Q^+(t)) \subset P(t) \tag{5.4}$$

holds for all $t \in \Gamma$, then inclusion (5.3) holds too. If the inclusion

$$E(a^-(t), Q^-(t)) \subset P(t) \tag{5.5}$$

is violated for some $t \in \Gamma$, then (5.3) does not hold either. We see that conditions (5.1), (5.4) and (5.2), (5.5) provide, respectively, sufficient and necessary conditions of solvability for our problems in terms of approximating ellipsoids. The closer are the inner and outer ellipsoids to each other, the smaller is the gap between the necessary and sufficient conditions. However, even if the gap is significant, the ellipsoidal estimates can be useful, because they can help to readily establish or refute the solvability of the problem under consideration.

5.3 Optimal Control Problems

Consider now possible applications of ellipsoidal bounds to optimal control problems [32]. First, we shall discuss the problem with a terminal cost functional and fixed time. Suppose we are to find an admissible control $u(t)$ satisfying the constraints (1.5) and an initial state vector $x(s) \in M$ such that the functional

$$J = F(x(T)), \qquad T > s \tag{5.6}$$

attains its minimal value for system (1.3). Here, T is a prescribed terminal instant of time, and $F(x)$ is a given scalar function. We have noted in Section 1.4 that the sought-for minimal value F^* of the functional (5.6) is equal to (see (1.10))

$$F^* = \min F(x), \quad x \in D(T, s, M) \tag{5.7}$$

Since inclusions (3.35) and (3.36) hold, we have

$$F^+ \leq F^* \leq F^-, \quad F^- = \min F(x), \quad x \in E(a^-(T), Q^-(T)) \tag{5.8}$$

$$F^+ = \min F(x), \quad x \in E(a^+(T), Q^+(T))$$

Thus, we have the two-sided bounds (5.8) on the minimal value F^* of the functional. To obtain F^- and F^+, we are to solve two problems: first, to find the approximating ellipsoids, second, to minimize the given function $F(x)$ over these ellipsoids at $t = T$. The first problem is reduced to integrating systems (3.29), (3.25), and (3.27) and depends on the right-hand sides of (1.3), on the constraints imposed on control, and on the initial set M. However, this problem is independent of the function $F(x)$ determining the cost functional (5.6). The second problem is an ordinary problem of nonlinear programming. Both problems can be solved by means of conventional numerical algorithms.

Remark 5.1. Suppose we are to solve a number of optimal control problems for system (1.3), (1.5) with different cost functionals (5.6). Then the first part of the problem (obtaining ellipsoids) should be solved only once, while the second part (calculating the minima in (5.8)) must be performed separately for each functional. Thus, we can save a considerable amount of computer time.

Consider the following example. For the system (4.20) we are to minimize the terminal functional

$$J = -\left\{[x(T)]^2 + [\dot{x}(T)]^2\right\}^{1/2}, \quad T = 2 \tag{5.9}$$

This problem implies maximization of the terminal length of the state vector in the $(x, \dot{x})$-plane. Reachable sets for system (4.20) as well as the corresponding approximating ellipses were obtained in Section 4.4. All these sets are centrally symmetric with respect to the origin of coordinates. Calculating the minimal value (5.7) of the functional (5.9) is reduced to finding the longest state vector contained in the reachable set. To obtain the minima in (5.8), we are to find the major semi-axes of the corresponding ellipses. Some numerical results are given in Table 5.1 where the values of the parameters α and k of system (4.20) as well as the corresponding values of F^-, F^*, and F^+ are presented.

k	α	F^-	F^*	F^+
0	0	0.61	1.12	1.40
1	0	1.07	1.69	1.86
1	1	0.61	0.68	0.94
-1	0	2.04	4.56	8.72
-1	1	0.96	2.14	4.25

Table 5.1, The optimal values of the functional (F^*) and the bounds $(F^-$ and $F^+)$.

Consider now a problem of minimal time as an example of optimal control problems with non-fixed time. We are to find an admissible control $u(t)$ and an initial state $x(s) \in M$ such that the system reaches the prescribed terminal set N in the minimal possible time $(T \to \min)$. We assume that the intersection $M \cap N$ is an empty set; then $T^* = \min T > s$. We have noted in Section 1.4 that the sought-for time T^* is the minimal one among all $T \geq s$ such that condition $N \cap D(T, s, M) \neq \emptyset$ holds. Using the ellipsoidal estimates (3.35) and (3.36), we obtain the following two-sided bounds on the optimal time T^*: $T^+ \leq T^* \leq T^-$. Here, T^- and T^+ are the least $t \geq s$ such that the respective ellipsoids $E(a^-(t), Q^-(t))$ and $E(a^+(t), Q^+(t))$ have non-empty intersections with the set N, i.e.

$$T^- = \min t, \quad E(a^-(t), Q^-(t)) \cap N \neq \emptyset \qquad (5.10)$$

$$T^+ = \min t, \quad E(a^+(t), Q^+(t)) \cap N \neq \emptyset$$

For example, let the terminal set N be a given point x^*, so that we are to bring our system (1.3), (1.5) to the state x^* in the minimal possible time. In this case relationships (5.10) can be rewritten as

$$T^- = \min t, \quad x^* \in E(a^-(t), Q^-(t)), \quad t \geq s$$

$$T^+ = \min t, \quad x^* \in E(a^+(t), Q^+(t)), \quad t \geq s$$

Here, T^- and T^+ are the earliest time instants such that the point x^* belongs to the boundaries of the respective ellipsoids. Hence, if we know the approximating ellipsoids, we can obtain the two-sided estimates on the minimal value of the cost functional. Other classes of optimal control problems can be considered in a similar manner.

5.4 Admissible Controls

Now we shall show how the approximating ellipsoids can be used for obtaining admissible controls bringing the system to the prescribed state. First, we consider the linear system (3.10) under the fixed initial condition $x(s) = x^0$. The matrix $B(t)$

is assumed to be symmetric and positive definite. The reachable set $D(t, s, x^0)$ of system (3.10) under the initial condition $x(s) = x^0$ satisfies the following inclusion (see (3.30))

$$D(t, s, x^0) \supset E(a(t), Q^-(t)) \qquad (5.11)$$

Here, the vector $a(t)$ and the symmetric matrix $Q^-(t)$ are the solutions of the initial value problems (3.29) and (3.27), i.e.

$$\dot{a} = Ca + f, \qquad a(s) = x^0 \qquad (5.12)$$

$$\dot{Q}^- = CQ^- + Q^-C^T + 2R, \qquad Q^-(s) = 0 \qquad (5.13)$$

$$R = B^{1/2}(B^{-1/2}Q^- B^{-1/2})^{1/2}B^{1/2} \qquad (5.14)$$

The solution of the initial value problem (5.13) under the degenerate initial condition is considered in Section 4.5. Let us fix the time instant $t = T > s$ and the state x^* belonging to ellipsoid (5.11) at $t = T$, i.e.

$$x^* \in E(a(T), Q^-(T)) \qquad (5.15)$$

Consider the following problem.

Problem 5.1. Find an admissible control $v(t), t \in [s, T]$, bringing system (3.10) from the initial state $x(s) = x^0$ to the terminal state x^* at $t = T$.

Let us introduce the n-dimensional vector-valued function $\psi(t)$ defined for $t \in [s, T]$ and satisfying the following linear differential equation and the initial condition

$$\dot{\psi} = -C^T\psi - (Q^-)^{-1}R\psi, \quad t \in [s, T] \qquad (5.16)$$

$$\psi(T) = [Q^-(T)]^{-1}[x^* - a(T)]$$

Here, a and Q^- are the solutions of the initial value problems (5.12) and (5.13), respectively, and the matrix R is defined by (5.14). The solution of Problem 5.1 is given by the following theorem [33].

Theorem 5.1. Let the vectors and matrices a, Q^-, R, x^* and ψ satisfy relationships (5.12)–(5.16). Then the control

$$v(t) = f(t) + R(t)\psi(t), \quad t \in [s, T] \qquad (5.17)$$

is admissible (it satisfies the constraint in (3.10)) and brings system (3.10) from the initial state $x(s) = x^0$ to the terminal state x^* at the time instant T.

Proof. At first, we prove that the quadratic form $(Q^-\psi, \psi)$ has a constant value for $t \in [s, T]$. Differentiating this form with respect to time and using equations (5.13) and (5.16), we obtain

$$d(Q^-\psi, \psi)/dt = (\dot{Q}^-\psi, \psi) + 2(Q^-\dot{\psi}, \psi)$$

$$= (CQ^-\psi, \psi) + (Q^-C^T\psi, \psi) + 2(R\psi, \psi) - 2(Q^-C^T\psi, \psi) - 2(R\psi, \psi) = 0$$

Therefore, the quantity $(Q^-\psi, \psi)$ is equal to its value at the terminal time instant $t = T$. Using (5.16), we get

$$(Q^-(t)\psi(t), \psi(t)) = (Q^-(T)\psi(T), \psi(T)) \tag{5.18}$$

$$= ([Q^-(T)]^{-1}[x^* - a(T)], [x^* - a(T)]) \le 1$$

The inequality in (5.18) follows from (5.15). Let us verify that the control (5.17) satisfies the constraint in (3.10). To do that, we are to prove that the inequality

$$(B^{-1}(v - f), (v - f)) \le 1 \tag{5.19}$$

holds for all $t \in [s, T]$. Let us calculate the quadratic form (5.19), using equations (5.17) and (5.14)

$$(B^{-1}(v - f), (v - f)) = (B^{-1}R\psi, R\psi) = (RB^{-1}R\psi, \psi)$$

$$= (B^{1/2}(B^{-1/2}Q^-B^{-1/2})^{1/2}(B^{-1/2}Q^-B^{-1/2})^{1/2}B^{1/2}\psi, \psi) = (Q^-\psi, \psi)$$

The obtained relationship and (5.18) entail (5.19). Thus, control (5.17) is admissible. Let us introduce the fundamental matrix $\Phi(t)$ of our system (3.10) by

$$\dot{\Phi} = C(t)\Phi, \quad \Phi(s) = I, \quad t \in [s, T] \tag{5.20}$$

The following identity stems from (5.20)

$$d\Phi^{-1}(t)/dt = -\Phi^{-1}(t)C(t) \tag{5.21}$$

Let us seek the solution of system (3.10) as follows

$$x(t) = \Phi(t)y(t) \tag{5.22}$$

where $y(t)$ is a new unknown function. Substituting (5.22) into equation (3.10), we obtain the equation for $y(t)$

$$\dot{y} = \Phi^{-1}(t)v(t) \tag{5.23}$$

We multiply the both sides of the matrix equation (5.13) by the vector ψ from the right, and the both sides of the vector equation (5.16) by the matrix Q^- from the left. Then we add the resulting equations and obtain

$$d(Q^-\psi)/dt = CQ^-\psi + R\psi \tag{5.24}$$

Let us express f and $R\psi$ by means of equations (5.12) and (5.24), respectively, and substitute them into (5.17). We have $v = d(Q^-\psi + a)/dt - C(Q^-\psi + a)$. Inserting the obtained control v into equation (5.23), we get

$$\dot{y} = \Phi^{-1}d(Q^-\psi + a)/dt - \Phi^{-1}C(Q^-\psi + a)$$

Transforming this equation by means of (5.21), we obtain

$$\dot{y} = \Phi^{-1}d(Q^-\psi + a)/dt + (d\Phi^{-1}/dt)(Q^-\psi + a) = d[\Phi^{-1}(Q^-\psi + a)]/dt$$

Therefore, we have $y = \Phi^{-1}(Q^-\psi + a) + c_1$ where c_1 is an arbitrary constant. Substituting the obtained expression for y into (5.22), we get

$$x(t) = Q^-(t)\psi(t) + a(t) + c_1\Phi(t)$$

Putting $t = s$ and using the initial conditions $x(s) = x^0$, (5.12), (5.13), and (5.20), we find the arbitrary constant c_1 to be zero. Hence, the state trajectory corresponding to control (5.17) is given by

$$x(t) = Q^-(t)\psi(t) + a(t) \tag{5.25}$$

Setting $t = T$ in (5.25) and taking account of (5.16), we obtain $x(T) = x^*$. Thus, control (5.17) brings our system to the prescribed state. The theorem is proved.

Remark 5.2. System (3.10) under condition $x(s) = x^0$ can be brought to any point of the reachable set at $t = T$. Theorem 5.1 provides a constructive way to obtain an admissible control bringing the system to any point of the inner approximating ellipsoid (5.11).

Remark 5.3. To obtain control (5.17), it is necessary to solve three initial value problems, namely, the problems (5.12) for $a(t)$, (5.13) for $Q^-(t)$, and (5.16) for $\psi(t)$. Two of these problems (for a and ψ) are linear, whereas the problem (5.13) for the matrix Q^- is nonlinear. Problems (5.12) and (5.13) are independent, while the problem for $\psi(t)$ depends on $a(t)$ and $Q^-(t)$ and can be solved only after problems (5.12) and (5.13). Note that only the problem for $\psi(t)$ depends on the given time T and the terminal vector x^*.

Remark 5.4. Consider the terminal problem of optimal control with functional (5.6) for the linear system (3.10) under the initial condition $x(s) = x^0$. Theorem 5.1 makes it possible to obtain the control providing the upper bound F^- on the functional. To find this control, we should first determine the vector x^* which provides the minimal value of the functional in the corresponding problem of nonlinear programming (5.8). After that, for the given T and obtained x^*, we can find the control $v(t)$ in accordance with (5.17). This control which can be called suboptimal brings our system to the state x^* at the time instant T. The corresponding value of the functional is $J = F^-$.

Remark 5.5. Consider the time-optimal control problem ($J = T \rightarrow \min$) for the linear system (3.10) with the fixed initial state $x(s) = x^0$ and terminal state $x(T) = x^*$. The upper bound T^- on the minimal time (see (5.10)) is defined as the earliest time when the point x^* lies on the boundary of the inner ellipsoid. Having found T^-, we can, using Theorem 5.1, construct the control bringing our system to the state x^* at the time instant T^-. This control can be called time-suboptimal.

5.5 Admissible Controls in Nonlinear Systems

Let us treat now a general case of the nonlinear system (1.3), (1.5) under the initial condition $x(s) = x^0$. Consider Problem 5.1 for this system. Suppose the first condition (3.33) holds. Thus, we can construct the auxiliary system (3.37). For this linear system which is analogous to (3.10), we can, on the strength of Theorem 5.1, find the control $v(t)$ bringing the system from the initial point $x(s) = x^0$ to the prescribed terminal point x^* at the instant T. The corresponding control $u(t)$ for the original nonlinear system (1.3) is determined by the condition

$$f(x(t), u(t), t) = C^-(t)x(t) + v(t), \quad t \in [s, T] \tag{5.26}$$

Here, $v(t)$ is given by (5.17), and $x(t)$ is determined by (5.25).

For each t, equation (5.26) is a transcendental equation for the vector $u(t)$. Since $X(t)$ is defined by (1.7) and the first inclusion (3.33) is true, equation (5.26) has for any t at least one solution $u(t)$ satisfying the constraint $u(t) \in U(x(t), t)$ from (1.5). If equation (5.26) is satisfied, then the trajectory of the original nonlinear system, starting at the initial point $x(s) = x^0$, coincides with the trajectory of the auxiliary linear system starting at the same point. Consequently, the terminal condition $x(T) = x^*$ is satisfied for the original system. The obtained result is summed up in the following theorem.

Theorem 5.2. Suppose the nonlinear system, defined by equations (1.3), (1.5) and the initial condition $x(s) = x^0$, satisfies the first inclusion (3.33). Let the vectors and matrices $a^-(t), Q^-(t), x^*, \psi(t), v(t)$, and $x(t)$ satisfy relationships (5.12)–(5.17) and (5.25), where C, f, B, and a are replaced by C^-, f^-, B^-, and a^-, respectively. Then, for each $t \in [s, T]$, there exists at least one vector $u(t)$ satisfying equation (5.26) and constraint $u(t) \in U(x(t), t)$. The control $u(t)$ brings system (1.3) from the initial state $x(s) = x^0$ to the state x^* at $t = T$ along the trajectory $x(t)$ given by (5.25).

Remark 5.6. In order to apply Theorem 5.2, we need to: 1) construct the auxiliary linear system; 2) construct the inner approximating ellipsoid for this system; 3) construct the control $v(t)$ and the trajectory for the auxiliary system; 4) find $u(t)$ from (5.26). The first two stages are the same as in Theorem 3.2; they were discussed in Section 3.5. The third stage implies the algorithm described in Theorem 5.1. The fourth stage depends on the right-hand sides of the original nonlinear system.

Remark 5.7. If all conditions of Theorem 5.2 are satisfied, we can obtain the suboptimal control (in the sense of Remarks 5.4 and 5.5) for the nonlinear system.

At first, we should construct the auxiliary linear system and find the suboptimal control for this system. Then, using (5.26), we can obtain the desired control for the nonlinear system.

Other results on application of ellipsoidal techniques to control problems for dynamical systems are given in [14].

5.6 Applications to Differential Games

Consider a differential game of two players X and Y described by (1.11)

$$X : \dot{x} = f(x, u, t), \quad u \in U, \quad x(s) = x^0 \tag{5.27}$$

$$Y : \dot{y} = g(y, v, t), \quad v \in V, \quad y(s) = y^0$$

Here, x and y are the state vectors of the players X and Y, respectively, u and v are their controls, and x^0 and y^0 are the respective initial states. The functions f and g as well as the closed sets U and V are assumed to be given. The cost functional J is a given scalar function of the terminal states of systems (5.27) at the prescribed terminal time instant T

$$J = \Phi(x(T), y(T)), \quad T > s \tag{5.28}$$

The player X seeks to minimize while the player Y seeks to maximize the functional (5.28). As is noted in Section 1.4, the optimal value J^* of the functional (5.28), corresponding to optimal strategies of the both players, satisfies the two-sided estimate (1.14), i.e.

$$\max_{y \in D_y} \min_{x \in D_x} \Phi(x, y) \leq J^* \leq \min_{x \in D_x} \max_{y \in D_y} \Phi(x, y) \tag{5.29}$$

Here, D_x and D_y are the reachable sets of the respective players X and Y. Suppose the nonlinear systems (5.27) satisfy the conditions of Theorem 3.2. Denote by a_x^-, Q_x^-, a_y^-, and Q_y^- the parameters of the inner approximating ellipsoids for the respective players X and Y, whereas by a_x^+, Q_x^+, a_y^+, and Q_y^+ the parameters of the corresponding outer ellipsoids. Then, by virtue of the estimates (3.35), (3.36), and (5.29), we have

$$\Phi_1 \leq J^* \leq \Phi_2, \quad \Phi_1 = \max_{y \in E_y^-} \min_{x \in E_x^+} \Phi(x, y) \quad \Phi_2 = \min_{x \in E_x^-} \max_{y \in E_y^+} \Phi(x, y) \tag{5.30}$$

$$E_x^\pm = E(a_x^\pm(T), Q_x^\pm(T)), \quad E_y^\pm = E(a_y^\pm(T), Q_y^\pm(T))$$

Suppose we have found the pairs of points x_1^*, y_1^* and x_2^*, y_2^* for which the maximin Φ_1 and minimax Φ_2 in (5.30) are attained, respectively. By means of Theorem 5.2 we can, under the corresponding conditions, find the control $u(t)$ bringing the player X to the state $x_2^* \in E_x^-$ at $t = T$, and also the control $v(t)$ bringing the player Y to the state $y_1^* \in E_y^-$ at $t = T$. If the player X applies the open-loop control $u(t)$, then the value of the functional J does not exceed Φ_2 from (5.30) under any admissible control of the player Y. On the other hand, if Y applies the control $v(t)$, the value of J is not less than Φ_1 from (5.30) under any admissible control of X. Thus, using the approximating ellipsoids, we can, in principle, obtain two-sided bounds on the value of the functional in our differential game and, moreover, find open-loop controls of the players ensuring these bounds.

Games of pursuit and evasion in which the terminal time T is not fixed were also considered using the method of ellipsoids. Here, the two-sided bounds $T_1 \leq T_0 \leq T_2$ on the time of pursuit T_0 can be obtained as well as the controls of the players which guarantee these bounds, namely, the lower bound T_1 for the evader and the upper bound T_2 for the pursuer. Since the procedure of obtaining these bounds uses certain specific methods of differential games, we shall not present it here. These results can be found in [14]. Let us present a numerical example.

Let the game of two similar players X and Y be described by the relationships

$$X : \dot{x} = Cx + u, \quad |u| \leq 1, \quad x(0) = 0 \tag{5.31}$$

$$Y : \dot{y} = Cy + v, \quad |v| \leq \nu, \quad y(0) = y^0$$

$$C = \begin{bmatrix} -1 & 0 \\ 0 & -\delta \end{bmatrix}, \quad \nu < 1, \quad \delta > 0$$

Here, $x, y, u,$ and v are two-dimensional vectors, δ and ν are positive constants, and the terminal condition of the game is given by

$$|x(T) - y(T)| = \ell, \quad \ell > 0 \tag{5.32}$$

where $\ell > 0$ is a given constant. For the value of the parameter $\delta = 1$, this game was investigated in [34] where the following value of the optimal terminal time T was found

$$T = \ln[(|y^0| + 1 - \nu)(\ell + 1 - \nu)] \tag{5.33}$$

y_1	y_2	δ	ν	ℓ	T_1	T_0	T_2
1.0	1.0	1.0	0.95	0.50	0.98	0.98	0.98
1.0	1.0	1.0	0.80	0.01	2.04	2.04	2.04
1.0	1.0	1.2	0.80	0.01	1.93	–	1.95
1.0	1.0	0.5	0.95	0.50	1.41	–	1.44
-1.5	0.5	0.8	0.85	0.02	2.35	–	2.37
1.0	1.0	0.8	0.80	0.01	2.18	–	2.20
1.5	0.5	0.8	0.80	0.02	2.12	–	2.13
0.8	-2.0	1.5	0.80	0.02	1.89	–	1.93
0.5	-2.0	0.9	0.80	0.02	2.48	–	2.49

Table 5.2, Numerical results for the differential game.

Some computational results obtained by the method of ellipsoids are presented in Table 5.2. Here, the following data are given: the components y_1 and y_2 of the initial vector y^0 (the initial state of the player X is zero, see (5.31)), the parameters δ and ν of the system (5.31), the parameter ℓ in (5.32), the exact time of pursuit T_0, and the bounds T_1 and T_2. The value of T_0 for $\delta = 1$ is given by (5.33).

5.7 Applications to Discrete Time Systems

We considered above dynamical systems described by ordinary differential equations. However, all obtained results can be easily extended to the case of finite-difference systems described by equations

$$x(t_{i+1}) = F(x(t_i), u(t_i), t_i), \quad t_0 < t_1 < \ldots \tag{5.34}$$

$$i = 0, 1, \ldots, \quad x(t_0) \in M, \quad u(t_i) \in U(x(t_i), T_i)$$

Here, x is again the vector of state, $x \in R^n$, u is the vector of control, $u \in R^m$, time takes discrete values, U is a closed set in R^m depending on state and time, and M is a closed initial set in R^n. Instead of differential equations, we obtain here finite-difference equations for the evolution in discrete time of ellipsoids approximating reachable sets for system (5.34). Inner and outer approximating ellipsoids optimal in the sense of volume can be obtained and used for the same purposes as in systems described by differential equations. These results are presented in [14].

6. HOW TO USE ELLIPSOIDAL MODELLING IN STATE ESTIMATION?

6.1 Problem of the Guaranteed State Estimation

Consider the dynamical system described by equations (1.3), (1.5), and (1.8)

$$\dot{x} = f(x, u, t), \quad u(t) \in U(x(t), t), \quad x(s) \in M \tag{6.1}$$

where the denotations are the same as in Section 1.2. Suppose the state $x(t)$ is observed at discrete instants of time $t_0, t_1, \ldots$, and the equation of observation is given by

$$y(t_i) = H(t_i)x(t_i) + \xi(t_i), \quad \xi(t_i) \in V_i, \tag{6.2}$$

$$s = t_0 < t_1 < t_2 < \ldots, \quad i = 0, 1, \ldots$$

Here, $y(t_i)$ is the ℓ-vector of observed parameters (results of observations), $H(t_i)$ is a given $\ell \times n$-matrix, and $\xi(t_i)$ is the vector of observation errors bounded by the given set V_i, $i = 0, 1, \ldots$. Let the following functions and sets be given beforehand: the function $f(x, u, t)$ and the sets $U(x, t)$ and M in (6.1), the matrices $H(t_i)$ and sets V_i in (6.2), $i = 0, 1, \ldots$. Instead of $f(x, u, t)$ and $U(x, t)$, the set $X(x, t)$ from (1.6) can be given. All these data comprise the a priori information.

At each instant t_i, the results of observations $y(t_i)$ are available, $i = 0, 1, \ldots$. The vector of disturbances $u(t)$ and the observation errors are unknown.

Denote by $D_x(t)$ the set of all possible values of the state vector $x(t)$ which are compatible with (do not contradict to) the observation results $y(t_i)$ for $t_i < t$. At the instant t_k, the set $D_x(t)$ is discontinuous and has a jump. To be definite, we assume that the set $D_x(t)$ is continuous from the left. Then $D_x(t_k)$ is the set of all $x(t_k)$ compatible with the observation results $y(t_i)$ for $i < k$, whereas $D_x(t_k + 0)$ is the set of all $x(t_k)$ compatible with the observation results $y(t_i)$ for $i \leq k$. In other words, the set $D_x(t_k + 0)$ takes into account the result of the observation made at $t = t_k$. Let us now state two problems.

Problem 6.1. Find the set $D_x(t)$ for $t \geq s$ using the a priori information and the results of observations $y(t_i)$, $i = 0, 1, \ldots, k$, where $t_k < t \leq t_{k+1}$; to obtain the set $D_x(t_k + 0)$, we should also use the observation result $y(t_k)$.

Problem 6.2. Find ellipsoids containing the set $D_x(t)$ for $t \geq s$:

$$D_x(t) \subset E(a(t), Q(t)), \quad t \geq s \tag{6.3}$$

using the a priori information and the observation results $y(t_i)$, $= 0, 1, \ldots, k$, where $t_k < t \leq t_{k+1}$; to obtain the parameters $a(t_k + 0)$ and $Q(t_k + 0)$, we should also use the observation result $y(t_k)$.

First, let us consider Problem 6.1 in a general case. After that, we shall present the algorithm solving Problem 6.2.

Denote by $D_y(y, t_i)$ the set of all state vectors $x(t_i)$ compatible with the observation result $y(t_i) = y$ made at $t = t_i$. We have

$$D_y(y, t_i) = \{x : y - H(t_i)x \in V_i\} \tag{6.4}$$

Denoting by $D(t, s, M)$ the reachable set of the system (6.1), we obtain the following relationships reflecting the process of observation

$$D_x(t_k + 0) = D_x(t_k) \cap D_y(y(t_k), t_k), \quad k = 0, 1, \ldots \tag{6.5}$$

$$D_x(t) = D(t, t_k, D_x(t_k + 0)), \quad t \in (t_k, t_{k+1}]$$

We have the initial condition

$$D_x(s) = M \qquad (s = t_0) \tag{6.6}$$

In principle, relationships (6.5) and (6.6) determine the sets $D_x(t)$ for all $t \geq s$. Practical computation of sets $D_x(t)$ presents great difficulties and requires finding reachable sets of the nonlinear system as well as intersections of sets. That is why we consider below Problem 6.2 which implies obtaining outer ellipsoidal bounds on the sets $D_x(t)$, see (6.3). Since we shall approximate all sets by ellipsoids, the operation of intersection of sets is reduced to the outer approximation of the intersection of two ellipsoids by means of an ellipsoid.

6.2 Intersection of Ellipsoids

Problem 6.3. Let $E(a_1, Q_1)$ and $E(a_2, Q_2)$ be two ellipsoids in R^n. Find an ellipsoid $E(a, Q)$ containing the intersection P of the given ellipsoids, i.e. such that

$$P = (E(a_1, Q_1) \cap E(a_2, Q_2)) \subset E(a, Q) \qquad (6.7)$$

Problem 6.3 admits the following interpretation. Suppose the vector $x \in R^n$ is observed (measured) twice, the results of observations being denoted by a_1 and a_2. The errors of both observations are unknown and bounded by ellipsoids $E(0, Q_1)$ and $E(0, Q_2)$. Therefore, the results of our observations mean that $x \in E(a_1, Q_1)$ and $x \in E(a_2, Q_2)$. These relationships entail $x \in P$ where P is given by (6.7). Problem 6.3 means that we seek an approximation a of the unknown vector x such that $x \in E(a, Q)$ where the matrix Q is also to be found. It is obvious that Problem 6.3 has an infinite number of solutions: any ellipsoid $E(a', Q')$ containing any solution $E(a, Q)$ of the problem is also a solution. It is desirable to find "the smallest" (in some sense) ellipsoid $E(a, Q)$ in order to diminish the information loss due to the approximation of the set P by an ellipsoid. We shall again take the volume of the resulting ellipsoid as a measure of its size, or as an optimality criterion. Thus, our approach is consistent with the principles stated above (see Section 2.7).

Problem 6.4. Let $E(a_1, Q_1)$ and $E(a_2, Q_2)$ be two ellipsoids in R^n. Find an ellipsoid $E(a, Q)$ of the minimal volume containing the intersection P of the given ellipsoids, i.e. an ellipsoid such that

$$P \subset E(a, Q), \qquad \det Q \to \min \qquad (6.8)$$

over all ellipsoids satisfying (6.7).

Problem 6.4 seems to be similar to Problems 2.1 and 2.2 dealing with the optimal approximations of the sum of ellipsoids. However, Problem 6.4 is more difficult. Note that the shape of the sum of two ellipsoids is independent of the location of the centers a_1 and a_2 of the initial ellipsoids. The center a^+ of the outer approximation of the sum is readily determined as $a^+ = a_1 + a_2$, whereas the matrix Q^+ depends only on the matrices Q_1 and Q_2 and is independent of a_1 and a_2 (see (2.19)). By contrast, the shape of the intersection P from (6.7) depends on the centers a_1 and a_2. Therefore, the center a and the matrix Q of the resulting ellipsoid depend on all parameters a_i and Q_i, $i = 1, 2$, of the initial ellipsoids.

In the one-dimensional case $n = 1$, Problem 6.4 is trivial. Here, the ellipsoids $E(a_1, Q_1)$ and $E(a_2, Q_2)$ as well as their intersection P are segments. Therefore, the exact approximation of the set P is possible: $E(a, Q) = P$.

It seems that Problem 6.4 cannot be solved explicitly in the general case $n > 1$. However, there are exact solutions of this problem for particular cases of ellipsoidal segment and layer. We propose approximate suboptimal solutions based on these exact solutions.

Let the first ellipsoid $E(a_1, Q_1)$ be non-degenerate (the matrix Q_1 is positive definite) and has the finite volume which is smaller than the volume of the second ellipsoid which may be infinite, $\det Q_1 \leq \det Q_2$. Let us define the set P as follows

$$P = E(a_1, Q_1) \cap E_2, \quad E_2 = \{x : (R(Hx - h), (Hx - h)) \leq 1\} \quad (6.9)$$

Representation (6.9) is a generalization of (6.7). Here, R is a symmetric positive definite $m \times m$-matrix, H is an $m \times n$-matrix, and h is an m-vector, where $1 \leq m \leq n$. The "generalized" ellipsoid E_2 in (6.9) has m finite semi-axes, whereas all its other $n - m$ semi-axes are of infinite length. In the case of $m = n$ and $H = I$, the domain E_2 becomes a non-degenerate ellipsoid, and relationships (6.9) are reduced to (6.7).

Note that the domain E_2 from (6.9) can be regarded as the uncertainty set corresponding to the observation of the m-dimensional vector Hx. In this case, h is the observation result, and R is the matrix of errors.

Let us describe some procedures for the outer ellipsoidal approximation of the set P from (6.9). We shall replace the "generalized" ellipsoid E_2 by a half-space or by a set bounded by two parallel hyperplanes, thus replacing P by an ellipsoidal segment or layer. After that, we shall use exact solutions of Problem 6.4 for ellipsoidal segment and layer and find the corresponding optimal ellipsoids.

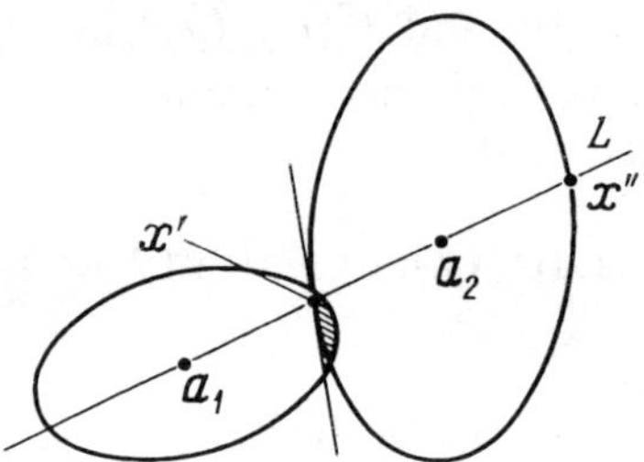

Fig. 6.1, Approximation of the intersection of two ellipsoids by a segment.

Denote by a_2 any vector satisfying the equation $Ha_2 = h$. If the ellipsoid E_2 is non-degenerate ($m = n$, $H = 1$), then a_2 is its center. In the general case, the vector a_2 is not unique, but the final result is independent of this non-uniqueness. Let us draw the line L through the points a_1 and a_2. This line intersects the boundary of E_2 at two points x' and x''. One of them (x') lies on the ray coming from a_2 through a_1, whereas the other (x'') belongs to the other half of the line L (see Fig. 6.1). The vectors introduced above are given by

$$x' = a_2 - \beta(a_2 - a_1), \quad x'' = a_2 + \beta(a_2 - a_1) \quad (6.10)$$

$$Ha_2 = h, \quad \beta = (R(Ha_1 - h), (Ha_1 - h))^{-1/2}$$

The vector ν normal to E_2 at the point x' and directed into E_2 is defined by

$$\nu = H^T R(h - Ha_1) \quad (6.11)$$

Let us replace E_2 by the half-space $(x, \nu) \geq (x', \nu)$ and, therefore, the set (6.9) by the corresponding segment (the shaded domain in Fig. 6.1). Omitting rather lengthy calculations, let us present the final results for this case [14]. The parameters of the suboptimal ellipsoid $E(a, Q)$ are

$$a = a_1, \quad Q = Q_1 \quad \text{if} \quad c \leq -n^{-1};$$

$$a = a_1 + \frac{(nc + 1)Q_1\nu}{(n + 1)(Q_1\nu, \nu)^{1/2}}, \tag{6.12}$$

$$Q = \left[\frac{(n^2 - 1)Q_1^{-1}}{n^2(1 - c^2)} + \frac{2(nc + 1)(n + 1)(\nu * \nu)}{n^2(1 - c^2)(1 - c)(Q_1\nu, \nu)} \right]^{-1}$$

$$\text{if} \quad -n^{-1} < c \leq 1; \quad c = \beta^{-2}(1 - \beta)(Q_1\nu, \nu)^{-1/2}$$

$$\beta = (R(Ha_1 - h), (Ha_1 - h))^{-1/2}, \quad \nu = H^T R(h - Ha_1)$$

Here, the symbol $*$ denotes the dyadic product of two vectors: $\nu * \nu$ is a matrix with elements $\nu_i\nu_j$; $i, j = 1, \ldots, n$.

Remark 6.1. If $E_2 = E(a_2, Q_2)$ is a non-degenerate ellipsoid, then we should set $R = Q_2^{-1}$, $H = I$, $h = a_2$, $\nu = Q_2^{-1}(a_2 - a_1)$ in (6.9)–(6.12).

Remark 6.2. Formulas (6.10)–(6.12) contain singularities, if $Ha_1 = h$ or $a_1 = a_2$. In these case we can take $a = a_1$ and $Q = Q_1$.

Several other procedures for approximation of the intersection of two ellipsoids, comparison and discussion of these procedures as well as a number of examples are presented in [14].

6.3 Procedure of the Guaranteed State Estimation

Let us return to Problem 6.2 and consider it for the linear system in the form given by (3.10)

$$\dot{x} = C(t)x + u, \quad u(t) \in E(f(t), B(t)) \tag{6.13}$$

$$x(s) \in E(a_0, Q_0), \quad t \geq s$$

Here, x is the n-dimensional state vector, u is the m-dimensional vector of disturbances, $C(t)$ is an $n \times n$-matrix, $f(t)$ is an n-dimensional vector, $B(t)$ is an $n \times m$-matrix, a_0 is an n-dimensional vector, and Q_0 is an $n \times n$-matrix. The matrices $B(t)$ and Q_0 are symmetric and non-negative definite, and at least one of them $(B(t)$ for all $t \geq s$ or $Q_0)$ is positive definite. Let us present (6.13) in the form of the differential inclusion

$$\dot{x} \in E(C(t)x + f(t), B(t)), \quad x(s) \in E(a_0, Q_0), \quad t \geq s \tag{6.14}$$

Let the equations of observations (6.2) have the form

$$y(t_i) = H_i x(t_i) + \xi_i, \quad \xi_i \in E(0, L_i), \tag{6.15}$$

$$i = 0, 1, \ldots, \qquad s = t_0 < t_1 < \ldots$$

Here, H_i are given $\ell \times n$-matrices, and L_i are given symmetric positive definite $\ell \times \ell$-matrices. The a priori information includes the matrices and vectors $C(t)$, $f(t)$, and $B(t)$ for $t \geq s$, matrices H_i and L_i for $i = 0, 1, \ldots$, a_0 and Q_0.

Due to the initial condition (6.13), we have

$$a(s) = a_0, \quad Q(s) = Q_0 \quad (s = t_0) \tag{6.16}$$

Let us describe the procedure of the guaranteed state estimation for system (6.13)–(6.15). Suppose the parameters of the ellipsoid $E(a(t_k)), Q(t_k))$ for some $k \geq 0$ are already determined. Let us first find $a(t_k + 0)$ and $Q(t_k + 0)$, and then obtain $a(t)$ and $Q(t)$ for $t \in (t_k, t_{k+1}]$.

At the observation instant $t = t_k$, the parameters $a(t)$ and $Q(t)$ are continuous from the left and have jumps which can be evaluated by means of the procedure for the outer ellipsoidal approximation of the intersection of two ellipsoids. Using the procedure (6.12), we should set

$$a_1 = a(t_k), \quad Q_1 = Q(t_k), \quad a = a(t_k + 0) \tag{6.17}$$

$$Q = Q(t_k + 0), \quad R = L_k^{-1}, \quad H = H_k, \quad h = y(t_k)$$

in equations (6.12). Denote the pair $\{a, Q\}$ by z. Then, equations (6.12) and (6.17) define the relationship

$$z(t_k + 0) = \psi_k(z(t_k), y(t_k)), \quad z = \{a, Q\}, \quad k = 0, 1, \ldots \tag{6.18}$$

Now let us obtain the parameters of the ellipsoid $E(a(t), Q(t))$ in the interval $(t_k, t_{k+1}]$ using the outer bound (3.30) for the reachable set of the linear system (6.13). Equations (3.29) and (3.25) for the vector $a(t)$ and matrix $Q(t)$ can be presented as follows

$$\dot{z} = Z(z, t), \quad z = \{a, Q\}, \quad Z = \{Z_a, Z_Q\} \tag{6.19}$$

Here the following notation is used

$$Z_a(a, t) = C(t)a + f(t)$$

$$Z_Q(Q, t) = C(t)Q + QC^T(t) + qQ + q^{-1}B(t) \tag{6.20}$$

$$q = \{n^{-1}\mathrm{Tr}[Q^{-1}B(t)]\}^{1/2}$$

The obtained results are summarized in the following theorem [14].

Theorem 6.1. A solution of Problem 6.2 for the linear system (6.13) or (6.14) with observations (6.15) is given by relationships (6.18)–(6.20) and initial conditions (6.16). The pair $z(t) = \{a(t), Q(t)\}$ determining the ellipsoid in (6.3) satisfies the system of differential equations (6.19), (6.20) between the observations, equations (6.18) at the instants of observations $t = t_k$, $k = 0, 1, \ldots$, and the initial conditions (6.16).

Remark 6.3. The right-hand sides (6.20) of equations (6.19) correspond to outer ellipsoidal approximations optimal in the sense of volume. Other equations given in Section 4.6 and corresponding to other optimality criteria can be also used, see Theorem 4.6.

Remark 6.4. The functions ψ_k in (6.18) correspond to the outer approximation of the intersection of two ellipsoids given by (6.12). Instead of it, other approximations of this intersection can be used [14].

Remark 6.5. The procedure of the guaranteed state estimation given by Theorem 6.1 can be generalized and extended to:

1) nonlinear systems described by equations (6.1) (here the same ideas as in Section 3.5 are used);

2) nonlinear equations of observations;

3) continuous observations described by the equation

$$y(t) = H(t)x(t) + \xi(t), \quad \xi(t) \in E(0, L(t))$$

where the matrices $H(t)$ and $L(t)$ similar to the respective matrices H_i and L_i in (6.15) are defined for all $t \geq s$;

4) systems governed by finite-difference equations given by (5.34).

All these results are presented in [14].

6.4 Examples

The computer software is developed which implement the guaranteed estimation procedure [14]. Let us restrict ourselves here with two examples belonging to A. M. Shmatkov. More examples can be found in [14].

1. Consider a system of two masses m_1 and m_2 connected by a spring with the stiffness k and subjected to bounded uncertain forces (disturbances) F_1 and F_2. The dynamics of the system is described by equations

$$m_1\ddot{x}_1 = F_1 + k(x_2 - x_1), \quad m_2\ddot{x}_2 = F_2 + k(x_1 - x_2)$$

where x_1 and x_2 are the coordinates of the masses. At the initial time instant $t = 0$ the coordinates and velocities of the both masses are known with certain bounded accuracy. During the motion, only the coordinate x_1 and velocity $\dot{x}_1$ of the first mass are observed with a certain bounded accuracy, whereas x_2 and $\dot{x}_2$ are not observed. Some results of simulation of the guaranteed estimation procedure are presented in Fig. 6.2. Here, the time histories of $x_1(t)$ and its obtained estimates for two cases are shown by solid lines. The estimate for each instant t lies in the middle of the vertical segment which reflects the obtained guaranteed bounds on $x_1(t)$. The upper and lower curves in Fig. 6.2 correspond to the disturbances equal to $F_1 = F_1^0$ and $F_1 = -F_1^0$, respectively, where F_1^0 is a positive constant. The force F_2 is zero in both cases. The observation errors are simulated by sinusoidal functions of time. The obtained results show that, starting from a certain instant of time, the estimation procedure allows to distinguish two processes corresponding to different forces.

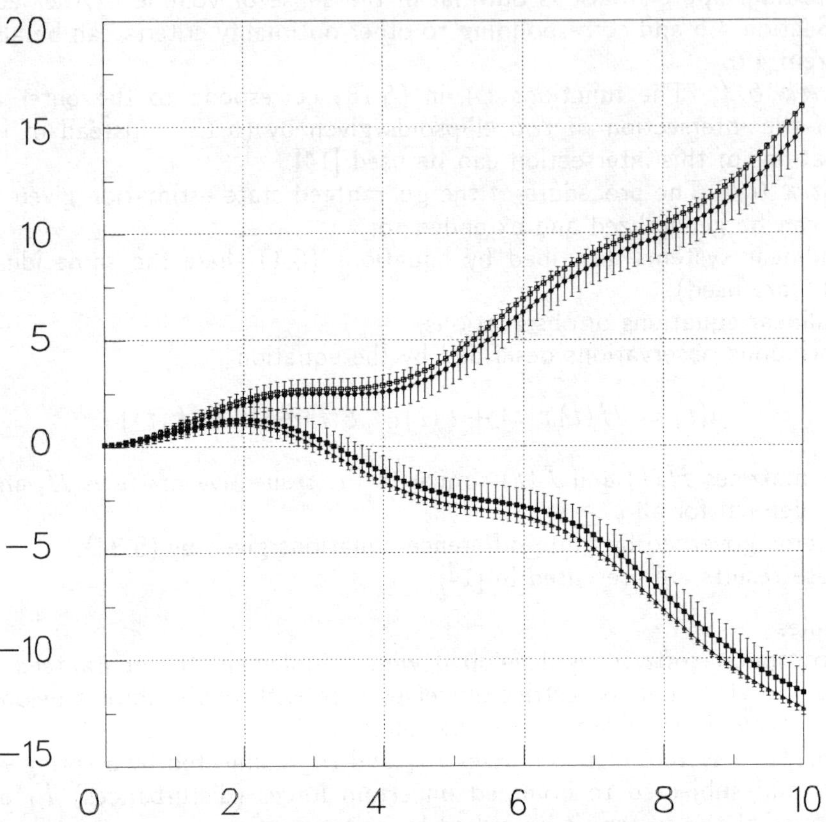

Fig. 6.2, Guaranteed state estimation for two masses connected by a spring.

2. Consider the motion of a unit mass in the vertical plane xy in the presence of the constant gravitational field and wind. We suppose that the velocity of the wind is horizontal (parallel to the x-axis) and grows linearly with height, being equal to $V_0 + \omega y$, where V_0 and ω are constants. The resistance force is assumed to be proportional to the velocity of the mass with respect to the air, the coefficient of proportionality being equal to k. The mass is subjected also to bounded but unknown disturbance. The equations of motion are

$$\ddot{x} = -k\dot{x} + \omega\dot{y} + k\omega y + kV_0 + F_x, \quad \ddot{y} = -k\dot{y} - g + Fy$$

Here, g is the gravity acceleration, F_x and F_y are the components of the disturbance. The coordinates x, y of the mass are being observed with some bounded observation errors, whereas the velocities are not observed. The time history of the vertical velocity $\dot{y}(t)$ as well as the results of the guaranteed estimation procedure are shown in Fig. 6.3 for two cases. The disturbance has the components $F_x = 0$, $F_y = F_y^0$ and $F_x = 0$, $F_y = -F_y^0$ for the upper and lower curves in Fig. 6.3, respectively. The denotations here are similar to those in Fig. 6.2. As in the first example, the results of simulation show that the guaranteed filtering makes it possible to distinguish the motions corresponding to different forces.

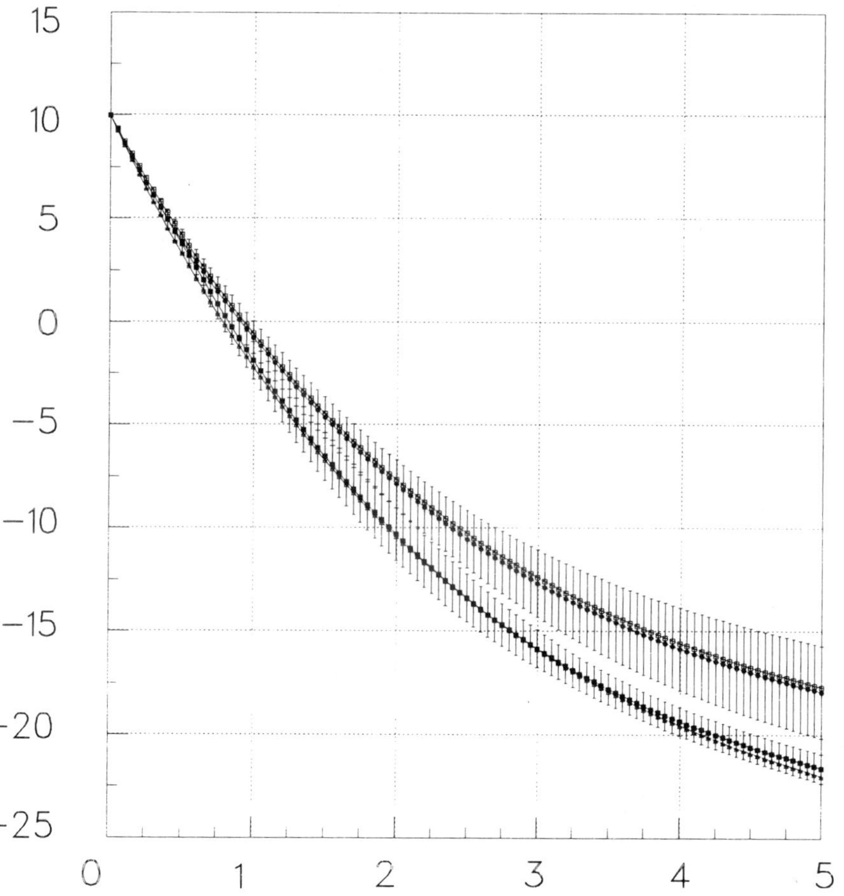

Fig. 6.3, Guaranteed state estimation for a mass in the vertical plane.

6.5 Discussion

Let us compare the algorithm of the guaranteed state estimation described above with the well-known stochastic (probabilistic) approach, namely, with the Kalman-Bucy filter. Although these algorithms show many similarities, they also exhibit a number of distinctions.

1) The both methods deal mostly worth linear dynamical systems with discrete or continuous time.

2) The observations in the both methods are described by linear relationships.

3) The equations of motion as well as the equations of observations include uncertain factors, namely, the external disturbances $(u(t)$ or $v(t))$ and the observation errors $\xi(t)$. The current observation results $y(t)$ are available.

4) In the stochastic approach, all uncertain factors are regarded as stochastic (random) processes with the given (known) parameters. Usually, these processes are assumed to be the Gaussian white noises. In the guaranteed approach, the uncertain factors are assumed to be arbitrary but bounded functions of time. In the method of ellipsoids considered here, the bounds are ellipsoidal and, in some sense, analogous to the Gaussian noises.

5) The goal of all state estimation procedures is to evaluate the state vector $x(t)$ using the observation data $y(\tau)$, $\tau \leq t$. In the stochastic approach, the probabilistic characteristics of the vector $x(t)$ are determined, namely, the average $m(t)$ and the correlation (covariance) matrix $K(t)$. In the guaranteed approach, we obtain the parameters of the ellipsoid $E(a(t), Q(t))$ containing the state $x(t)$.

6) In the both approaches, the solution of the estimation problem at the instant t is the pair z comprising the vector $(m$ or $a)$ and the symmetric positive definite matrix $(K$ or $Q)$. For the both approaches, the equations for the evolution of the pair $z(t)$ are obtained. These equations (finite-difference or differential) depend on the observation results $y(t)$ and are called the equations of the filter.

7) There are essential distinctions between the two approaches. The equations for the vector m are linear and depend on the observation results. The nonlinear matrix Riccatti equation for $K(t)$ is independent of m and y and thus can be integrated beforehand. In the guaranteed approach, the equations for a and Q are nonlinear and coupled; therefore, they cannot be solved separately.

8) The both approaches include optimization of the estimation results. In the Kalman filter, the obtained estimates are optimal in the average, whereas the ellipsoidal estimates are optimal with respect to the volume or other criteria.

9) The stochastic approach needs the statistical information about the external noises and observation errors. This information is often not available in practical problems. Advantages and shortcomings of probabilistic methods are discussed also in [35]. By contrast, the guaranteed approach uses only the bounds on the uncertain factors.

10) The stochastic approach provides results applicable to the whole ensemble of processes. The ellipsoidal approach can be applied to any particular process and gives the reliable guaranteed bounds which may be rather wide compared to the probabilistic estimates.

Remark 6.6. New results on ellipsoidal approximation of reachable sets for linear systems of differential equations with uncertain matrices of coefficients were recently obtained in [36]. These results can be applied to mechanical systems with unknown, uncertain, or varying parameters.

ACKNOWLEDGEMENTS
The paper is based on research sponsored by the Russian Foundation for Basic Research (Grant 96-01-01137). The author thanks Dr. A.M. Shmatkov, R.P. Soldatova, and D. Ya. Rockityanskii for their great help in preparing the manuscript.

REFERENCES

1. Bakhshiyan, B.Ts., Nazirov, R.R. and P.E. Elyasberg: Determination and Correction of Motion, Nauka, Moscow, 1980 (in Russian).
2. Elishakoff, I., Lin Y.K. and L.P. Zhu: Probabilistic and Convex Modeling of Acoustically Excited Structures, Elsevier, Amsterdam, 1994.
3. Kalman, R.E.: On the general theory of control systems, in: Proc. 1st IFAC Congress, 1, Butterworths, London 1960.
4. Bryson, A.E. and Yu-Chi Ho: Applied Optimal Control, Blaisdell, New York 1969.
5. Krasovskii, N.N.: Theory of Control of Motion, Nauka, Moscow 1968 (in Russian).
6. Lee, E.B. and L.Markus: Foundations of Optimal Control Theory, Wiley, New York 1967.
7. Pontryagin, L.S., Boltyanski, V.G., Gamkrelidze, R.V. and E.F.Mischenko: Mathematical Theory of Optimal Processes, Wiley-Interscience, New York 1962.
8. Isaacs, R.: Differential Games, John Wiley, New York 1965.
9. Krasovskii, N.N. and A.I.Subbotin: Positional Differential Games, Springer-Verlag, Berlin 1988.
10. Schweppe, F.C.: Uncertain Dynamic Systems, Prentice-Hall, Englewood Cliffs 1973.
11. Chernousko, F.L. and A.A.Melikyan: Game Problems of Control and Search, Nauka, Moscow 1978 (in Russian).
12. Kurzhanski, A.B.: Control and Observation under Conditions of Uncertainty, Nauka, Moscow 1977 (in Russian).
13. Chernousko, F.L.: Estimation of the Phase State of Dynamic Systems, Nauka, Moscow 1988 (in Russian).
14. Chernousko, F.L.: State Estimation for Dynamic Systems, CRC Press, Boca Raton 1994.
15. Kurzhanski, A. and I.Valyi: Ellipsoidal Calculus for Estimation and Control, Birkhäuser, Boston 1997.
16. Dunford N. and J.T.Schwartz: Linear Operators, Part 1, Wiley-Interscience, New York 1967.
17. Rockafellar, R.T.: Convex Analysis, Princeton University Press, Princeton 1970.
18. Gantmacher, F.R.: Matrix Theory, Chelsea Publishing Co., New York 1960.
19. Berger, M.: Géométrie, Part 3, CEDIC/Fernand Nathan, Paris 1978.

20. Grünbaum, B.: Convex Polytopes, Interscience, London 1967.
21. Leichtweiss, K.: Konvexe Mengen, VEB Deutscher Verlag der Wissenschaft, Berlin 1980.
22. John, F.: Extremum problems with inequalities as subsidiary conditions, in: Studies and Essays Presented to R. Courant on His 60th Birthday (Eds. K. Friedrichs, O. Neugebauer, J. J. Stoker), John Wiley & Sons Inc., New York 1948.
23. Zaguskin, V.L.: On circumscribed and inscribed ellipsoids of the extremal volume, Uspekhi Mat. Nauk, 13, 6 (1958), 89–93 (in Russian).
24. Chernousko, F.L.: Optimal guaranteed estimates of uncertainties by means of ellipsoids, I–III, Izvestiya Akad. Nauk SSSR, Tekhnicheskaya kibernetika (1980), I: N 3, 3–11, II: N 4, 3–11, III: N 5, 5–11 (in Russian), translated into English as Engineering Cybernetics.
25. Chernousko, F.L.: Guaranteed estimates of undetermined quantities by means of ellipsoids, Soviet Math. Doklady, 21 (1980), 396–399.
26. Chernousko, F.L.: Ellipsoidal estimates of a controlled system's attainability domain, J. Applied Mathematics and Mechanics, 45 (1981), 7–12.
27. Chernousko, F.L.: On equations of ellipsoids approximating reachable sets, Problems of Control and Information Theory, 12 (1983), 97–110.
28. Ovseevich, A.I. and F.L.Chernousko: Two-sided estimates on the attainability domains of controlled systems, J.Applied Mathematics and Mechanics, 46 (1982), 590–595.
29. Ovseevich, A.I.: Limit behaviour of attainable and superattainable sets, in: Modelling, Estimation and Control Systems with Uncertainty, Proc. Conf. (Eds. G. B. Di Mazi, A. Gombani, and A. B. Kurzhanski), Birkhäuser, Boston 1991, 324–333.
30. Ovseevich, A.I. and Yu.N.Reshetnyak: Asymptotic behaviour of the ellipsoidal estimates for reachable sets, Izvestiya AN SSSR. Tekhnicheskaya kibernetika, 1 (1993), 90–100 (in Russian).
31. Reshetnyak, Yu.N.: Summation of ellipsoids in the guaranteed estimation problem, J. Applied Mathematics and Mechanics, 53 (1989), 193–197.
32. Chernousko, F.L.: Ellipsoidal bounds for sets of attainability and uncertainty in control problems, Optimal Control Applications and Methods, 3 (1982), 187–202.
33. Komarov, V.A.: Estimates on reachable sets for linear systems, Izvestiya Akad. Nauk SSSR, Ser. Mat., 48 (1984), 865–879 (in Russian).
34. Melikyan, A.A.: On the minimal number of observation instants in the model game of approach, Izvestiya Akad. Nauk ArmSSR, Mat., 9 (1974), 242–247 (in Russian).
35. Ben-Haim, Y. and I.Elishakoff: Convex Models of Uncertainty in Applied Mechanics, Elsevier, Amsterdam, 1990.
36. Chernousko, F.L.: Ellipsoidal approximation of attainability sets of a linear system with indeterminate matrix, J. Applied Mathematics and Mechanics, 60 (1996), 921-931.

CHAPTER 4

WHAT IS DIFFERENTIAL STOCHASTIC CALCULUS?

M. Di Paola
University of Palermo, Palermo, Italy

WHAT ARE THE BASIC CONCEPTS OF THE STOCHASTIC DIFFERENTIAL CALCULUS ?

M. Di Paola
Dipartimento di Ingegneria Strutturale e Geotecnica,
Università degli Studi di Palermo,
Viale delle Scienze, Palermo, Italy

ABSTRACT

Some well known concepts of stochastic differential calculus of non linear systems corrupted by parametric normal white noise are here outlined. Ito and Stratonovich integrals concepts as well as Ito differential rule are discussed. Applications to the statistics of the response of some linear and non linear systems is also presented.

1. INTRODUCTION

Mechanical systems are studied in an analytical way through the identification of a mathematical model and the solution of its governing equations. A satisfactory level of generality is achieved by writing a set of nonlinear ordinary differential equations in the form:

$$\dot{x}(t) = f(x(t),t) \tag{1.1}$$

where $x(t)$ is an n-vector of state variables, t is the time and the superimposed dot means time derivative. Often Eq. (1.1) is obtained by a transformation of r second order Lagrange equations relative to a mechanical system with r-degree of freedom and thus $n = 2r$.

If some noises of forcing vector are present, then Eq. (1.1) is rewritten in more explicit form

$$\dot{x}(t) = f(x(t),w(t),t) \tag{1.2}$$

where $w(t)$ is an m vector of excitations. A particular class, that will be studied in detail in what follows, is constituted by the equation

$$\dot{x} = f(x(t),t) + G(x(t),t)\, w(t) \tag{1.3}$$

where $G(x(t),t)$ is an nxm matrix of non-linear functions of $x(t)$ and t. In Eq. (1.3) the excitation vector $w(t)$ is called *parametric* because its intensity depends on the response $x(t)$ itself. In the case of Eq. (1.3) the vector $w(t)$ of excitation is called *multiplicative*. In the particular case in which the matrix G depends only on t, the excitation $w(t)$ is called *additive*.

Our study is confined to the case in which $f(x(t),t)$ and $G(x(t),t)$ are deterministic non linear functions of x and t and $w(t)$ is a vector of stochastic processes.

As customary in the following we shall denote stochastic processes with capital letters and the corresponding domains with the corresponding lower case roman letters. Then if the excitation vector is a stochastic vector process the corresponding response vector is a stochastic process too and equation (1.3) is written in the form:

$$\dot{X} = f(X(t),t) + G(X(t),t)\, W(t) \tag{1.4}$$

Here we confine ourselves to the case in which $W(t)$ is a normal or non normal white noise process or the case in which the stochastic vector of excitation is a filtered normal or non-normal white noise. Appendix A.1 summarizes some well known concepts and definition on the probabilistic description of random variables and stochastic processes.

2. NORMAL WHITE NOISE AND WIENER PROCESS

The basic process for the analytical treatment of stochastic differential equations is the *Wiener* or *Wiener-Levy* process $B(t)$, (also called Brownian motion). The process $B(t)$ is a normal process characterized by the first two moments as follows

$$E[B(t)] = 0 \tag{2.1}$$

$$E[B(t_1)B(t_2)] = q_2 \, min \, (t_1, t_2) \tag{2.2}$$

where $E[\cdot]$ means stochastic average and q_2 is a deterministic function whose physical signifiance will be explained later on. Because of the normality of the process $B(t)$ all higher order cumulant than the second one are exactly zero, that is:

$$K_3[B(t_1)B(t_2)B(t_3)] = 0 \; ; \ldots ; K_j[B(t_1)B(t_2)\ldots B(t_j)] = 0 \tag{2.3}$$

Furthermore, it was shown that, if $0 \le t_1 \le t_2 \le \ldots \le t_n$, the increments of $B(t)$, $\Delta B(t_1, t_2) = B(t_2) - B(t_1), \ldots, \Delta B(t_{j,j+1}) = B(t_{j+1}) - B(t_j), \ldots, \Delta B(t_{n-1}, t_n) = B(t_n) - B(t_{n-1})$, are mutually independent random variables, then

$$E[dB(t_1)\,dB(t_2)] = E\Big[\big(B(t_1+dt_1) - B(t_1)\big)\big(B(t_2+dt_2) - B(t_2)\big)\Big] =$$

$$= q_2 \, \delta(t_1 - t_2)\,dt_1\,dt_2 \tag{2.4}$$

where $\delta(t_1 - t_2)$ is the Dirac's delta. In virtue of Eq. (2.4) it can be shown that

$$E[dB(t)] = 0$$

$$E[(dB(t))^2] = q_2 \, dt \tag{2.5}$$

Moreover due to the normality of the process $dB(t)$

$$E[(dB(t))^{2j+1}] = 0$$

$$E[(dB(t))^{2j}] = (2j-1)!!\left(E[(dB(t))^2]\right)^j = (2j-1)!! \, q_2^j \, (dt)^j \qquad j = 1, 2, \ldots, \infty \tag{2.6}$$

Eqs. (2.5) and (2.6) show that an increment of the Wiener process $B(t)$ is of order $(dt)^{1/2}$. This is one of the keys for understanding the stochastic differential calculus, how it will be shown later. It can be also shown that sample functions of the Wiener process $B(t)$ are continuous with probability one, but, in accordance with a theorem of Wiener, they are nowhere differentiable due to the extreme irregularity of the Wiener process. In Fig. 1 a sample function of the Wiener process is depicted, showing a peculiar aspect of these sample functions, that is they exhibit unbounded variations in infinitesimal intervals.

Because $B(t)$ is nowhere differentiable, then $\dot{B}(t)$ does not exist in the frame of ordinary differential calculus. However, the derivative of the process $B(t)$ can be framed in the context of distribution theory, at the moment one may only say that $\dot{B}(t)$ is a *formal* derivative of the stochastic process $B(t)$. This derivative is denoted as $W^0(t)$ and it is called *normal white noise* process. By indicating $W^0(t) = dB(t)/dt$ and using Eq. (2.4) one can write

$$E[W^0(t_1)\, W^0(t_2)] = E[dB(t_1)\, dB(t_2)]/dt_1\, dt_2 = q_2\, \delta(t_1 - t_2) \tag{2.7}$$

it follows that $W^0(t)$ is a stationary process having zero mean and it is uncorrelated. That means that it is constituted by Dirac's deltas independent of each other having normal zero mean random amplitudes. A sample function of such a process is depicted in Fig. 2.

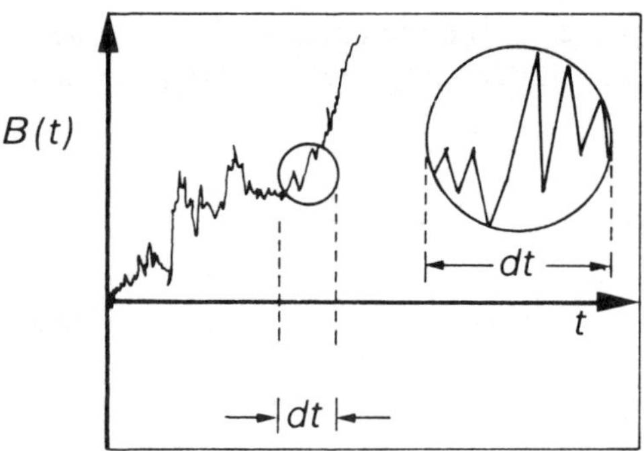

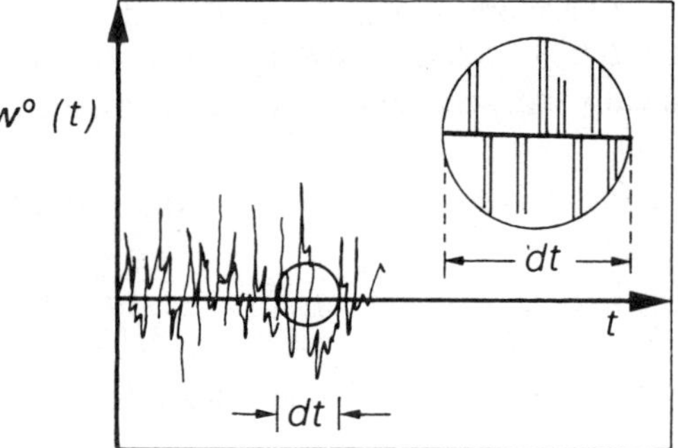

Fig. 1 - Sample function of the Wiener Process   Fig. 2 - Sample function of normal white noise process

Moreover, from Eq. (2.7) one recognizes that the variance, being the correlation evaluated at zero, attains an infinite value.

The Power Spectral Density (PSD) of $W^0(t)$ is given as

$$S_{W^0}(\omega) = \frac{q_2}{2\pi} \int_{-\infty}^{\infty} e^{-i\omega\tau}\, \delta(\tau)\, d\tau = \frac{q_2}{2\pi} \tag{2.8}$$

that is the PSD of the process $W^0(t)$ takes a constant value at all frequencies and this justifies the nomenclature of white noise; in the optic the white noise colour possesses constant frequency power.

Further hints on the subject of this section can be found in the papers of Wiener [1] and Lévi [2], or in the boocks of Doob [3], Loéve [4] and Soong and Grigoriu [5].

3. STOCHASTIC DIFFERENTIAL CALCULUS

In this section some well known results of the classical stochastic differential calculus are outlined.

Let the equation of motion be given in the form

$$\dot{X} = f(X(t),t) + g(X(t),t)\, W^0(t) \; ; \; X(t_0) = X_0 \tag{3.1}$$

where $f(\cdot)$ and $g(\cdot)$ are deterministic non linear functions of the response $X(t)$, X_0 is the (deterministic or random) initial condition, while $W^0(t)$ is a normal white noise. Equation (4.1) can be also written in the incremental form as follows

$$dX(t) = f(X(t),t)dt + g(X(t),t)dB(t) \; ; \quad X(t_0) = X_0 \tag{3.2}$$

in which $dB(t)$ is an increment of the Wiener process $B(t)$. Integration of Eq. (3.2) yelds to

$$X(t) - X(t_0) = \int_{t_0}^{t} f(X(\tau),\tau)d\tau + \int_{t_0}^{t} g(X(\tau),\tau)dB(\tau) \tag{3.3}$$

The first integral on the rigth hand side of Eq. (3.3) is a mean square Riemann integral, while the second one is not a Riemann nor a Fourier-Stieltjes integral, as in fact the limit of the partial sum does not achieve a unique value but it depends on the choice of the intermediate point selected (see e.g. [8], [9]). In order to show this let the time interval $[t_0, t]$ be subdivided into n subintervals by means of partitioning points t_j, let Δn be the maximum length of these intervals, then the Riemann sum is given as

$$\int_{t_0}^{t} g(X(\tau),\tau)dB(\tau) = \underset{\substack{n\to\infty \\ \Delta n\to 0}}{l.i.m.} \sum_{s=1}^{n} g(X(\bar{t}_s),\bar{t}_s)\,[(B(t_s) - B(t_{s-1}))] \tag{3.4}$$

$\bar{t}_s$ being an intermediate point such that $t_{s-1} \leq \bar{t}_s \leq t_s$, and $l.i.m.$ means limit in mean square (see Appendix A).

3.1 Itô integral

In order to show that Eq. (3.4) does not attain unique solution a very simple example is examined, let us consider integrals of the type

$$\int_{t_0}^{t} B(t)\,dB(t) = \underset{\substack{n\to\infty \\ \Delta n\to 0}}{l.i.m.} \sum_{s=1}^{n} B(\bar{t}_s)\,[(B(t_s)-B(t_{s-1})] \tag{3.5}$$

If for all the intervals one selects the initial point, $\bar{t}_s = t_{s-1}$, then the integral is the *Itô stochastic integral* [8], that is

$$(I)\int_{t_0}^{t} g(X(\tau),\tau)\,dB(\tau) = \underset{\substack{n\to\infty \\ \Delta n\to 0}}{l.i.m.} \sum_{s=1}^{n} g(X(t_{s-1}),t_{s-1})\,[(B(t_s)-B(t_{s-1})] \tag{3.6}$$

where the symbol *(I)* before the integral means that the integration is performed in Itô sense. If the definition of the integral follows Eq. (3.6), the integral (3.5) (see e.g. for proof Gardiner [9]) gives

$$(I)\int_{t_0}^{t} B(t)\,dB(t) = \frac{1}{2}\left[B^2(t)-B^2(t_0)-(t-t_0)\right] \tag{3.7}$$

From this equation it can be seen that Itô integrals does not follow the classical rules of integrals. As in fact for the ordinary Riemann-Streltjes integral the term $(t-t_0)$ would be absent.

If the integrals of the differential equations are performed in Itô sense Eq. (3.2) are called *Itô stochastic differential equation.*

3.2 Stratonovich integral

If for all intervals one selects the midpoint point that is $\bar{t}_s = (t_{s-1}+t_s)/2$, then the mean square integrals of the type (3.5) attains an unique value, and one refers to this as *Stratonovich integral* [10]

$$(S)\int_{t_0}^{t} B(\tau)\,dB(\tau) = \underset{\substack{n\to\infty \\ \Delta n\to 0}}{l.i.m.} \sum_{s=1}^{n} B\left(\frac{t_{s-1}+t_s}{2}\right)[(B(t_s)-B(t_{s-1})] =$$

$$= \left[B^2(t)-B^2(t_0)\right]/2 \tag{3.8}$$

It can be easily seen that using $(S)\int$ also all other rules of ordinary differential calculus are preserved.

It can be also monstrated that

$$(S) \int_{t_0}^{t} \psi(\tau)\,dB(\tau) = (I) \int_{t_0}^{t} \psi(\tau)\,dB(\tau) \tag{3.9}$$

in the case in which $\psi(t)$ is a deterministic smooth function. If the $(S)\!\int$ is used in Eq. (3.3) one refers the differential equation (3.3) as a *Stratonovich differential equation,* in the sense that the last term on the rigth hand side is performed in Stratonovich sense.

At this stage some comments and questions arise: i) integrals in Itô sense or in Stratonovich sense give quite different results; ii) using Itô integrals the rules of the ordinary differential calculus and integrations are not preserved; iii) in view of ii) why the Itô integral has been introduced?; iiii) what is the true solution of Eq. (3.2)?: iv) if $W^0(t)$ is a band limited white noise, do physical white noise $(I)\!\int$ and $(S)\!\int$ lead to the same results?

All these questions are of primary importance and have been subject of discussions and different interpretations by people working in the random vibration theory. All these questions will be examined in the next sections.

3.3 Non-anticipating property

An important concept, which is the key of the Itô calculus is that of *non-anticipating function.* A deterministic or stochastic function $\phi(t)$ is said to be non-anticipating, if for every $t_{j+1} > t_j$, $\phi(t_j)$ is independent of the increment of the Wiener process $B(t_{j+1}) - B(t_j)$.

Example of non anticipating functions are the solution $X(t)$ of Eq. (3.2), or every function $\phi(X(t),t)$ of the solution. Moreover, every deterministic function possesses this property. For any non anticipating function (see Gardiner for proofs [9])

$$(I) \int_{t_0}^{t} \phi(t)\,dB(t) = 0$$

$$(I) \int_{t_0}^{t} \phi(t)\,(dB(t))^2 = q_2 \int_{t_0}^{t} \phi(t)\,dt \tag{3.10}$$

$$(I) \int_{t_0}^{t} \phi(t)\,(dB(t))^N = 0 \qquad \forall N > 2$$

If Stratonovich integrals where adopted these relationships cannot be used. In this case extraterms in Eq. (3.10) would appear due to the fact that in the Stratonovich integral interpretation the intermediate point selected is the mid point so that the independence between $dB(t)$ and $\phi(t)$ cannot be invoked. At this stage it is evident that, using Itô calculus, one can manipulate the differential equation involving product of non linear functions of the response and increments of the Wiener process such as in Eq. (3.2) with simplicity. This is the main motivation in the choice of Itô integral concept, but the price to be payed for this simplicity is that one has to treat with care these integrals since the usual rules of differentiation and integration fail.

3.4 Relationship between $(I)\int$ and $(S)\int$.

In general no relationships between the Itô integral and the Stratonovich integral exist. However, if the stochastic process $X(t)$ is the solution of Eq. (3.3) there is a connection between these two integrals given as

$$(S) \int_{t_0}^{t} \phi(X(\tau),\tau) \, dB(\tau) = (I) \int_{t_0}^{t} \phi(X(\tau),\tau) \, dB(\tau) +$$

$$+ \frac{1}{2} \int_{t_0}^{t} \phi(X(\tau),\tau) \frac{\partial \phi(X(\tau),\tau)}{\partial X} q_2 \, d\tau \tag{3.11}$$

$\phi(X(t),t)$ being any non-linear function of the response, the second term on the right hand side of equation (3.11) is the so-called *Wong-Zakaj* or *Stratonovich (WZ or S) correction term* [11].

Equation (4.11) can be rewritten in differential form as follows

$$(S) \, \phi(X(t),t) dB(t) = (I) \, \phi(X(t),t) dB(t) + \frac{1}{2} \, \phi(X(t),t) \frac{\partial \phi(X(t),t)}{\partial X(t)} q_2 \, dt \tag{3.12}$$

in which the symbols (S) or (I) means Stratonovich and Itô differential, respectively, in the sense that the corresponding integrals are performed in Stratonovich or in Itô sense.

Letting $\phi(X(t),t) = g(X(t),t)$ in Eq. (3.12) and substituting in Eq. (3.2) leads to

$$(S) \, dX(t) = f(X(t),t) dt + (I) \, g(X(t),t) dB(t) + \frac{1}{2} \, g(X(t),t) \frac{\partial g(X(t),t)}{\partial X(t)} q_2 \, dt \tag{3.13}$$

or in the more compact form

$$(S) \, dX(t) = m(X(t),t) dt + (I) \, g(X(t),t) dB(t) \tag{3.14}$$

where $m(X(t),t)$ is the so-called *drift term* given as

$$m(X(t),t) = f(X(t),t) + \frac{q_2}{2} g(X(t),t) \frac{\partial g(X(t),t)}{\partial X(t)} \tag{3.15}$$

In the case of external noise, that is $g(X(t),t)$ is independent on $X(t)$, the *WZ* or *S* correction term vanishes and the Itô and Stratonovich integrals lead to the same result. On the contrary if parametric excitation appears the two integrals give entirely different results.

At this stage people working in the area of stochastic dynamics can be disconcerted because a main question arises: what is the correct integral that one has to performe in order to obtain the solution in terms of statistics of the physical system under white noise? The answer to this question cannot be obtained by numerical or experimental test since it is impossible to generate true white noise, but only band limited or physical white noise. It will shown that the correct integral is the Stratonovich one.

In order to retain the fundamental non-anticipating property of Itô integral, in virtue of Eq. (3.14); the most common way to handle with stochastic differential equations forced by white noise processes is that the differential equation is transformed in Eq. (3.14) by modifying the drift term and then by performing the integrals in Ito sense. Since on the right hand side of equation (3.14) Itô differential often appears, Eq. (3.14) is sometimes called *Itô differential equation.*

The usual way to handle stochastic differential equations of the type (3.2) is that of rewriting Eq. (3.2) into an Itô form by inserting the *WZ* or *S* correction term into the drift coefficient and then performing integral in Itô sense. In this way the non-anticipating property is retained and the output is a reponse evaluated in Stratonovich sense.

3.5 Ito differential rule

Another fundamental concept for evaluating the response in statistical sense of Eq. (3.1) is the so-called *differential rule.*

Let $\phi(X(t),t)$ be any scalar real valued function of the process $X(t)$ and t continuously differentiable on t and twice differentiable on $X(t)$, then an increment of $\phi(X(t),t)$, can be written in the form [9], [12]

$$d\phi(X(t),t) = \frac{\partial\phi(X(t),t)\,dt}{\partial t} + \frac{\partial\phi(X(t),t)}{\partial X}\,dX(t) + \frac{1}{2}\frac{\partial^2\phi(X(t),t)}{\partial X^2}(dX(t))^2 \qquad (3.16)$$

Eq. (3.16) is a Taylor expansion truncated up to the second order term. This is another main difference with the classical differential calculus, in the latter case, in fact, the second term is an infinitesimal of the second order $((dt)^2)$ and then it can be neglected. In the stochastic differential calculus $dX(t)$ is composed by two terms the first one is of order dt while the second one is of order $dB(t)$ $(0\,(dB) = dt^{1/2})$ then the second term in the Taylor expansion is of the same order than $[\partial\phi(X(t),t)\,dt/\partial t]$, and $[\partial\phi(X(t),t)/\partial X(t)]dX(t)$

The Itô differential rule is a fundamental tool for practical applications how it will be shown in the following examples.

4. SOME EXAMPLES

In this sub-section some examples are reported in order to show the way for manipulating the concepts already discussed.

4.1 Linear system (external excitation)
  Let the differential equation of motion be given in the form

$$\dot{X} = f(t)X(t) + g(t)W(t); \quad X(0) = X_0 \tag{4.1}$$

$f(t)$ and $g(t)$ being deterministic (smooth) function, X_0 is the initial condition (independent of the white noise $W(t)$).
  In this case, since the equation is forced by external excitation the Itô differential equation and the Stratonovich one are exactly the same

$$dX(t) = f(t)X(t)dt + g(t)dB \tag{4.2}$$

The moments of every order of X(t) can be found letting $\phi(X(t),t) = X^k$ in Eq. (3.16), then

$$dX^k = kX^{k-1}dX + \frac{k(k-1)}{2}X^{k-2}(dX)^2 \tag{4.3}$$

inserting Eq. (4.2) in (4.3), leads to

$$dX^k = kf(t)X^k dt + kg(t)X^{k-1}dB + \frac{k(k-1)}{2}X^k f^2(t)(dt)^2 +$$

$$+ \frac{k(k-1)}{2}X^{k-2}g^2(t)(dB)^2 + k(k-1)f(t)g(t)X(t)dB\,dt \tag{4.4}$$

making stochastic average, dividing by dt, using the non-anticipating property $(E[X^r(t)dB(t)] = E[X^r(t)]E[dB(t)] = 0)$ and neglecting higher order differentials than dt Eq. (4.4) leads to

$$\dot{E}[X^k] = kf(t)E[X^k] + \frac{k(k-1)}{2}E[X^{k-2}]g^2(t)q_2 \tag{4.5}$$

This is an ordinary differential equation (deterministic) that can be solved by imposing the initial condition

$$E[X_0^k] = \int_{\infty}^{\infty} p_{X_0}(x_0)\, x_0^k\, dx_0 \tag{4.6}$$

$p_{X_0}(x_0)$ being the probability density function of X_0. It will be noted that, if X_0 is deterministic or a normal random variable then $X(t)$ is also a normal process. It follows that once the first two moments are evaluated the higher order moments can be immediately found by imposing that higher order cumulants than the second one of X are zeros.

Once the moments of the response are evaluated by solving Eq. (4.5) higher order statistics can be evaluated by multiplying Eq. (4.2) by $X(t_1)$, $t_1 \leq t$, in this way one can write

$$X(t_1)\,dX(t) = f(t)\,X(t_1)X(t)\,dt + g(t)X(t_1)\,dB(t) \tag{4.7}$$

using the non-anticipating property, making stochastic average and dividing by dt one obtains

$$\frac{d}{dt} E[X(t_1)X(t)] = f(t)E[X(t_1)X(t)]; \qquad t_1 \leq t \tag{4.8}$$

this is an ordinary differential equation whose initial condition (obtained by setting $t_1 = t$) is $E[X^2(t)]$ already evaluated by solving Eq. (4.5) for $k = 2$.

Higher order statistics can be easily found by multiplying Eq. (4.2) by $X(t_1)X(t_2)...X(t_r)$, being $t_1 \leq t$, $t_2 \leq t,...,$ $t_r \leq t$. As an example the third order statistics can be evaluated as

$$\frac{d}{dt} E[X(t_1)X(t_2)X(t)] = f(t)\,E[X(t_1)X(t_2)X(t)] \tag{4.9}$$

The appropriate initial condition (obtained by setting $t_1 = t_2 = t$) can be derived at every time instant t by solving the differential Eq. (4.5) particularized for $k = 3$.

If $f(t) = f$ and $g(t) = g$ are constants, and the system operates from $t = -\infty$, then the solution $X(t)$ reaches a stationary state. The second order moment can be evaluated from Eq. (4.5), due to the stationarity, as follows

$$E[X] = 0\,; \quad E[X^2] = -\frac{g^2}{2f}\,q_2 \tag{4.10}$$

Moreover the correlation function depends only on the difference $\tau = t - t_1$, then from Eq. (4.8)

$$\frac{d}{d\tau} R_X(\tau) = f R_X(\tau); \quad R_X(0) = -\frac{g^2}{2f} q_2 \tag{4.11}$$

and the correlation function is given as

$$R_X(\tau) = -\frac{g^2 q_2}{2f} \exp(f\tau); \tau \geq 0 \quad \text{or} \quad R_X(\tau) = -\frac{g^2 q_2}{2f} \exp(f|\tau|); \tau > 0 \tag{4.12}$$

4.2 Quasi-linear system

Let the equation of motion be given in the form

$$\dot{X} = f(t) X(t) + (g(t) X(t) + h(t)) W(t); \quad X(0) = X_0 \tag{4.13}$$

In this case the system is quasilinear (also called bilinear, or simply linear) and a parametric excitation appears, then the equation (4.13) has to be transformed into an Itô type stochastic differential equation by inserting the WZ or S correction term in Eq. (4.13), that is

$$dX = f(t) X(t) dt + \frac{1}{2} g(t) (g(t) X(t) + h(t)) q_2 dt + (g(t) X(t) + h(t)) dB(t) \tag{4.14}$$

Once the differential equation has been transformed into an Itô type differential equation the Itô rule can be applied, setting in Eq. (3.16) $\phi(X(t),t) = X^k$ one can write

$$dX^k = k m(t) X^k dt + k g(t) X^k dB + k h(t) X^{k-1} dB(t) +$$

$$+ \frac{k(k-1)}{2} X^{k-2} (g(t) X(t) + h(t))^2 (dB(t))^2 \tag{4.15}$$

$m(t)$ being the drift term $(m(t) = fX + q_2 g(gX + h)/2)$. By making stochastic average and dividing by dt one obtains

$$\frac{1}{k} \dot{E}[X^k] = \left[m(t) + \frac{(k-1)}{2} g^2(t) q_2 \right] E[X^k] + (k-1) g(t) h(t) q_2 E[X^{k-1}] +$$

$$+ \frac{k-1}{2} h^2(t) q_2 E[X^{k-2}] \tag{4.16}$$

it can be easily seen that if $g(t) \neq 0$ the stochastic process X(t) is non normal.

Also in this case the higher order statistics can be evaluated by multiplying Eq. (4.14) by $X(t_1) X(t_2) \dots X(t_r)$, $t_1 \leq t_2 \leq \dots \leq t_r$, and making stochastic average. As an example the second order statistics is given as

$$\frac{d}{dt} E[X(t_1) X(t)] = \left(f + \frac{q_2}{2} g^2 \right) E[X(t_1) X(t)] + \frac{1}{2} g h q_2 \quad t_1 \le t \tag{4.17}$$

the appropriate initial conditions are simply obtained by solving eq. (4.16) for $k = 2$.

4.3 Bernoulli equation

Let the equation of motion be given in the form

$$\dot{X} = f(t)X + g(t) X^s W(t) \, ; \; X(0) = X_0 \tag{4.18}$$

this equation is a non linear (for $s \ne 1$). The Ito equation is given in the form

$$dX = m(X) \, dt + g X^s dB \tag{4.19}$$

where the drift term $m(X)$ in this case is given as

$$m(X) = f(t)X + \frac{s}{2} g^2(t) X^{2s-1} q_2 \tag{4.20}$$

Setting $\phi(X(t), t) = X^k$ in Eq. (3.16) one obtains

$$dX^k = k X^{k-1} m(X) \, dt + k g(t) X^{k+s-1} dB + \frac{k(k-1)}{2} g^2 X^{2s+k-2} (dB)^2 \tag{4.21}$$

making stochastic average and dividing by dt one obtains

$$\dot{E}[X^k] = k f E[X^k] + \frac{k}{2} q_2 g^2 E[X^{k+2s-2}](s+k-1) \tag{4.22}$$

Equation (4.22) is an infinite hierarchy in the sense that the equation of moments of a fixed order involves moments of higher order, then no exact solution in terms of moments can be achieved.

Now, another approach is presented for the Bernoulli equation, putting

$$Z = X^{1-s} \, ; \quad Z(0) = X_0^{1-s} \tag{4.23}$$

differentiation yields to

$$\dot{Z} = (1-s) X^{-s} \dot{X} \tag{4.24}$$

inserting in Eq. (4.18) one obtains

$$\dot{Z} = (1-s)fZ + (1-s)g\,W(t)\,; \qquad Z_0 = X_0^{1-s} \tag{4.25}$$

this is a linear differantial equation forced by external input, then the Itô equation is simply writen in the form

$$dZ = (1-s)fZ\,dt + (1-s)g\,dB(t) \tag{4.26}$$

for this equation the moments of every order are given as

$$\dot{E}[Z^r] = r(1-s)f\,E[Z^r] + \frac{r(r-1)}{2}(1-s)^2\,g^2\,q_2\,E[Z^{r-2}] \tag{4.27}$$

now putting $r = k\,/\,1-s$ one exactly obtain (4.22). This is a very interesting result, one can in fact approach directly Eq. (4.18) writing the Itô stochatic differential equation by modifying the drift term and then applying Itô differential rule, or for the Bernoulli equation, by means of the transformation (4.23), is reduced into a linear one (excited by external excitation). For the latter the Itô differential equation does not require the modification of the drift term, and at least the two approaches lead in mean identical results.

5. FOKKER-PLANCK EQUATION

Let the equation of motion be given in the form (4.1), then one can write this equation in the Ito form

$$dX = m(X(t),t)\,dt + g(X(t),t)\,dB\,; \quad X(t_0) = X_0 \tag{5.1}$$

letting $\phi(X,t) = exp(-i\,\vartheta\,X)$, ϑ being a real parameter, by inserting in the Itô differential rule (3.16) one obtains

$$de^{-i\vartheta X} = -i\,\vartheta\,e^{-i\vartheta X}(m\,dt + g\,dB) - \frac{\vartheta^2}{2}\,q_2\,g^2\,e^{-i\vartheta X}\,dt \tag{5.2}$$

making stochastic average and a dividision by dt leads to

$$\dot{M}_X(\vartheta) = -i\,\vartheta\int_{-\infty}^{\infty} e^{-i\vartheta X}\,p_X(x,t)\,m(X,t)\,dX -$$

$$- \frac{\vartheta^2 q_2}{2}\int_{-\infty}^{\infty} e^{-i\vartheta X}\,g^2(X,t)\,p_X(x,t)\,dx \tag{5.3}$$

where $M_X(\vartheta)$ is the characteristic function (see Appendix A) and $p_X(x,t)$ is the probability density function of the process $X(t)$. By making the inverse Fourier transform of Eq. (5.3) one obtains

$$\dot{p}_X(x,t) = -\frac{\partial}{\partial x}(p_X(x,t)m(x,t)) + \frac{q_2}{2}\frac{\partial^2}{\partial x^2}[p_X(x,t)\,g^2(X,t)] \tag{5.4}$$

This equation, that gives the evolution of the probability density function, is known as *Fokker-Planck (FPK)* equation.

The stationary solution (if exists) of this equation can be obtained by setting $\dot{p}_X(x,t) = 0$. The appropriate initial condition of Eq. (5.4) is $p_{X_0}(x_0,t_0)$.

Extension of the FPK equation to multidegree of freedom systems will be discussed in the comparison paper [13].

CONCLUSIONS

In this paper the statistics of the response of linear or non linear systems corrupted by parametric normal white nose has been presented. It has been shown that, due to the extreme irregularity of the input, the usual rule of differentiation and integration of the stochastic differential equation fail.

The main strategy for the mathematical treatment of such equations is to transform the stochastic differential equations into Itô type equations. This can be done by modifying the drift term and accounting for the so-called Wong and Zakay or Stratonovich correction term. Then by using the non anticipating property of the response and increments of the Wiener process, the statistics of the response can be easily evaluated.

Examples of linear, quasi-linear and non linear equations have been presented.

REFERENCES

1. Wiener, N.: Differential space, Journal Math. and Phys., 2, 1923, 131-174.

2. Levy, P.: Processes stochastic et mouvement Brownien, Gauthier-Villars, Paris, 1948

3. Doob, J.L.: Stochastic processes, Wiley, New York, 1953.

4. Loéve, M.: Probability theory, Van Nostrand Reinhold, Princeton, N.J., 1965.

5. Soong, T.T. and Grigoriu, M.: Random vibration of mechanical and structural systems, Prentice Hall, 1993.

6. Stratonovich, R.L.: Topics in the theory of random noise, Gordon and Breach, New York, N.Y., 1963.

7. Lin, Y.K.: Probabilistic theory of structural dynamics, Mcgraw-Hill, New York, N.Y., 1967.

8. Itô, K.: On a formula concerning stochastic differentials, Nagoya Mathematical Journal, 3, 55-65, 1951.

9. Gardiner, C.W.: Handbook of stochastic methods, Springer verlag, Berlin, 1990.

10. Stratonovich, R.L.: A new form of representing stochastic integral and equation, J. SIAM Control, 4, 1966, 362-371.

11. Wong, E. and Zakai, M.: On the relation between ordinary and stochastic differential equations, International Journal of Engineering Sciences, 3, 1965, 215-229.

12. Arnold, L.: Stochastic differential equations: theory and applications, John Wiley & Sons, New York, N.Y., 1973.

13. Di Paola, M.: Stochastic differential calculus: Non linear systems under non normal white noise processes, CISM, Udine, 1998.

APPENDIX A

In this appendix some few fundamentals of random variables and stochastic process are summarized just to fix the symbol list adopted.

A.1 Random variables

A continuous variable X assuming values in a given domain χ is characterized by the probability density functions $p_X(x)$. It is a non-negative valued function whose integral in a given subspace $\Omega \in \chi$ gives the probability that X belongs to that interval, and the integral over χ must be one.

Let ϑ be a real parameter, then the *characteristic function* is written as

$$M_X(\vartheta) = E[exp(-i\vartheta X)] = \int_{-\infty}^{\infty} exp(-i\vartheta x)\, p_X(x)\, dx \tag{A.1}$$

Since $p_X(x)$ is a real function $M_X(\vartheta)$ is a complex function having even real part and odd imaginary part. Once the characteristic function is known, the $p_X(x)$ can be evaluated by an inverse Fourier transform, that is

$$p_X(x) = \frac{1}{2\pi} \int_{-\infty}^{\infty} exp(i\vartheta x) M_X(\vartheta)\, d\vartheta \tag{A.2}$$

The Taylor expansion of the characteristic function can be written in the form

$$M_X(\vartheta) = \sum_{j=0}^{\infty} \frac{(-i)^j}{j!} E[X^j]\, \vartheta^j \tag{A.3}$$

where the *moment* of order j is defined as

$$E[X^j] = \int_{-\infty}^{\infty} x^j p_X(x)\, dx \tag{A.4}$$

Another important function in the theory of random variables is the *log-characteristic function*, that is $ln M_X(\vartheta)$, its Taylor expansion is given in the form

$$ln M_X(\vartheta) = \sum_{j=1}^{\infty} \frac{(-i)^j}{j!} K_j[X]\, \vartheta^j \tag{A.5}$$

where $K_j[X]$ are the so-called *cumulants* of order j of the random variable X. Moments and cumulants are related each others by nonlinear algebraic relationships

$$E[X] = K_1[X]$$

$$E[X^2] = K_2[X] + K_1[X]E[X]$$

$$E[X^3] = K_3[X] + 2K_2[X]E[X] + K_1[X]E[X^2] \qquad\qquad (A.6)$$

$$\vdots$$

A random variable is said to be Gaussian (or normal) if all cumulants of order $j > 2$, are exactly normal. In this case the log characteristic function write

$$ln\, M_X(\vartheta) = -iK_1[X]\vartheta + \frac{(-i)^2}{2} K_2[X]\vartheta^2 \qquad\qquad (A.7)$$

and hence

$$p_X(x) = \frac{1}{\sqrt{2\pi K_2[X]}} exp\left\{ -\frac{(x - E[X])^2}{2K_2[X]} \right\} \qquad\qquad (A.8)$$

Extension to multivariate random vectors X can be easily made starting from the *joint probability function* $p_{X_1 X_2 \ldots X_n}(x_1, x_2, \ldots, x_n)$ in the following denoted as $p_X(x)$, where the random vector X assumes value in a domain χ^n. $p_X(x)$ is a non-negative valued function whose integral over a cell in χ^n gives the probability that X belongs to that cell. The integral in all the space χ^n must be one.

The characteristic function $M_X(\vartheta)$ is written as

$$M_X(\vartheta) = E[exp(-i\,\vartheta^T x)] = \int_{-\infty}^{\infty} \ldots \int_{-\infty}^{\infty} exp(-i\,\vartheta^T x)\, p_X(x)\, dx \qquad\qquad (A.9)$$

ϑ being an n-vector of real parameters.

The Taylor expansions of $M_X(\vartheta)$ and $ln\, M_X(\vartheta)$ can be written in the form

$$M_X(\vartheta) = 1 + (-i) \sum_{j=0}^{n} E[X_j]\vartheta_j + \frac{(-i)^2}{2!} \sum_{j=1}^{n} \sum_{k=1}^{n} E[X_j X_k]\vartheta_j \vartheta_k + \ldots \qquad (A.10)$$

and

$$ln M_X(\boldsymbol{\vartheta}) = (-i) \sum_{j=0}^{n} K_1[X_j]\vartheta_j + \frac{(-i)^2}{2!} \sum_{j=1}^{n} \sum_{k=1}^{n} K_2[X_j X_k]\vartheta_j \vartheta_k + \dots \qquad (A.11)$$

where $K_j[X_r^{m_1},\dots,X_s^{m_j}]$ $(m_1+\dots+m_j = j)$ is the cumulant of order j of the random variables $X_r^{m_1},\dots,X_s^{m_j}$, and $E[X_r^{m_1},\dots,X_s^{m_j}]$ is the corresponding moment of order j.

If all cumulants of order greather than two wre exactly zero, the n-random vector X is said to be Gaussian or normal, and the log-characteristic function is given in the form

$$ln M_X(\boldsymbol{\vartheta}) = (-i) \sum_{j=1}^{n} K_1[X_j]\vartheta_j + \frac{(-i)^2}{2!} \sum_{j=1}^{n} \sum_{k=1}^{n} K_2[X_j X_k]\vartheta_j \vartheta_k \qquad (A.12)$$

while the corresponding probability density function is given in the form

$$p_X(x) = \frac{1}{\sqrt{(2\pi)^n\, Det(\Sigma_X)}}\, exp\left[-\frac{1}{2}(x - E[X])^T \Sigma_X^{-1}(x - E[X])\right] \qquad (A.13)$$

Σ_X being the so-called *covariance matrix* $(\Sigma_X = E[X X^T] - E[X]E[X^T])$ containing the cumulants of the second order of the random variables X_j.

A.2 Random processes

If the random variable X depends upon a parameter t, then $X(t)$ is a random or stochastic process. At the fixed time t_0, $X(t_0) = X_0$ is a random variable then it can be defined from a probabilistic point of view by the probability density function $p_{X_0}(x_0;t_0)$ or by its Fourier transform that is the characteristic function

$$M_{X_0}(\vartheta_0;t_0) = \int_{-\infty}^{\infty} exp(-i\vartheta_0 x_0)\, p_{X_0}(x_0;t_0)\, dx_0 \qquad (A.14)$$

or by moments and cumulants of every order $E[X(t_0)^j], K_j[X(t_0)]$.

The functions $p_{X_0}(x_0;t_0)$ or $M_{X_0}(\vartheta_0;t_0)$ or the quantity $E[X(t_0)^j], K_j[X(t_0)]$ are *statistics of first order* of the random process. More detailed informations on the stochastic process can be obtained by the random variables $X(t_0)$ and $X(t_1)$, for these two random variables the functions $p_{X_0 X_1}(x_0,x_1;t_0,t_1)$ or $M_{X_0 X_1}(\vartheta_0,\vartheta_1;t_0,t_1)$ or the quantities $E[X_0^m X_1^s], K_{m+s}[X_0^m X_1^s]$ have to be defined. In particular

$$K_2[X_0 X_1] = E[X(t_0) X(t_1)] - E[X(t_0)] E[X(t_1)] \qquad (A.15)$$

is called *correlation function* and is usually denoted as $R_X(t_0,t_1)$. If each couple of random variables $X(t_0)$, $X(t_1)$ are jointly Gaussian or normal the stochasti process is a *normal stochastic process* and is fully characterized in probabilistic sense by the mean and by correlation function $R_X(t_0,t_1)$. The functions $p_{X_0 X_1}(x_0,x_1;t_0,t_1)$, $M_{X_0 X_1}(\vartheta_0,\vartheta_1;t_0,t_1)$, or the quantities $E[X_0^m X_1^s]$, $K_{m+s}[X_0^m X_1^s]$ are called *statistics of second order*. Similarly higher order statistics can be defined for p values of the paramater t. If the process is normal, then higher order statistics can be constructed by the second order statistics.

A.3 Operation on stochastic processes

Consider a sequence of values $X(t_n)=X_n$ and let $E[X_n^2]$ be finite: one says that X is the limit, in a mean square sense, of X_n and writes

$$l.i.m._{n\to\infty} X_n = X \tag{A.16}$$

provided that

$$\lim_{n\to\infty} E[|x_n - x|^2]=0 \tag{A.17}$$

This operator and the expectation commutes.

If $R_X(t_1,t_2)$ exists and is finite, then the process $X(t)$ is continuous and its derivative is defined as

$$X(t) = l.i.m._{\tau\to 0} \frac{X(t+\tau)-X(t)}{\tau} \tag{A.18}$$

The following properties hold

$$\frac{d}{dt} E[X(t)] = E[\dot{X}(t)] \tag{A.19}$$

$$\frac{d}{dt} E[\dot{X}_1(t) X_2(t)] = E[X_1(t) X_2(t)] + E[X_1(t) \dot{X}_2(t)] \tag{A.20}$$

$$\frac{\partial}{\partial t_1} R_{XX}(t_1,t_2) = R_{\dot{X}X}(t_1,t_2) \tag{A.21}$$

Given a deterministic function $h(t, \tau)$, one introduces the mean square Riemann integral as the stochastic processes $Y(t)$

$$Y(t) = \int_a^b X(\tau) h(t,\tau)d\tau = l.i.m._{\substack{n\to\infty \\ \Delta n\to 0}} \sum_{j=1}^{n} h(t,\bar{t}_j) X(\bar{t}_j)(t_j - t_{j-1}) \tag{A.22}$$

where $a = t_0 < t_1 < \ldots < t_n = b$, $\Delta n = max_j\,(t_j - t_{j-1})$ and $\bar{t}_j$ is a time instant in the interval (t_{j-1}, t_j). The concept of Fourier-Stieltjes integral replaces $X(\bar{t}_j)\,(t_j - t_{j-1})$ with $(X(t_j) - X(t_{j-1}))$.

Put

$$\bar{t}_j = \alpha\, t_j + (1 - \alpha)\, t_{j-1} \tag{A.23}$$

then one distinguishes:

- the *Itô integral* which corresponds to the adoption of $\alpha = 0$
- the *Stratonovich integral* which corresponds to the adoption of $\alpha = 1/2$.

HOW TO ANALYZE THE NON-LINEAR SYSTEMS UNDER POISSON WHITE NOISE PROCESSES ?

M. Di Paola
Dipartimento di Ingegneria Strutturale e Geotecnica,
Università degli Studi di Palermo,
Viale delle Scienze, Palermo, Italy

ABSTRACT

In this paper the extension of stochastic differential calculus to non-normal white noise processes (Poisson white noise) is handled by starting from the case of a single parametric impulse. Moment equations and Fokker-Planck equation are derived for both external and parametric Poisson processes.

1. POISSON WHITE NOISE AND COMPOUND POISSON PROCESSES

The Poisson white noise process, in the following denoted as $W(t)$, is defined in the form

$$W(t) = \sum_{k=1}^{N(t)} Y_k \, \delta(t - T_k)$$ (1.1)

This function is a train of Dirac's delta impulses $\delta(t - T_k)$ occurring at Poisson distributed random times T_k, $N(t)$ is the number of impulses in the time interval $[0, t)$ with initial condition $N(0) = 0$ with probability one. The random variable Y_k, that constitute the amplitude of the spike at the corresponding time T_k, are assumed to be mutually independent and independent of the random instants T_k.

The Poisson process $W(t)$ defined in Eq. (1.1) is completely characterized on the expected rate $\lambda(t)$ of events (impulse occurrences) that is the mean number of impulses per unit time and the assigned probability distribution of Y. The mean number of impulses in the time interval $t \div t+dt$, as $N(dt)$, is given as

$$E[N(dt)] = \lambda(t)dt$$ (1.2)

The entire probabilistic description of the process W(t) defined in Eq. (3.1) is defined as follows

$$R_W^{(s)}(t_1, t_2, \ldots, t_s) = \lambda(t) \, E[Y^s] \, \delta(t_1 - t_s) \ldots \delta(t_1 - t_s)$$ (1.3)

where $R_W^{(s)}(t_1, t_2, \ldots, t_s)$ is the cumulant of order s of the random variables $W(t_1)$, $W(t_2)$, $\ldots, W(t_s)$. Following the nomenclature of Stratonovich (see ref. [6] of [1]), $R_W^{(s)}(t_1, t_2, \ldots, t_s)$ is the so called *correlation function* of order s. Since it is constituted by product of Dirac's deltas such a process is also called *delta correlated*. If $\lambda(t)$ keeps a constant value, the Poisson white noise is a stationary one. It will be emphasized that, since higher order cumulants are different from zero, the Poisson white noise is a non-normal one also in the case in which Y is normally distributed.

The nomenclature of white noise is due to the fact that, if $\lambda(t)$ is a constant and Y is a zero mean process, then the second correlation coincide with $E[W(t_1) W(t_2)]$, in this case

$$R_W^{(2)}(t_1, t_2) = \lambda \, E[Y^2] \delta(t_1 - t_2)$$ (1.4)

then $W(t)$ is a (non-normal) stationary process, that at least in second order exactly coincides with a normal white noise having an intensity $q_2 = \lambda E[Y^2]$. Then the PSD function is also in this case a constant

$$S_W(\omega) = \frac{\lambda \, E[Y^2]}{2\pi} \tag{1.5}$$

explaining the nomenclature of white noise process, since it has a constant power at the overall frequency range. The difference between $W_0(t)$ and $W(t)$ is that for normal white noise $W^0(t)$, $R_{W^0}^{(s)}(t_1,t_2,\ldots,t_s)=0$, $\forall s \geq 2$ while for the Poisson white noise higher order correlations given as products of Dirac's deltas are different from zero.

The Poisson white noise $W(t)$ can be considered as the formal derivative of the so-called *homogeneous compound Poisson process* $C(t)$, defined as

$$C(t) = \sum_{k=1}^{N(t)} Y_k \, H(t-t_k) \tag{1.6}$$

$H(t)$ being the unit step function $(H(X)=0 \; \forall X < 0; \; H(X)=1 \; \forall X \geq 0)$.

It is easy to show that the correlation functions of $C(t)$ can be expressed as

$$R_C^{(1)}(t) = \lambda \, t \, E[Y]$$
$$R_C^{(s)}(t_1,t_2,\ldots,t_s) = \lambda \, E[Y^s] \, min \, (t_1,t_2,\ldots,t_s); \; \forall s \geq 2 \tag{1.7}$$

Differentiating Eq. (1.7) with respect $t_1,t_2,\ldots,t_s$, leads to

$$\frac{\partial^s R_C^{(s)}(t_1,t_2,\ldots,t_s)}{\partial t_1 \partial t_2 \ldots \partial t_s} = R_W^{(s)}(t_1,t_2,\ldots t_s) \tag{1.8}$$

This equation shows that the correlation function of a Poisson white noise coincides with the derivative of the correlations of the corresponding homogeneous compound Poisson process. Then the formal derivative of $C(t)$ is $W(t)$, $(\dot{C}(t)=W(t))$, in Figs. 1 and 2 the sample functions of $C(t)$ and $W(t)$ are plotted, respectively.

It can be easily seen that the correlation functions of an increment of the compound Poisson process $dC(t) = C(t + dt) - C(t)$ are given as

$$R_{dC}^{(s)}(t_1,t_2,\ldots,t_s) = \lambda \, E[Y^s] \delta(t_1 - t_2) \, \delta(t_1 - t_3) \ldots \delta(t_1 - t_s) dt_1 \, dt_2 \ldots dt_s \tag{1.9}$$

taking into account the relationships between moments and cumulants, one obtains

$$E[dC(t_1)] = \lambda\, E[Y]dt$$

$$E[dC(t_1)\,dC(t_2)] = \lambda\, E[Y^2]\delta(t_1 - t_2)dt_1\,dt_2 + \lambda^2 E[Y]^2 dt_1\,dt_2$$

$$E[dC(t_1)\,dC(t_2)\,dC(t_3)] = \lambda\, E[Y^3]\delta(t_1 - t_2)\delta(t_1 - t_3)dt_1\,dt_2\,dt_3 + \qquad (1.10)$$

$$+ \lambda^2\, E[Y]E[Y^2]\,(\delta(t_2 - t_1) + \delta(t_3 - t_1) + \delta(t_3 - t_2))dt_1\,dt_2\,dt_3 +$$

$$+ \lambda^3 E[Y^3]dt_1\,dt_2\,dt_3$$

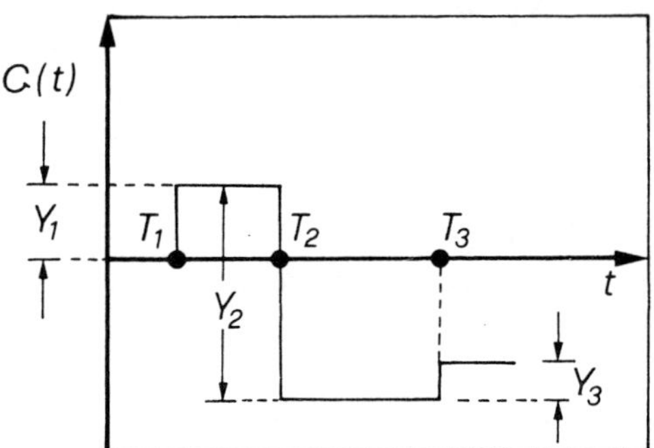

Fig. 1 - Sample function of compound
Poisson process

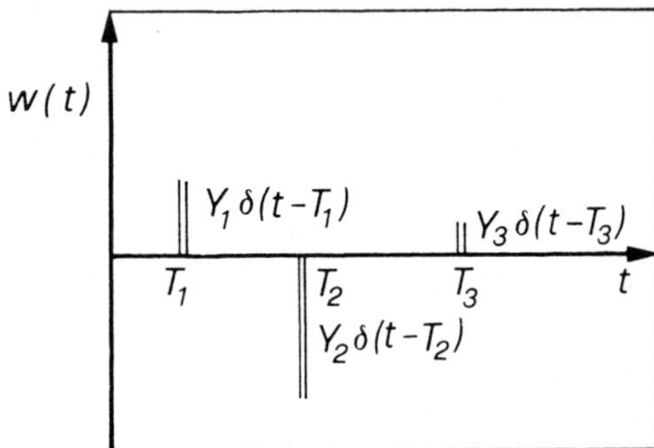

Fig. 2 - Sample function of Poisson
white noise process

Moreover, when t_i approaches t_j, the Dirac's deltas $\delta(t_i - t_j)$ attains an infinite value of the same order of $1/dt$, by imposing $t_1 = t_2 = \ldots = t_s$ in equation (1.10) and neglecting higher order infinitesimal than dt, one obtains the very remarkable result

$$E[(dC(t))^r] = K_r[dC] = \lambda\, E[Y^r]dt \qquad (1.11)$$

Equation (1.11) shows that cumulants and moments of the same order of an increment of the Compound Poisson process coincide. Moreover, it can be seen that moments (or cumulants) of order r of the increment $dC(t)$ is of order dt. This is the main difference between dB and dC. In the former case dB is of order $(dt)^{1/2}$ $(0\,(dB) = dt^{1/2})$, while in the latter one cannot describe the order of dC but one has to keep in mind that $0(dC(t))^r = dt$ $(r = 1, 2, \ldots)$. At least one can see that if $\lambda \to \infty$ and at the same time $\lambda\, E[Y^2]$ keeps a constant value then $\lambda\, E[Y^r] \to 0, r > 2$ and the Poisson white noise $W(t)$ tends towards the normal white noise $W^0(t)$.

In Table I the main features of increments of the Wiener process, increments of the compound Poisson process, normal and Poisson white noise are reported.

Further hint on the Poisson white noise processes can be found in the books of Stratonovich and Lin (Ref. [6] and [7] of [1]).

$X(t)$	$dB(t)$	$W^0(t)$	dC	$W(t)$
$E[X(t)]$	0	0	$\lambda E[Y]dt$	$\lambda E[Y]$
$R_X(t_1,t_2)$	$q_2\,\delta(t_1-t_2)dt_1\,dt_2$	$q_2\,\delta(t_1-t_2)$	$\lambda E[Y^2]\delta(t_1-t_2)dt_1\,dt_2$	$\lambda E[Y^2]\delta(t_1-t_2)$
$R_X^{(s)}(t_1,t_2,\ldots,t_s)$	0	0	$\lambda E[Y^s]\delta(t_1-t_2)\ldots \ldots\delta(t_1-t_s)dt_1\,dt_2\ldots dt_s$	$\lambda E[Y^s]\delta(t_1-t_2)\ldots \ldots\delta(t_1-t_s)$
$K_2(X(t))$	$q_2\,dt$	∞	$\lambda E[Y^2]dt$	∞
$K_s(X(t))$	0	0	$\lambda E[Y^s]dt$	∞

Table I, Statistics of generating processes

2. SINGLE PARAMETRIC IMPULSE

How it has been stated in the previous Section, the Poisson white noise is constututed by a train of Dirac's delta then in order to provide extension of Ito calculus to Poisson white noise processes one the simpler deterministic problem have to be solved

$$\dot{x} = f(x,t) + Y\,g(x,t)\,\delta(t-\bar{t}); \quad x(t_0) = x_0 \tag{2.1}$$

Let $g(x,t)$ be a deterministic non linear function of x and t belonging to the class C^∞ (infinite times differentiable on x and t). In equation (2.1) δ is the Dirac's delta and Y is the amplitude of the parametric impulse.

In order to evaluate the response $x(t)$, one can subdivide the time interval $[t_0, \infty)$ in three parts: the first one is $t_0 \leq t < \bar{t}^-$; the second one is $\bar{t}^-, \bar{t}^+$; the third one is $t > \bar{t}^+$, where $\bar{t}^-$ and $\bar{t}^+$ indicate the time instants immediately before and after the impulse, respectively.

In the first and third interval, the differential equation (2.1) can be written in the form

$$\dot{x} = f(x,t) \qquad \forall\ t : t_0 \leq t < \bar{t}^- \tag{2.2a}$$

$$\dot{x} = f(x,t) \qquad \forall\ t : t > \bar{t}^+ \tag{2.2b}$$

The appropriate initial conditions are $x(t_0) = x_0$ for Eq. (6.2a) and $x(\bar{t}^+)$ for Eq. (2.2b). In order to find the initial conditions for Eq. (2.2b), it is necessary to solve the differential equation in the second interval $(\bar{t}^-,\bar{t}^+)$. Because of the impulse in $\bar{t}$, the response $x(\bar{t}^+)$ is different from $x(\bar{t}^-)$, that is the response exhibits a jump $j(\bar{t})$ given as

$$j(\bar{t}) = x(\bar{t}^+) - x(\bar{t}^-) \tag{2.3}$$

In order to evaluate the jump, let $\bar{t} - \varepsilon; \bar{t} + \varepsilon$ a small interval centered in $\bar{t}$, in which Eq. (2.1) may be expressed in the integral form as follows

$$x(\bar{t} + \varepsilon) - x(\bar{t} - \varepsilon) = \int_{\bar{t}-\varepsilon}^{\bar{t}+\varepsilon} f(x,t)dt + Y \int_{\bar{t}-\varepsilon}^{\bar{t}+\varepsilon} g(x,t)dH(t - \bar{t}) \tag{2.4}$$

being $H(\cdot)$ the unit step function. The first integral on the right hand side of Eq. (6.4) is a Riemann integral, the second one is not a Riemann or a Fourier-Stieltjes integral, in fact the limit of the Riemann sum does not achieve unique solution, but it depends on the intermediate point selected [2].

As a consequence the classical differential calculus cannot predict the jump $j(\bar{t})$. The key to find the jump for equation (2.1) is in writing the Taylor expansion of an increment Δ of a real valued function $\phi(x,t)$, belonging to the class C_∞, as follows

$$\Delta \phi(x,t) = d\phi(x,t) + \frac{1}{2!} d^2\phi(x,t) + \ldots = \sum_{j=1}^{\infty} \frac{d^j\phi(x,t)}{j!} \tag{2.5}$$

By neglecting infinitesimals of higher order than dt the first few differentials appearing in Eq. (2.5) can be written in the form (with argument omitted)

$$\begin{aligned}
d\phi &= \frac{\partial \phi}{\partial x} dx + \frac{\partial \phi}{\partial t} dt \\[2mm]
d^2\phi &= \frac{\partial}{\partial x}(d\phi)dx = \frac{\partial^2 \phi}{\partial x^2}(dx)^2 + \frac{\partial \phi}{\partial x} d^2x \\[2mm]
d^3\phi &= \frac{\partial}{\partial x}(d^2\phi)dx = \frac{\partial^3 \phi}{\partial x^3}(dx)^3 + 3\frac{\partial^2 \phi}{\partial x^2} d^2x\, dx + \frac{\partial \phi}{\partial x} d^3x \\
&\quad\ \vdots
\end{aligned} \tag{2.6}$$

The numerical series (2.5) proves that in general increment $\Delta\phi$ and first order differential of $\phi(x,t)$ do not coincide. If $\phi(x,t)$ is not a singular function then $\Delta\phi(x,t) \equiv d\phi(x,t)$ since higher order differentials of $\phi(x,t)$ are infinitesimals of higher order that dt, but if $x(t)$ experiences a finite jump as in the case of Eq. (2.1) at $\bar{t}$ the higher order differentials have the same order of the first one and then they cannot be neglected. As in fact, selecting $\phi(x,t) = x$ and using Eq. (2.5) it leads to

$$\Delta x = dx + \frac{1}{2!} d^2x + \frac{1}{3!} d^3x + \ldots = \sum_{j=1}^{\infty} \frac{d^j x(t)}{j!} \tag{2.7}$$

If $x(t)$ is the response of Eq. (2.1) one has two possibilities, $t \neq \bar{t}$ or $t = \bar{t}$, in the first case

$$dx = f(x,t)\, dt \tag{2.8}$$

then $\Delta x \equiv dx$ since higher order differentials of $x(t)$ are infinitesimals of higher order than dt, while in $\bar{t}$ one can write

$$dx = f(x,t)dt + Y\, g(x,t)\, dH(t - \bar{t}) \tag{2.9}$$

then using Eq. (2.9) it follows

$$d^2x = Y^2 g^{(2)}(x,t)\,(dH(t-\bar{t}))^2;\ldots; d^j x = Y^j g^{(j)}(x,t)\,(dH(t-\bar{t}))^j \tag{2.10}$$

where $g^{(j)}(x,t)$ can be evaluated by the recurrence relationship

$$g^{(j)}(x,t) = \frac{\partial g^{(j-1)}}{\partial x}\, g^{(1)}(x,t);\quad g^{(1)}(x,t) = g(x,t) \tag{2.11}$$

$H(t-\bar{t})$ is defined as

$$H(t-\bar{t}) = 1 \quad \forall t \geq \bar{t};\quad H(t-\bar{t}) = 0 \quad \text{elsewhere} \tag{2.12}$$

$(dH(t-\bar{t}))^j = 0\ \forall t \neq \bar{t}^-$, and then the jump is given in the form

$$j(\bar{t}) = \Delta x = \sum_{j=1}^{\infty} \frac{Y^j}{j!}\, g^{(j)}(x(\bar{t}^-),\bar{t}) \tag{2.13}$$

from Eq. (2.13) one recognizes that the jump of the differential Eq. (2.1) depends on the value of the state variable immediately before the impulse as well as on the intensity of the jump. If ∞ terms are retained in the series (2.13) the jump prediction is exact. Once the jump is evaluated the initial condition of Eq. (2.2) is known.

2.1 Incremental rule

The incremental rule already presented in Eq. (2.5) can be also rewritten in a more expressive form

$$\Delta \phi(x,t) = \frac{\partial \phi(x,t)}{\partial t}\, dt + \frac{\partial \phi(x,t)}{\partial x}\left(\sum_{j=1}^{\infty} \frac{d^j x}{j!} \right) + \frac{1}{2!}\frac{\partial^2 \phi(x,t)}{\partial x^2}\left(\sum_{j=1}^{\infty} \frac{d^j x}{j!} \right)^2 + \ldots$$

$$\ldots + \frac{1}{k!}\frac{\partial^k \phi(x,t)}{\partial x^k}\left(\sum_{j=1}^{\infty} \frac{d^j x}{j!} \right)^k + \ldots \tag{2.14}$$

substituting Eq. (2.7) in Eq. (2.14) leads to the incremental rule

$$\Delta\phi(x,t) = \frac{\partial\phi(x,t)}{\partial t}\,dt + \frac{\partial\phi(x,t)}{\partial x}\,\Delta x + \frac{1}{2!}\frac{\partial^2\phi(x,t)}{\partial x^2}(\Delta x)^2 + \ldots$$

$$\ldots + \frac{1}{k!}\frac{\partial^k\phi(x,t)}{\partial x^k}(\Delta x)^k + \ldots \tag{2.15}$$

If $t \neq \bar{t}$, then $\Delta x \equiv dx$, that is an increment is an infinitesimal of first order, it follows that, by neglecting infinitesimals of higher order than dt, $\Delta\phi(x,t) = d\phi(x,t)$, and equation (2.15) reduces to the classical rule of differentiation of composite functions, that is

$$\Delta\phi(x,t) = \frac{\partial\phi(x,t)}{\partial t}\,dt + \frac{\partial\phi(x,t)}{\partial x}\,dx\,, \qquad \forall t \neq \bar{t} \tag{2.16}$$

In contrast, if the increment Δx is evaluated at $\bar{t}$, then Δx is a finite quantity, the term $(\partial\phi(x,t)/\partial t)dt$ can be neglected since it is an infinitesimal quantity with respect to all higher order terms (finite quantities). In corrispondence of the impulse, that is at $t = \bar{t}$, the jump of $\phi(x,t)$ is given as

$$\Delta\phi(x,\bar{t}) = \phi(x(\bar{t}^+),\bar{t}) - \phi(x(\bar{t}^-),\bar{t}) = \sum_{k=1}^{\infty}\frac{1}{k!}\frac{\partial^k\phi(x(\bar{t}^-),\bar{t})}{\partial x^k}(\Delta x)^k\,; \quad t = \bar{t} \tag{2.17}$$

From Eq. (2.17) one recognizes that all one needs for evaluating the jump of any scalar real valued function $\phi(x,t)$ belonging to the class C_∞, is the value of the function immediately before the impulse as well as the jump of the state variable x.

As an example the Bernoulli equation given in the form

$$\dot{x} = ax + Y x^3 \delta(t - \bar{t})\,; \quad x(0) = x_0\,, \quad \bar{t} > 0 \tag{2.18}$$

will be studied. For every time $0 \leq t < \bar{t}$ the equation of motion is

$$\dot{x} = ax\,; \qquad x(0) = x_0 \tag{2.19}$$

The value of x immediately before the impulse occurrence is $x(\bar{t}^-) = x_0 e^{a\bar{t}}$. By using Eq. (2.13), the jump is given in the form

$$j(\bar{t}) = x(\bar{t}^+) - x(\bar{t}^-) = \sum_{j=1}^{\infty}\frac{Y^j}{j!}x^{2j+1}(\bar{t}^-)\,\Gamma_{2j-1} = \sum_{j=1}^{\infty}\frac{Y^j}{j!}(x_0 e^{a\bar{t}})^{2j+1}\,\Gamma_{2j-1} \tag{2.20}$$

where $\Gamma_{2j-1} = [(2j-1)\cdot(2j-3)...5\cdot3\cdot1]$. Once the jump is evaluated, the value $x(\bar{t}^{+})$ is immediately evaluated and is the initial condition for $t > \bar{t}$.

Now the same system will be studied in a different form, due to the fact that Eq. (2.18) is a Bernoulli equation, it can be exactly solved, in order to do this the following positions can be made

$$y = x^{-2} \; ; \; \dot{y} = -2x^{-3} \; ; \; y(0) = x_0^{-2} \tag{2.21}$$

By substituting Eq. (2.21) in Eq. (2.18) one obtains

$$\dot{y} = -2ay - 2\gamma\delta(t-\bar{t}) \tag{2.22}$$

Now, Eq. (6.22) is forced by an external impulse, then one can write

$$y(t) = y_0 e^{-2at} \; ; \; \forall t < \bar{t} \tag{2.23}$$

the jump in $\bar{t}$, on the basis of elementary consideration, for this simple case is exactly given as

$$y(\bar{t}^{+}) - y(\bar{t}^{-}) = y(\bar{t}^{-})(-2\gamma) \tag{2.24}$$

by substituting Eq. (2.23) particularized in $\bar{t}$ in Eq. (2.24) one obtains

$$y(\bar{t}^{+}) = y_0 e^{-2a\bar{t}}(1-2\gamma) \tag{2.25}$$

this is the initial condition for $t > \bar{t}$, then the response for $t > \bar{t}$ write

$$y(t) = y_0 e^{-a(t-\bar{t})} e^{-2a\bar{t}}(1-2\gamma) \tag{2.26}$$

In order to test the validity of Eq. (2.17) put $x = y^{-1/2}$, using the incremental rule given in Eq. (2.17) one obtains exactly Eq. (2.20). From this example one can conclude that the jump evaluated by the numerical series (2.13) is exact.

3. MULTIDEGREE OF FREEDOM SYSTEMS

A dynamical non linear system forced by parametric impulsive input, including structural one, can be cast in the form

$$\dot{x} = f(x,t) + Y g(x,t)\delta(t-\bar{t}) \; ; \; x(t_0) = x_0 \tag{3.1}$$

where x is an n-vector of state variables, $f(x,t)$ and $g(x,t)$ are non linear n-vector functions of x and t.

In the time intervals $[0;\bar{t}^-)$ and $(\bar{t}^+;\infty)$ the solution of Eq. (3.1) can be reached by solving the ordinary differential equation

$$\dot{x} = f(x,t); \quad x(t_0) = x_0 \tag{3.2}$$

$$\dot{x} = f(x,t); \quad x(\bar{t}^+) \tag{3.3}$$

while the jump in correspondence of the Dirac's delta can be evaluated in the form

$$j(\bar{t}) = x(\bar{t}^+) - x(\bar{t}^-) = \sum_{j=1}^{\infty} \frac{Y^j}{j!} g^{(j)}(x(\bar{t}^-),\bar{t}) \tag{3.4}$$

where $g^{(j)}(x;t)$ can be evaluated in recursive form as follows:

$$g^{(j)}(x;t) = (\nabla g^{(j-1)}(x,t)\, g(x,t)); \quad g^{(1)}(x,t) = g(x,t) \tag{3.5}$$

$\nabla g^{(j)}(x,t)$ being now the gradient operator, that is

$$\nabla g^{(j)}(x,t) = \begin{bmatrix} \dfrac{\partial g_1^{(j)}}{\partial x_1} & \dfrac{\partial g_1^{(j)}}{\partial x_2} & \cdots & \dfrac{\partial g_1^{(j)}}{\partial x_n} \\[2mm] \dfrac{\partial g_2^{(j)}}{\partial x_1} & \dfrac{\partial g_2^{(j)}}{\partial x_2} & & \dfrac{\partial g_n^{(j)}}{\partial x_n} \\[2mm] \cdots & \cdots & & \cdots \\[2mm] \dfrac{\partial g_n^{(j)}}{\partial x_1} & \dfrac{\partial g_n^{(j)}}{\partial x_2} & \cdots & \dfrac{\partial g_n^{(j)}}{\partial x_n} \end{bmatrix} \tag{3.6}$$

where $g_i^{(j)}(x,t)$ is the i-th component of the vector $g^{(j)}(x,t)$.

Once the jump is evaluated by the summation in Eq. (3.4), the value of the state variables immediately after the impulse is already evaluated and the Eq. (3.3) can be solved.

The incremental rule for an n-degree of freedom system can be obtained by using the relationship

$$\Delta\phi(x,t) = \frac{\partial\phi}{\partial t}\,dt + \sum_{j=1}^{n}\frac{\partial\phi}{\partial x_j}\,\Delta x_j + \frac{1}{2!}\sum_{j=1}^{n}\sum_{j=1}^{n}\frac{\partial^2\phi}{\partial x_j\,\partial x_k}\,\Delta x_j\,\Delta k_k + \ldots \tag{3.7}$$

where Δx_j is the jump of the j-th component of the vector x.

4. DELTA CORRELATED INPUT

Let now the differential Eq. (2.1) be driven by a Poisson white noise process, then it has the form

$$\dot{X} = f(X(t),t) + g(X(t),t)\,W(t) \tag{4.1}$$

where, according to Eq. (1.1) $W(t)$ is a train of impulses having random amplitude Y_k and distributed in time according to Poisson law, that is

$$W(t) = \sum_{k=1}^{N(t)} Y_k\,\delta(t - t_k) \tag{4.2}$$

In the infinitesimal time lag $t \div t + dt$, for each sample function in the following denoted as $\tilde{X}$, according to Eq. (2.13) one can write

$$\Delta\tilde{X} = \tilde{X}(t+dt) - \tilde{X}(t) = f(\tilde{X}(t),t)dt + \sum_{j=1}^{\infty} \frac{g^{(j)}(\tilde{X}(t),t)}{j!}\left(\sum_{k=N(t)}^{N(t+dt)} \tilde{Y}_k\right)^j \tag{4.3}$$

where $\Sigma_{k=N(t)}^{N(t+dt)}\tilde{Y}_k$ represent the summation of the spikes in the time interval $t \div t+dt$, then it is the sample function of the increment of the compound Poisson process $(\Sigma_{k=N(t)}^{N(t+dt)}\tilde{Y}_k = d\tilde{C}(t))$.

The second term on the right hand side of Eq. (4.3), in each sample function is present or not if in the time interval $t;\ t+dt$, there is a spike or not. In the latter case $\Delta\tilde{X}$ is simply $f(\tilde{X}(t),t)dt$ and $\Delta\tilde{X} \equiv d\tilde{X}$. When one or more Dirac's deltas occur in the interval $t;\ t+dt$, then the summation in Eq. (4.3) is present, but is necessary to keep in mind that since $g^{(j)}(\tilde{V}(t),t)$ is evaluated at the beginning of the interval and the spike starts in the subsequent time interval, the function $g^{(j)}(\tilde{X}(t),t)$ is independent of the process $dC(t)$, that is the non anticipating property can be invoked. Using the properties of the process dC in Eq. (1.11), the mean value of ΔX is then gives as

$$E[\Delta X] = E[f(X(t),t)]dt + \lambda\sum_{j=1}^{\infty} \frac{E[g^{(j)}(X(t),t)]}{j!}E[Y^j]dt \tag{4.4}$$

where the non-anticipating property has been accounted for, since $g^{(j)}(X(t),t)$ is evaluated at the time instant t while the impulse occurs immediately after. Eq. (7.4) is exactly the same already found in [3] and [4] framed in the context of Ito calculus.

It will be noted that in Eq. (4.4) there is a summation of two infinitesimal of first order and then $E[\Delta X] \equiv E[dX]$, it follows that

$$\dot{E}[X] = E[f(X(t),t)] + \lambda \sum_{j=1}^{\infty} \frac{E[g^{(j)}(X(t),t)]}{j!} E[Y^j]$$ (4.5)

In order to find the differential equations of moments of any order of the response process $X(t)$, the incremental rule (2.15) have to be adopted. Let $\phi(X(t),t)$ be a scalar real valued function of X and t, belonging to the class C_∞, then an increment of the stochastic process $\phi(X(t),t)$ is given in the form

$$\Delta\phi(X(t),t) = \frac{\partial\phi(X(t),t)}{\partial t} + \sum_{r=r!}^{\infty} \frac{1}{r!} \frac{\partial^r\phi(X(t),t)}{\partial X^r}(\Delta X)^r$$ (4.6)

where ΔX is given in Eq. (4.3). Equation (4.6) is the rule for increments of functions of X and t. The latter can be rewritten in the form

$$\Delta X = f(X(t),t)dt + \sum_{j=1}^{\infty} \frac{g^{(j)}(X(t),t)}{j!}(dC(t))^j$$ (4.7)

an other hand, neglecting higher order infinitesimal than dt

$$(\Delta X)^2 = \sum_{j=1}^{\infty} \sum_{\ell=1}^{\infty} \frac{g^{(j)}(X(t),t)\,g^{(\ell)}(X(t),t)}{j!\,\ell!}(dC(t))^{j+\ell}$$

$$\vdots$$ (4.8)

$$(\Delta X)^k = \underbrace{\sum_{j=1}^{\infty} \sum_{\ell=1}^{\infty} \cdots \sum_{p=1}^{\infty}}_{k-times} \frac{g^{(j)}(X(t),t)g^{(\ell)}(X(t),t)\ldots g^{(k)}(X(t),t)}{j!\,\ell!\ldots p!}(dC(t))^{j+\ell+\ldots+p}$$

Putting $\phi(X,t) = X^k$ in equation (4.6), making stochastic average and dividing by dt the differential equation of moments of every order of X are given as

$$\dot{E}[X^k] = k\,E[X^{k-1}f(X,t)] + k\lambda \sum_{j=1}^{\infty} \frac{E[X^{k-1}g^{(j)}(X,t)]}{j!}E[Y^j] +$$

$$+ \lambda \frac{k(k-1)}{2!} \sum_{j=1}^{\infty} \sum_{\ell=1}^{\infty} \frac{E[X^{k-2}g^{(j)}(X,t)\,g^{(\ell)}(X,t)]}{j!\,\ell!}E[Y^{j+\ell}] + \ldots$$

$$\ldots + \lambda \sum_{j=1}^{\infty} \sum_{\ell=1}^{\infty} \cdots \sum_{p=1}^{\infty} \frac{E[g^{(j)}(X,t)\,g^{(\ell)}(X,t)\ldots g^{(p)}(X,t)]}{j!\,\ell!\ldots p!}E[Y^{j+\ell+\ldots+p}]$$ (4.9)

In this equation the non-anticipating property of increments $dC(t)$ and any scalar function of the response has been used, that is

$$E[g(X)(dC)^r] = E[g(X)]E[dC^r] \tag{4.10}$$

If $\lambda \to \infty$ and $\lambda E[Y^2]$ keep a constant value, then the Poisson white noise becomes a normal white noise, in this case $dC \to dB$ and dB is of order $(dt)^{1/2}$, then equation (4.3) becomes

$$\Delta X = f(X(t),t)dt + g(X(t),t)dB(t) + \frac{g}{2}(X(t),t)\frac{\partial g(X(t),t)}{\partial X}[dB(t)]^2 \tag{4.11}$$

and this equation coincides with the Ito equation (see Eq. (3.14) of [1]). Moreover, the differential rule given in Eq. (4.6) is simply written as

$$\Delta\phi(X,t) = \frac{\partial\phi(X,t)}{\partial t} + \frac{\partial\phi}{\partial X}\Delta X + \frac{1}{2!}\frac{\partial^2\phi}{\partial X^2}(\Delta X)^2 \tag{4.12}$$

that coincides with the classical differential rule given in Eq. (3.16) of [1]. By comparing Eq. (4.11) with Eq. (3.13) of [1] one can conclude that in the case of normal white noise input $\Delta X \equiv S(dX)$.

4.1 Moment equation approach for external delta correlated input

In the case of external delta correlated input, that is when $g(X(t),t)$ is independent of $X(t)$, $(g(X(t),t)=g(t))$, from equation (4.7) it may be recognized that $g^{(j)}(X,t)=0 \;\forall\, j>1$, then

$$\Delta X(t) \equiv dX(t) = f(X(t),t)\,dt + g(t)\,dC(t) \tag{4.13}$$

and equation (4.6) becomes

$$d\phi(X(t),t) = \frac{\partial\phi(X(t),t)}{\partial t} + \sum_{r=1}^{\infty}\frac{1}{r!}\frac{\partial^r\phi(X(t),t)}{\partial X^r}(dX)^r = \frac{\partial\phi(X(t),t)}{\partial t}dt +$$

$$+ \frac{\partial\phi(X(t),t)}{\partial X}f(X(t),t)dt + \sum_{r=1}^{\infty}\frac{1}{r!}\frac{\partial^r\phi(X(t),t)}{\partial X^r}g(t)^r(dC)^r \tag{4.14}$$

Equation (4.14) represents the differential rule for external delta correlated input.

Putting $\phi(X(t),t)=X^k$ in Eq. (4.14) the differential moment equation is written in the form

$$\dot{E}[X^k] = k\,E[X^{k-1}f(X(t),t)] + \lambda \sum_{r=1}^{k} \frac{k(k-1)\ldots(k-r+1)}{r!} E[X^{k-r}]g(t)^r E[Y^r] \quad (4.15)$$

This equation can be proved to be equivalent to that obtained by Iwankievicz et al. [5].

4.2 Some examples

In this section some examples are reported in order to elucidate the concepts previously discussed.

4.2.1 Linear systems

Let the equation of motion be given in the form

$$\dot{X} = f(t)X + g(t)W(t); \quad X(0) = X_0 \qquad (4.16)$$

Equation (4.16) is a linear equation and can be rewritten in the equivalent form

$$dX(t) = f(t)X(t)\,dt + g(t)\,dC(t) \qquad (4.17)$$

since the system is forced by external excitation, $\Delta X \equiv dX$, the differential rule given in Eq. (4.14) can be used, and the differential moment equations of any order can be obtained as follows

$$\dot{E}[X^k] = kf\,E[X^k] + \lambda \sum_{r=1}^{k} \frac{k(k-1)\ldots(k-r+1)}{r!} E[X^{k-r}]g^r(t) E[Y^r] \qquad (4.18)$$

If $f(t) = \alpha < 0$ and $g(t) = \beta$ are constant, and the system operates from $t = -\infty$, then the response reaches the stationary state and $\dot{E}[X^k] = 0$, then Eq. (4.18) simply write

$$k\alpha E[X^k] = -\lambda \sum_{r=1}^{k} \frac{k(k-1)\ldots(k-r+1)}{r!} E[X^{k-r}]\beta^r E[Y^r] \qquad (4.19)$$

By putting $k = 1, k = 2, \ldots$ in Eq. (4.19) one obtains

$$E[X] = -\frac{\lambda \beta}{\alpha} E[Y] \qquad (4.20)$$

$$E[X^2] = (E[X])^2 - \frac{\lambda}{2\alpha} \beta^2 E[Y^2] \qquad (4.21)$$

then the variance of the response is given as

$$\sigma_X^2 = -\frac{\lambda}{2\alpha}\,\beta^2 E[Y^2] \tag{4.22}$$

This expression is exactly the same as that obtained by considering a linear system excited by a normal white noise whose strength is $q_2 = \lambda\,E[Y^2]$.

However, for higher order moments, the response of linear system under normal white noise input is also a normal process while the response of the linear system under delta correlated input is non normal. In order to see this, let $k = 3$ in Eq. (4.19), the third moment for $E[Y] = 0$ evaluated is given as

$$E[X^3] = -\lambda\,\frac{\beta^3 E[Y^3]}{\alpha}\left[\frac{3\beta^2 + \alpha^2}{3\alpha^2}\right] \tag{4.23}$$

while in the case of normal white noise $E[X^3] = 0$.

Once the moments of every order are evaluated by Eq. (4.18), higher order statistics can be evaluated using the approach described in [6].

4.2.2 Quasi linear system

Let the equation of motion be given in the form

$$\dot{X} = f(t)X + (r(t)X + h(t))\,W(t) \tag{4.24}$$

This equation is now forced by parametric excitation and increments ΔX and differentials dX do not coincide. According to Eq. (4.6) particularized for $\phi(X(t),t) = X(t)$, the increment is given as

$$\Delta X = f(t)X\,dt + \sum_{j=1}^{\infty}\frac{g^{(j)}(X,t)}{j!}(dC)^j \tag{4.25}$$

in this case

$$g^{(1)}(X,t) = r(t)X + h\,;\ldots;\,g^{(j)}(X,t) = \frac{\partial g^{(j-1)}(X,t)}{\partial X}\,g^{(1)}(X,t) =$$

$$= r^j(t)X + h(t)\,r^{j-1}(t) \tag{4.26}$$

then

$$\Delta X = f(t)X\,dt + X\sum_{j=1}^{\infty}\frac{r^j(t)}{j!}\,dC^j + h(t)\sum_{j=1}^{\infty}\frac{r^{j-1}(t)}{j!}\,dC^j \tag{4.27}$$

Making stochastic average, using the non-anticipating property $(E[X\, dC^j] = E[X]\, E[(dC)^j])$ and dividing by dt the differential equation of motion of first order moment is easily derived in the form

$$\dot{E}[X] = \left[f(t) + \lambda \sum_{j=1}^{\infty} \frac{r^j(t) E[Y^j]}{j!} \right] E[X] + \lambda\, h(t) \sum_{j=1}^{\infty} \frac{r^{j-1}(t)}{j!} E[Y^j] \qquad (4.28)$$

the differential equation of moments of higher order can be obtained by Eq. (4.6), as an example the second order moment is given as

$$\dot{E}[X^2] = 2E[X^2] f(t) + 2\lambda\, E[X^2] \left(\sum_{j=1}^{\infty} \frac{r^j(t)}{j!} E[Y^j] \right) +$$

$$+ 2\lambda\, E[X]\, h(t) \left(\sum_{j=1}^{\infty} \frac{r^{j-1}(t)}{j!} E[Y^j] \right) + \lambda\, E[X^2] \left(\sum_{j=1}^{\infty} \sum_{\ell=1}^{\infty} \frac{r^j(t)\, r^{\ell}(t)}{j!\, \ell!} E[Y^{j+\ell}] \right) +$$

$$+ 2\lambda\, E[X] \left(\sum_{j=1}^{\infty} \sum_{\ell=1}^{\infty} \frac{r^j(t)\, r^{j-1}(t)}{j!\, \ell!} E[Y^{j+\ell}] \right) h(t) \qquad (4.29)$$

An inspection of equation (4.28) and (4.29) reveals that the moment equations do not constitute an infinite hyerarchy then they can be exactly solved. Once the moments of every order are evaluated the statistics of higher order can be evaluated. As an example the second order statistics can be evaluated by multiplying Eq. (4.27) by $X(t_1)$ $(t_1 \leq t)$, by making stochastic average and dividing by dt, one obtains

$$\frac{\partial}{\partial t} E[X(t) X(t_1)] = E[X(t) X(t_1)] \left(f(t) + \lambda \sum_{j=1}^{\infty} \frac{r^j(t)}{j!} E[Y^j] \right) +$$

$$+ \lambda\, h(t)\, E[X(t_1)] \sum_{j=1}^{\infty} \frac{r^{j-1}(t)}{j!} E[Y^j] \qquad (4.30)$$

in which the non-anticipating property has been taken into account. Eq. (4.30) is a differential equation that can be solved since the initial condition is $E[X^2(t)]$ already known by solving Eqs. (4.28) and (4.29).

4.2.3 Bernoulli equation

Let the equation of motion be given in the form

$$\dot{X} = \gamma X^2\, W(t); \quad X(0) = X_0 \qquad (4.31)$$

this is a Bernoulli equation forced by a parametric Poisson white noise, then

$$\Delta X = \sum_{j=1}^{\infty} \frac{g^{(j)}}{j!} (dC)^j \tag{4.32}$$

in this case

$$g^{(1)}(X) = \gamma X^2 \; ; \; g^{(2)}(X) = 2\gamma^2 X^3 \; ; \dots ; g^{(j)}(X) = j! \, \gamma^j X^{j+1} \tag{4.33}$$

then an increment of X is given in the form

$$\Delta X = \sum_{j=1}^{\infty} \gamma^j X^{j+1} (dC)^j \tag{4.34}$$

the differential rule for any scalar real valued function $\phi(X)$ belonging to C_∞, write

$$\Delta \phi(X) = \sum_{i=1}^{\infty} \frac{1}{i!} \frac{\partial^2 \phi(X)}{\partial X^i} (\Delta X)^i \tag{4.35}$$

Letting $\phi(X) = X^k$ in Eq. (4.35) leads to

$$\Delta X^k = k \sum_{j=1}^{\infty} \gamma^j X^{k+1} (dC)^j + \frac{k(k-1)}{2!} \sum_{j=1}^{\infty} \sum_{\ell=1}^{\infty} \gamma^{j+\ell} X^{k+j+\ell} (dC)^{j+\ell} + \dots$$

$$+ \dots + \frac{k!}{k!} \sum_{j=1}^{\infty} \sum_{\ell=1}^{\infty} \dots \sum_{p=1}^{\infty} \gamma^{j+\ell+\dots+p} X^{k+j+\ell+\dots+p} (dC)^{j+\ell+\dots+p} \tag{4.36}$$

Making stochastic average and dividing by dt the moments of every order of X can be easily evaluated. As an example the first two differential equations of moments are given in the form

$$\dot{E}[X] = \lambda \sum_{j=1}^{\infty} \gamma^j E[X^{j+1}] E[Y^j] \tag{4.37}$$

$$\dot{E}[X^2] = 2\lambda \sum_{j=1}^{\infty} \gamma^j E[X^{k+j}] E[Y^j] + \lambda \sum_{j=1}^{\infty} \sum_{\ell=1}^{\infty} \gamma^{j+\ell} E[X^{2+j+\ell}] E[Y^{j+\ell}] \tag{4.38}$$

from these equations one recognizes that the system of moment equations is an infinite hierarchy whose solution can be obtained by using some closure scheme [7-9].

On the other hand Eq. (4.31) can be transformed into a linear differential equation by means of the transformation

$$X^{-1} = Z \; ; \quad Z(0) = X_0^{-1} \tag{4.39}$$

then

$$\dot{Z} = -X^{-2}\dot{X} \tag{4.40}$$

by inserting Eq. (4.40) in (4.31) one obtains

$$\dot{Z} = -\gamma W(t) \tag{4.41}$$

now this equation, in terms of Z is a linear differential equation forced by external excitation, then $\Delta Z \equiv dZ$ and Eq. (4.41) can be rewritten in the form

$$\Delta Z = dZ = -\gamma \, dC \tag{4.42}$$

By applying the differential rule one obtains

$$\Delta \phi(Z) = \sum_{j=1}^{\infty} \frac{1}{j!} \frac{\partial^j \phi(Z)}{\partial Z^j} (dZ)^j \tag{4.43}$$

then the moments of every order of Z can be easily found letting $\phi(Z) = Z^k$ $(k = 1, 2, \ldots)$

$$\dot{E}[Z] = -\gamma \lambda E[Y]$$
$$\dot{E}[Z^2] = -2\gamma \lambda E[Z] E[Y] + \gamma^2 \lambda E[Y^2]$$
$$\vdots$$
$$\dot{E}[Z^k] = -k\gamma \lambda E[Z^{k-1}] E[Y] + \frac{k(k-1)}{2} \gamma^2 \lambda E[Z^{k-2}] E[Y^2] + \ldots$$
$$\ldots + \frac{(-1)^k}{k!} \gamma^k \lambda E[Y^k] \tag{4.44}$$

Putting now $\phi(Z) = Z^{-1}$ in Eq. (4.43) one obtains

$$\Delta Z^{-1} = \sum_{j=1}^{\infty} \gamma^j \frac{j!}{j!} Z^{-j-1} (dC)^j \tag{4.45}$$

then ΔZ^{-1} exactly coincides with ΔX evaluated directly by means of Eq. (4.34). It can be seen that by applying the above described procedure the moment equations in terms of Z^{-s}

by using Eq. (4.43) exactly coincide with the moment equation in terms of X evaluated using the rule (4.35).

5. FOKKER-PLANCK EQUATION FOR DELTA CORRELATED INPUT

Let the equation of motion be given in the form

$$\dot{X} = f(X(t),t) + g(X(t),t)\,W(t) \tag{5.1}$$

where $W(t)$ is a Poisson white noise process. Let

$$\phi(X(t),t) = exp(-i\vartheta X(t)) \tag{5.2}$$

ϑ bing a real parameter, then by using the differential rule given in Eq. (4.6) one can write

$$\Delta\,exp(-i\vartheta X) = \sum_{k=1}^{\infty} \frac{1}{k!} \frac{\partial^k\,exp(-i\vartheta X)}{\partial\vartheta^k}(\Delta X)^k \tag{5.3}$$

where

$$\Delta X = f(X)dt + \sum_{j=1}^{\infty} \frac{g^{(j)}(X(t),t)}{j!}\,dC^j \tag{5.4}$$

by inserting equation (5.4) in Eq. (5.3), making stochastic average, one obtains the differential equation of the characteristic function $M_X(\vartheta,t)$ that is

$$\dot{M}_X(\vartheta,t) = (-i\vartheta)\,E[exp(-i\vartheta X)f(X)] + \tag{5.5}$$

$$+\lambda \sum_{k=1}^{\infty} \frac{(-i\vartheta)^k}{k!} \underbrace{\sum_{j=1}^{\infty}\sum_{\ell=1}^{\infty}\cdots\sum_{p=1}^{\infty}}_{k-fold} \left\{ E\left[exp(-i\vartheta X)\frac{g^{(i)}(X)g^{(\ell)}(X)\ldots g^{(p)}(X)}{j!\,\ell!\ldots p!} \right] E[Y^{j+\ell+\ldots p}] \right\}$$

Making an inverse Fourier transform one obtains

$$\dot{p}_X(x,t) = -\frac{\partial}{\partial x}(p_X(x,t)f(x)) +$$

$$+\lambda \sum_{k=1}^{\infty} \frac{(-1)^k}{k!} \frac{\partial^k}{\partial x^k} \left[\underbrace{\sum_{j=1}^{\infty}\sum_{\ell=1}^{\infty}\cdots\sum_{p=1}^{\infty}}_{k-fold} p_X(x,t)\frac{g^{(i)}(x)g^{(\ell)}(x)\ldots g^{(p)}(x)}{j!\,\ell!\ldots p!} E[Y^{j+\ell+\ldots p}] \right] \tag{5.6}$$

Equation (5.6) represents the differential equation in terms of probability density function. For the case of zero mean normal white noise process $(\lambda \to \infty,\ \lambda E[Y^2]=q_2)$ equation (5.6) becomes

$$\dot{p}_X(x,t)=-\frac{\partial}{\partial x}\left(p_X(x,t)f(x)\right)+\frac{q_2}{2}\frac{\partial^2}{\partial x^2}\left[p_X(x)g^2(x)\right]-\frac{q_2}{2}\frac{\partial}{\partial x}\left[p_X(x)g^{(2)}(x)\right] \quad (5.7)$$

this equation exactly coincides with the FPK equation (see [1]).

In the case of external excitation Eq. (4.6) reduces to

$$\dot{p}_X(x,t)=-\frac{\partial}{\partial x}\left(p_X(x)f(x)\right)+\sum_{k=1}^{\infty}\frac{(-1)^k}{k!}\,g^k(t)\frac{\partial^k}{\partial x^k}\left[p_X(x)\right]\lambda\,E[Y^k] \quad (5.8)$$

This equation proves to be identical to that obtained by Roberts [10]. Exact solutions for some classes of non linear systems driven by parametric excitation have been found by Vasta [11].

6. EXTENSION TO MDOF SYSTEMS

The equation of motion of non-linear n-DOF system driven by parametric delta correlated process can be cast in the form

$$\dot{X}=f(X,t)+g(X,t)W(t);\quad X(0)=X_0 \quad (6.1)$$

where X is the n-vector of state variables, $f(X,t)$ and $g(X,t)$ are nonlinear n-vector deterministic functions of the stochastic vector process $X(t)$ and t, $W(t)$ is a Poisson white noise process whose correlations have already been defined in Eq. (1.3), X_0 is the vector of deterministic or random initial conditions. An increment of the vector $X(t)$ can be evaluated in the form

$$\Delta X=f(X,t)\,dt+\sum_{u=1}^{\infty}\frac{1}{u!}\,g^{(u)}(X,t)\,(dC)^u \quad (6.2)$$

where, according to Eq. (3.5), $g^{(u)}(X,t)$ is given in recursive form as follows

$$g^{(u)}(X,t)=(\nabla g^{(u-1)}(X,t))\,g(X,t);\quad g^{(1)}(X,t)=g(X,t) \quad (6.3)$$

where ∇ is the gradient operator already defined in Eq. (3.6).

The incremental rule for an a n-degree of freedom system can by obtained in the form

$$\Delta\phi(X,t)=\frac{\partial\phi}{\partial t}\,dt+\sum_{j=1}^{n}\frac{\partial\phi}{\partial X_j}\,\Delta X_j+\frac{1}{2!}\sum_{j=1}^{n}\sum_{\ell=1}^{n}\frac{\partial^2\phi}{\partial X_j\,\partial X_\ell}\,\Delta X_j\,\Delta X_\ell+\dots \tag{6.4}$$

where $\phi(X,t)$ is a scalar function of the $n+1$ variables belonging to the class C_∞; ΔX_j is the j-th component of the vector (6.2). By selecting $\phi(X,t)=X_p$; $\phi(X,t)=X_pX_q$;... the moment equation of first, second, etc. of various order can be easily obtained

$$\dot{E}[X_p]=E[f_p(X,t)]+\lambda\sum_{u=1}^{\infty}\frac{1}{u!}E[g_p^{(u)}(X,t)]\,E[Y^u] \tag{6.5}$$

$$\dot{E}[X_pX_q]=E[f_p(X,t)X_q]+E[f_q(X,t)X_p]+$$

$$+\lambda\sum_{j=1}^{\infty}\sum_{\ell=1}^{\infty}\frac{E[g_p^{(j)}(X,t)\,g_q^{(\ell)}(X,t)]}{j!\,\ell!}\,E[Y^{j+\ell}] \tag{6.6}$$

$$\vdots$$

Examples of solution of linear and non linear MDOF systems can be found in [12], [13].

If the input process is a normal white noise then the summation in Eq. (6.2) will be extended to the second term and the incremental rule will contain only the first three terms in Eq. (6.4).

Let the equation of motion be given in the form

$$\dot{X}=f(X,t)+g(X,t)\,W^0(t) \tag{6.7}$$

where $W^0(t)$ is zero mean normal white noise. An increment ΔX can be evaluated by truncating the series (6.2) up to the second term since the higher order terms are infinitesimals of higher order

$$\Delta X=f(X,t)dt+\sum_{u=1}^{n}\frac{1}{u!}g^{(u)}(X,t)\,(dB)^u \tag{6.8}$$

$$g^{(1)}(X,t)=g(X,t);\quad g^{(2)}(X,t)=(\nabla g(X,t))\,g(X,t) \tag{6.9}$$

the incremental rule is now given as

$$\Delta\phi(X,t)=\frac{\partial\phi}{\partial t}\,dt+\sum_{j=1}^{n}\frac{\partial\phi}{\partial X_j}\,\Delta X_j+\frac{1}{2!}\sum_{j=1}^{n}\sum_{\ell=1}^{n}\frac{\partial^2\phi}{\partial X_i\,\partial X_j}\,\Delta X_j\,\Delta X_\ell \tag{6.10}$$

by letting $\phi(X,t)=X_p$; $\phi(X,t)=X_p X_q$;... the moment equations can be obtained in the form

$$\dot{E}[X_p]= E[f_p(X,t)]+\frac{q_2}{2}E[g_p^{(2)}(X,t)] \tag{6.11}$$

$$\dot{E}[X_p X_q]= E[f_p(X,t) X_q]+ E[f_q(X,t) X_p]+ q_2 E[g_p(X,t) g_q(X,t)]+$$

$$+\frac{q_2}{2}\left\{E[X_q g_p^{(2)}(X,t)]+ E[X_p g_q^{(2)}(X,t)]\right\} \tag{6.12}$$

$$\vdots$$

Compact expressions for evaluating moments of every order can be obtained by using Kronecker algebra whose fundamentals are reported in Appendix A.

Since $E[X \otimes X]= E[X^{[2]}],...,E[X \otimes X \otimes...\otimes X]= E[X^{[j]}]$ contains all moments of order $2, ... , j$ of the state variables X, then an extension of the incremental rule for every vector function $\phi(X,t)$ is necessary for obtaining moment equations in compact form, to do this Eq. (6.4) can be rewritten in the form

$$\Delta \phi(X,t) = \frac{\partial}{\partial t} \phi(X,t)dt +\left[\nabla_X^T \otimes \phi(X,t)]\right]\Delta X + \frac{1}{2!}\left[\nabla_X^{[2]^T} \otimes \phi(X,t)\right](\Delta X)^{[2]} +...=$$

$$=\frac{\partial}{\partial t}\phi(X,t)dt + \sum_{j=1}^{\infty} \frac{1}{j!}\left[\nabla_X^{[j]^T} \otimes \phi(X,t)\right](\Delta X)^{[j]} \tag{6.13}$$

∇_X being the gradient vector $(\nabla_X^T =[\partial/\partial X_1 \ \partial/\partial X_2 \ ... \ \partial/\partial X_n])$.
By selecting $\phi(X,t)= X^{[k]}$, the incremental rule (6.13) write

$$\Delta X^{[k]} = \left(\nabla_X^T \otimes X^{[k]}\right)f(X,t)dt + \sum_{j=1}^{k} \frac{1}{j!}\left(\nabla_X^T \otimes X^{[k]}\right) g^{(j)}(X,t)(dC)^j \tag{6.14}$$

making stochastic average and dividing by dt the moment equation $[X^{[k]}]$ are obtained in the form

$$\dot{E}[X^{[k]}]= E\left[\left(\nabla_X^T \otimes X^{[k]}\right)f(X,t)\right]+ \lambda \sum_{j=1}^{k} \frac{E[Y^j]}{j!} E\left[\left(\nabla_X^{[j]^T} \otimes X^{[k]}\right)g^{(j)}(X,t)\right] \tag{6.15}$$

Explicit expressions of moments can be obtained by performing the derivatives in Eq. (6.15) (see [12] and [13]).

7. FOKKER-PLANCK EQUATION: MULTIDIMENSIONAL CASE

An extension of the FPK equation for multidimensional systems can be easily obtained by setting $\phi(X,t) = exp(-i\,\vartheta^T X(t))$, where ϑ is an n-vector of real parameters, inserting this expression in Eq. (6.17), making stochastic average and dividing by dt one obtains the differential equation of the characteristic function in the form [12, 13]:

$$dM_X(\vartheta;t) = \sum_{j=1}^{\infty} \frac{(-i)^j}{j!}\,\vartheta^{[j]^T} E\left[exp(-i\,\vartheta^T X)(\Delta X)^{[j]}\right] \tag{7.1}$$

or in explicit form

$$\dot{M}_X(\vartheta;t) = -i\,\vartheta^T E[f(X,t)exp(-i\,\vartheta^T X) - i\lambda\,\vartheta^T \sum_{\ell=1}^{\infty} E[exp(-i\,\vartheta^T X)g^{(\ell)}(X,t)]\frac{E[Y^\ell]}{\ell!} -$$

$$+ \frac{\lambda\,\vartheta^{[2]^T}}{2!} \sum_{\ell=1}^{\infty}\sum_{r=1}^{\infty} E[exp(-i\,\vartheta^T X)g^{(\ell)}(X,t)\otimes g^{(r)}(X,t)]\frac{E[Y^{\ell+r}]}{\ell!\,r!} + ... \tag{7.2}$$

making an inverse Fourier transform the differential equation of the joint probability density function is written as

$$\frac{\partial}{\partial t} p_X(x,t) = -\nabla_x^T(p_X(x,t)f(x,t)) - \lambda \sum_{\ell=1}^{\infty} \frac{E[Y^\ell]}{\ell!}\,\nabla_x^T(p_X(x,t)g^{(\ell)}(x,t)) +$$

$$+ \frac{\lambda}{2!}\sum_{\ell=1}^{\infty}\sum_{r=1}^{\infty} \frac{E[Y^{\ell+r}]}{\ell!\,r!}\,\nabla_x^{[2]^T}(p_X(x,t)g^{(\ell)}(x,t)\otimes g^{(r)}(x,t)) + ... \tag{7.3}$$

in the case of purely external loads, Eq. (7.3) reduces to

$$\frac{\partial}{\partial t} p_X(x,t) = -\nabla_x^T(p_X(x,t)f(x,t)) + \lambda \sum_{j=1}^{\infty} \frac{E[Y^j](-i)^j}{j!}\left[\nabla_x^{[j]^T} p_X(x,t)\right] g^{[j]}(t) \tag{7.4}$$

8. MONTE CARLO SIMULATION OF NON LINEAR SYSTEMS UNDER WHITE NOISE PROCESS

The Monte Carlo simulation approach consists in generating sample functions having prescribed probabilistic characteristics, then for each sample function the deterministic analysis is performed and on the sample functions so obtained of the response the statistics are evaluated.

In the case of delta correlated input one first generates a sequence of time instants T_k according to Poisson law having a mean number of occurrences λ. Moreover, realizations of random variables Y_k is generated with assigned probability density function. At each instant T_k in which the Dirac's delta occours one associates the corresponding strength Y_k. Between two subsequent Dirac's deltas the differential equation of motion is given in the form

$$\dot{X} = f(X,t) \tag{8.1}$$

this equation can be easily integrated by means of the usual integration schemes (as central difference scheme). In corrispondence of the Dirac's occurrence the jump can be evaluated by means of Eq. (3.4). The value of the response immediately after the Dirac's delta becomes initial condition for equation (8.1).

In the case of normal white noise, some additional comments have to be made. As afore mentioned a normal white noise can be seen as an impulsive Poisson process when λ tends to infinity and at the same time $\lambda E[P^2] = q_2$ attains a constant value. In order to simulate a normal white noise, let t^* be a fictitious reference time such that $\lambda = 1$ that means there is one impulse per second in mean and let $\lambda E[P^2] = q_2$. By selecting a time step of integration Δt, and replacing the original time t^* with t, in such a way that when in the second case the time is Δt in the first case $t = 1$ sec, the point location of the impulses in the time t^* can be obtained by stretching the effective time t of $1/\Delta t$. In the new time axis t, since λ is $1/\Delta t$, any amplitude of the impulse Y_k must be multiplied by $\sqrt{\Delta t}$. In this way the two processes have the same value of $\lambda E[Y^2]$ and, at the limit when $\Delta t \to 0$, the process $W(t)$ tends to the normal white noise $W^0(t)$ and this is independent on the original distribution of Y. For Δt very small, by using Eq. (3.4), it follows that:

$$X(t_{k+1}) = X(t_k) + f(X,t_k)\,\Delta t + \sum_{j=1}^{\infty} \frac{Y_k^j (\Delta t)^{1/2}}{j!}\, g^{(j)}(X(t_k),t_k) \tag{8.2}$$

where Y_k is simply the sum of all pulses amplitude occurring in the time step $t_k \div t_{k+1}$.

In equation (8.2) the coefficients of the series decreases with law $(\Delta t)^{j/2}/j!$ so that if Δt is very small, then the first two terms in the summation (8.2) give contributions of the same order of $f(X,t_k)\Delta t$. As $\Delta t \to 0$, only the first two terms of the series in Eq. (3.4) give the contribution of order dt and Eq. (8.2) reduces to the Ito equation. If Δt is small but finite quantity, the results are affected by some errors. The first one is connected with the central difference scheme adopted, the other one is connected with the number of terms of the series (8.2) included in the analysis. In order to minimize the latter error one can assume as Y a normal distribution with zero mean and standard deviation $\sqrt{q_2}$ in this way $E[Y^3] = 0$, $E[Y^4] = 3q_2^2$, meaning that the third term in the summation (8.2) is zero and the fourth term is of order $3\Delta t^2/4!$ so that for zero mean normal white noise, the central scheme difference simply is modified in:

$$X(t_{k+1}) = X(t_k) + f(X(t_k),t_k)\,\Delta t + Y_k\,g^{(j)}(X(t_k),t_k)\,\Delta t^{1/2} +$$

$$+ \frac{Y_k^2}{2!}\,g^{(2)}(X(t_k),t_k)\Delta t \tag{8.3}$$

Another source of error is connected with the fact that the equation (8.3) does not account for the exact location of the impulse in the time step Δt, because $g^{(j)}(X(t_k),t_k)$ is evaluated at the beginning of the step t_k independently of the location of the total impulse Y_k. However this is a source of minor error since Caddemi and Di Paola [2] showed that the jump in a very small interval is independent of the shape of the impulse but depends only of the total intensity of the impulse occurring in the time interval $t_k \div t_{k+1}$. It follows that for the generation of the normal white noise and response one can define the time step Δt and then generates a random variable Y with assigned distribution law (as an example normal, with variance $q_2\,\Delta t$) and then one can perform integration at each time step by using Eq. (8.3) retaining only the first two terms of the summation.

9. PHYSICAL AND IDEAL WHITE NOISE

In this section a last question is examined: in the case of parametric physical white noise, do increments and differentials coincide? A paper has been devoted to this subject [14], the main results can be summarized as follows.

In the case of deterministic impulse the differential equation

$$\dot{X} = f(X,t) + Y\,g(X,t)\,\delta(t - t_0) \tag{9.1}$$

has a jump at to given as

$$J = \sum_{j=1}^{\infty} \frac{Y^j}{j!}\left[g^{(j)}(X,t)\right]_{X=X(t_0)} \tag{9.2}$$

If one has a physical impulse, that is an impulse of finite duration u having an amplitude $1/\Delta$, then into the time interval $t_0 \div t_0 + u$ one can write

$$\dot{X} = f(X,t) + \frac{Y}{u}\,g(X,t), \quad \forall t : t_0 < t \le t + u \tag{9.3}$$

in this interval the differential equation is given as

$$dX = f(X,t)dt + \frac{Y}{u} g(X,t)\, dt \tag{9.4}$$

that is increment and differential coincide. The solution of equation (9.4) in the time interval $[t_0 \div t_0 + u]$ can be obtained by using the central integration scheme by subdividing the time interval $[t_0 \div t_0 + u]$ in N parts of equal length Δu and evaluating the response in the form

$$X(t_{k+1}) = X(t_k) + f(X(t_k),t_k)\Delta u + \frac{Y}{u} g(X(t_k),t_k)\Delta u \tag{9.5}$$

from this equation it seems that for physical impulse no correction terms are needed in passing from differentials to increments. However the output at the end of the interval has to be evaluated recursively by Eq. (9.5), that is

$$X(t_0 + \Delta u) = X(t_0) + f(X(t_0),t_0)\Delta u + \frac{Y}{u} g(X(t_0),t_0)\Delta u$$

$$X(t_0 + 2\Delta u) = X(t_0 + \Delta u) + f(X(t_0 + \Delta u),t_0 + \Delta u)\Delta u +$$

$$+ \frac{Y}{u} g(X(t_0 + \Delta u),t_0 + \Delta u)\Delta u$$

$$\vdots \tag{9.6}$$

$$X(t_0 + N\Delta u) = X(t_0 + (N-1)\Delta u) + f(X(t_0 + (N-1)\Delta u),t_0 + (N-1)\Delta u)\Delta u +$$

$$+ \frac{Y}{u} g(X(t_0 + (N-1)\Delta u),t_0 + (N-1)\Delta u)\Delta u$$

By performing such a procedure at the end of the interval, if u is small enough, the result coincides with that evaluated by the series (2.13). Numerical tests can be found in [14]. As a conclusion by considering a physical impulse, that is a load modelled in such a way that it possesses a finite duration (and unchanged total area), also in the time instants in which the physical impulse is present no corrective terms are necessary and the response can be evaluated by means of a classical incremental analysis. But if the physical impulse has a duration sufficiently small the value at the end of the interval $[t_0 \div t_0 + u]$ evaluated by equation (9.6) coincide with that evaluated by equation (2.13). That means that all terms of the series expansion are necessary for integration in a single step for parametric Dirac delta. In [14] also the Monte Carlo simulation for physical impulses distributed in time according to Poisson law and physical normal (band limited) white noise are examined, and the results performed using Eq. (9.6) always give the same result as those obtained by making the jump prediction by means of the series (2.13). In the case of band limited white noise only the first two terms of the series (2.13) are necessary for obtaining the response in terms of moments.

REFERENCES

1. Di Paola, M.: Stochastic differential calculus non linear systems under normal white noise process, CISM, Udine 1998 (companion paper).

2. Caddemi, S. and Di Paola, M.: Nonlinear systems response for impulsive parametric input, Journal of Applied Mechanics, ASME, Vol. 64, 642-648.

3. Di Paola, M. and Falsone, G.: Stochastic dynamics of nonlinear systems driven by non-normal delta correlated processes, Journal of Applied Mechanics (ASME), 60, 1663, 407-423.

4. Di Paola, M. and Falsone, G.: Ito and Stratonovich integral for delta correlated processes, Probabilistic Engineering Mechanics, 8, 1663, 167-208.

5. Iwankiewicz, R., Nielsen, S.R.K. and Christensen, P.: Dynamic response of nonlinear systems to Poisson distributed pulse train: Markov approach, Nonlinear Structural Systems under random Condition, (Eds. Casciati, I, Elishakoff, I. and Roberts, J.B.), 1660, 223-238.

6. Falsone, G.: Cumulants and correlations for linear systems under non-stationary delta correlated processes, Probabilistic Engineering Mechanics, 6, 1664, 157-165.

7. Ibrahim, R.A.: Parametric random vibration, research Studies Press, Letchworth, England, 1685.

8. Ibrahim, R.A., Soundrarajan, A. and Heo, H.: Stochastic response of nonlinear dynamics systems based on a non Gaussian closure, Journal of Applied Mechanics, (ASME), 52, 1685, 665-670.

9. Wu, W.F. and Lin, Y.K.: Cumulant closure for nonlinear oscillators under random parametric and external excitation, International Journal of Nonlinear Mechanics, 16, 1684, 346-362.

10. Roberts, J.B.: System response to random impulses, Journal of Sound and Vibratin, 24, 1672, 23-24.

11. Vasta, M.: Exact stationary: solution for a class of nonlinear systems driven by nonnormal delta correlated processes, International Journal of Nonlinear Mechanics, 30, 1665, 407-418.

12. Casciati, F. and Di Paola, M.: Stochastic process models, Mathematical models for structural realiability analysis, CRC Mathematical Modelling series, (Casciati F. and Roberts B., Eds.), 1666.

13. Di Paola, M.: Stochastic differential calculus, CISM Courses and Lectures, Casciati F. (Ed.), Dynamic motion: Chaotic and stochastic behaviour, Springer-verlag, Wien, 1663, 26-62.

14. Caddemi, S. and Di Paola, M.: Ideal and physical white noise in stochastic analysis, International Journal of Nonlinear Mechanics, 31, 5, 1666, 581-560.

APPENDIX A

In this appendix some fundamentals of Kronecker algebra are listed. Let A and B be two given matrices of size $p \times q$ and $s \times t$, respectively. The Kronecker product of these two matrices, denoted by $A \otimes B$, is the matrix of size $(ps \times qt)$ obtained by multiplying each element a_{ij} of A by whole matrix B, that is

$$A \otimes B = \begin{bmatrix} a_{11}B & a_{12}B & \cdots & a_{1q}B \\ a_{21}B & a_{22}B & \cdots & a_{2q}B \\ \vdots & \vdots & \ddots & \vdots \\ a_{p1}B & a_{p2}B & \cdots & a_{pq}B \end{bmatrix} \tag{A.1}$$

The Kronecker product has the following properties

$$A \otimes (B \otimes C) = (A \otimes B) \otimes C \tag{A.2}$$

$$(A + B) \otimes (C + D) = A \otimes C + A \otimes D + B \otimes C + B \otimes D \tag{A.3}$$

$$(A \otimes B)(C \otimes D) = (A C) \otimes (B D) \tag{A.4}$$

$$(A \otimes B)^T = (A^T \otimes B^T) \tag{A.5}$$

For two square matrices

$$(A \otimes B)^{-1} = (A^{-1} \otimes B^{-1}) \tag{A.6}$$

The Kronecker power of a matrix A, denoted as $A^{[k]}$, is defined as follows

$$A^{[1]} = A \; ; \; A^{[k+1]} = A \otimes A^{[k]} = A^{[k]} \otimes A \tag{A.7}$$

The Kronecker sum of two matrices A and B of order $p \times p$ and $s \times s$, respectively, is defined as

$$A \oplus B = A \otimes I_s + I_p \otimes B \tag{A.8}$$

I_s and I_p being identity matrices of order s and p, respectively.
Let C be a matrix of order $r \times r$, then

$$(A \oplus B) \oplus C = A \oplus (B \oplus C) \tag{A.9}$$

Note that

$$A \oplus A = A \otimes I_p + I_p \otimes A \tag{A.10}$$

$$A \oplus A \oplus A = A \otimes I_p^{[2]} + I_p \otimes A \otimes I_p + I_p^{[2]} \otimes A \tag{A.11}$$

Let $E_{k,j}$ denote the perturbation matrix of order $(kj \times kj)$, consisting of $j \times k$ array of elementary submatrices E^{mn} of order $(k \times j)$

$$E_{k,j} = \begin{bmatrix} E^{11} & E^{21} & \dots & E^{k1} \\ E^{12} & E^{22} & \dots & E^{k2} \\ \vdots & \vdots & \ddots & \vdots \\ E^{1j} & E^{2j} & \dots & E^{kj} \end{bmatrix} \tag{A.12}$$

where each entry of the matrix E^{mn} has the value one in the (m, n)-th position and zero in all other positions. The perturbation matrices have the following properties

$$E_{k,1} = E_{1,k} = I_k \tag{A.13}$$

$$E_{k,j}^{T} = E_{k,j}^{-1} = E_{j,k} \tag{A.14}$$

I_k being the identity matrix of order k.

Let A and B be two matrices of size $(p \times q)$ and $(s \times t)$, respectively, then

$$B \otimes A = E_{p,s}(A \otimes B)E_{t,q} \tag{A.15}$$

Let V be a differential operator of order $j \times k$, that is

$$V = \begin{bmatrix} \partial/\partial x_{11} & \partial/\partial x_{12} & \dots & \partial/\partial x_{1k} \\ \partial/\partial x_{21} & \partial/\partial x_{22} & \dots & \partial/\partial x_{2k} \\ \vdots & \vdots & \ddots & \vdots \\ \partial/\partial x_{j1} & \partial/\partial x_{j2} & \dots & \partial/\partial x_{jk} \end{bmatrix} \tag{A.16}$$

and let the product AB exist. Then

$$V \otimes (AB) = (V \otimes A)(I_k \otimes B) + (I_j \otimes A)(V \otimes B) \tag{A.17}$$

Moreover

$$V \otimes (A \otimes B) = (V \otimes A) \otimes B + (I_j \otimes E_{s,p})[(V \otimes B) \otimes A](I_k \otimes E_{q,t}) \tag{A.18}$$

For two vectors a and b having u and v components, respectively, the previous equation simplifies into

$$V \otimes (a \otimes b) = (V \otimes a) \otimes b_v + (I_k \otimes E_{v,u})[(V \otimes b) \otimes a] \tag{A.19}$$

HOW TO APPLY STOCHASTIC DIFFERENTIAL CALCULUS TO EARTHQUAKE ENGINEERING ?

M. Di Paola
Dipartimento di Ingegneria Strutturale e Geotecnica,
Università degli Studi di Palermo,
Viale delle Scienze, Palermo, Italy

ABSTRACT

In this paper the applications of the stochastic differential calculus to earthquake engineering are presented. Moment equations for linear systems and approximation for calculating the moments of the response of non linear systems excited by earthquake ground accelerations are also presented.

1. INTRODUCTION

Stochastic models for characterization and simulation of earthquake ground motions have been of interest for a long time. From study of recorded accelerograms it is well understood that earthquake ground motions are non-stationary normal stochastic processes in which intensity as well as frequency content changes with time. Early attempts on stochastic modeling of such phenomena were based on filtered stationary processes (see e.g. [1] ÷ [4]). In order to take into account for the non-stationarity of the earthquake ground motion, models belonging to the class of uniformly modulated processes have been widely used (see e.g. [4] ÷ [6]). More realistic models are those belonging to the class of evolutionary models that account also for the non-stationarity in the frequency content (see e.g. [9] ÷ [11]).

Once the earthquake ground motion is characterized in probabilistic sense the statistics linear or non linear analysis can be performed by using the proper tools of stochastic dynamics.

In the following the probabilistic characterization of earthquake ground motion and the response analysis will be discussed in detail.

2. STOCHASTIC MODELLING OF EARTHQUAKE GROUND MOTION

The stochastic modelling of the acceleration $\ddot{Z}_g(t)$ at a fixed point has been faced with different level of accuracy.

2.1 Stationary models

A refined model, widely used by people working in seismic area, is the Kanai-Tajimi [1] model. In order to understand the physics of this model, let be the ground modelled as a single oscillator, M_g, C_g and K_g are the mass, damping and stiffness, respectively, as shown in Fig. 1.

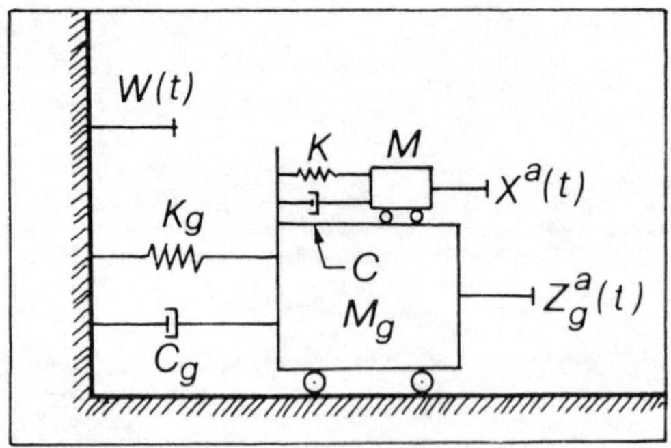

Fig. 1 - The Kanai-Tajimi model

Moreover, let M, C, K the parameters of the superimposed structure. Let $W(t)$, $Z_g^{(a)}(t)$ and $X^{(a)}(t)$ be the absolute displacements of the support (bedrock) of the mass M_g and of the mass M, respectively. Moreover, let $Z_g(t)$, $X(t)$ be the relative displacements between the support and of the superimposed mass M with respect to M_g, respectively. The equations of motion, for a such system write:

$$M\ddot{X} + C\dot{X} + KX = -M\ddot{W} - M\ddot{Z}_g \tag{2.1a}$$

$$M_g\ddot{Z}_g + C_g\dot{Z}_g + K_g Z_g = C\dot{X} + KX - M_g\ddot{W} \tag{2.1b}$$

by dividing equation (2.1a) by M, and equation (2.1b) by M_g for $M_g \to \infty$, equation (2.1) can be rewritten in the form

$$\ddot{X} + 2\zeta\omega_0\dot{X} + \omega_0^2 X = 2\zeta_g\omega_g\dot{Z}_g + \omega_g^2 Z_g \tag{2.2a}$$

$$\ddot{Z}_g + 2\zeta_g\omega_g\dot{Z}_g + \omega_g^2 Z_g = -\ddot{W} \tag{2.2b}$$

with the positions

$$\omega_0 = \sqrt{K/M} \ ; \quad \omega_g = \sqrt{K_g/M_g} \ ; \quad \zeta_g = C_g\big/2\sqrt{K_g M_g} \ ; \quad \zeta = C\big/2\sqrt{KM} \tag{2.3}$$

Equations (2.2) and (2.3) are a set of differential equations forced by the bedrock acceleration $\ddot{W}(t)$. Assuming that $\ddot{W}(t)$ is a normal white noise process, the input for the superimposed structure has a PSD of Kanai-Tajimi form [1]

$$S(\omega) = S_0 \frac{4\zeta_g^2\omega_g^2\omega^2 + \omega_g^4}{(\omega_g^2 - \omega^2)^2 + 4\zeta_g^2\omega_g^2\omega^2} \tag{2.4}$$

S_0 being the (constant) PSD of the bedrock acceleration that controls the ground acceleration peacks. From equation (2.4) one can see that ω_g and ζ_g may be thought of as characteristic ground frequency and damping, respectively; it will be noted that the total area of the PSD $S(\omega)$, that is the variance of the input is infinite. In Figs. 2, 3, the Kanai-Tajimi PSD is plotted for typical values of ζ_g and ω_g, in particular ω_g depends on the soil type (F = Firm, M = Medium, S = Soft).

Another refined model for stationary seismic ground motion is that proposed in [2], it mainly consists in adding to equations (2.2) another filter in order to assure finite power for the PSD, the equation of motion are

$$\ddot{X} + 2\zeta\omega_0\dot{X} + \omega_0^2 X = 2\zeta_g\omega_g\dot{Z}_g + \omega_g^2 Z_g \tag{2.5a}$$

$$\ddot{Z}_g + 2\zeta_g\omega_g\dot{Z}_g + \omega_g^2 Z_g = -\ddot{Z}_f \tag{2.5b}$$

$$\ddot{Z}_f + 2\,\zeta_f\,\omega_f\,\dot{Z}_f + \omega_f^2\,Z_f = \ddot{W}(t) \tag{2.5c}$$

from equations (2.5b,c) it can be recognized that the support motion $W(t)$ of the Kanai-Tajimi model has been substituted by $Z_f(t)$, that is a filtered white noise process obtained by filtered passing the white noise $\ddot{W}(t)$ by a linear high band filter represented by equation (2.5c), removing the low frequency content. In this way the PSD of the input of the system writes

$$S(\omega) = S_0 \frac{\omega_g^4 + 4\,\zeta_g^2\,\omega_g^2\,\omega^2}{(\omega_g^2 - \omega^2)^2 + 4\,\zeta_g^2\,\omega_g^2\,\omega^2} \cdot \frac{\omega^4}{(\omega_f^2 - \omega^2)^2 + 4\,\zeta_f^2\,\omega_g^2\,\omega^2} \tag{2.6}$$

ω_g and ζ_g may be thought of as characteristic ground motion frequency and damping, and ω_f and ζ_f are the parameters of an additional filter, introduced to assure finite power for the PSD.

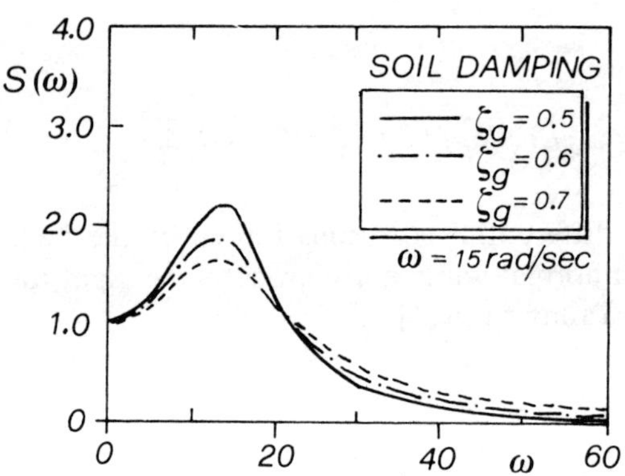

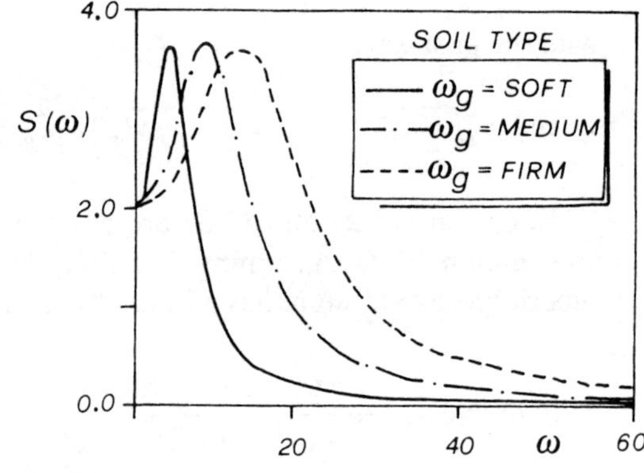

Fig. 2 - PSD of Kanai-Tajimi model for different values of soil damping

Fig. 3 - PSD of Kanai-Tajimi for different values of soil frequency

Also for this model the characteristic ground frequency ω_g depends on the soil type as follows

$$\omega_g(F) = 15\,rad/sec\,;\,\omega_g(M) = 10\,rad/sec\,;\quad \omega_g(S) = 5\,rad/sec \tag{2.7}$$

where (F), (M), (S) means Firm, Medium and Soft, respectively, while the other parameters of the filter can be determined as follows

$$\zeta_g = \omega_g/25\,;\quad \omega_f = \omega_g/10\,;\quad \zeta_f = 0.6 \tag{2.8}$$

In Fig. 4, the PSD $S(\omega)$ for different values of ω_g are plotted for $S_0 = 1\,cm^2/sec^3$.

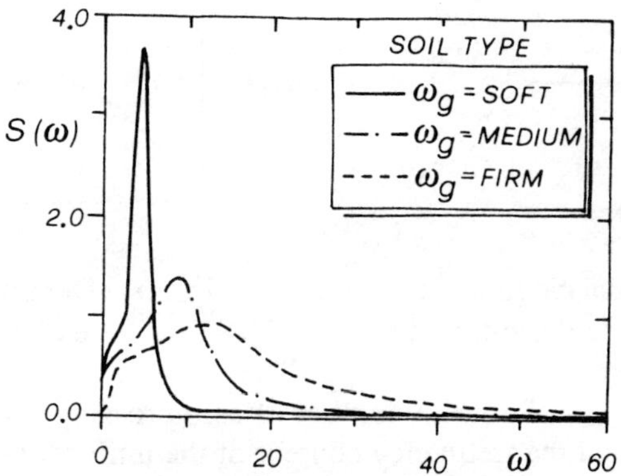

Fig. 4 - Clough and Penzien PSD, for different soil conditions

Another modification of the Kanai-Tajimi spectrum is that proposed by Shinozuka [3], it mainly consists in multiplying the Kanai-Tajimi PSD by a $sin^2 \omega T/2$ for $\omega < \pi/T$, in this way the stationary velocity process exhibit a finite variance.

2.2 Non stationary models of seismic ground motion

Attempts to model the earthquake ground motion as a non-stationary process have been made by a large number of investigators. The simplest model of non-stationary earthquake ground motion is the uniformly modulated one. It mainly consists in multiplying the stationary process having an assigned PSD by a deterministic modulating function ([5] + [8]), that is

$$\ddot{Z}_g(t) = a(t)Q(t) \tag{2.9}$$

where $a(t)$ is a deterministic modulating function and $Q(t)$ is the stationary acceleration having a PSD in the form of equations (2.4) or (2.6), or other spectra. The deterministic modulating function $a(t)$ is essentially of two types: a piecewise linear function how is depicted in Fig. 5a); or of exponential type how is depicted in Fig. 5b).

The latter is given in two different form

$$a(t) = \alpha\, t\, exp(-\beta t)\,; \quad a(t) = \gamma(e^{-\alpha t} - e^{-\beta t}) \tag{2.10a,b}$$

for different values of the parameters α, β and γ, a variety of modulating functions can be obtained. Such an example, for $a(t)$ defined in (2.10a), the parameter β could be chosen in such a way that the maximum of $a(t)$ is reached at a time instants t^*, then $\beta = 1/t^*$, and the value of the maximum is one, then $\alpha = exp\,(1)$.

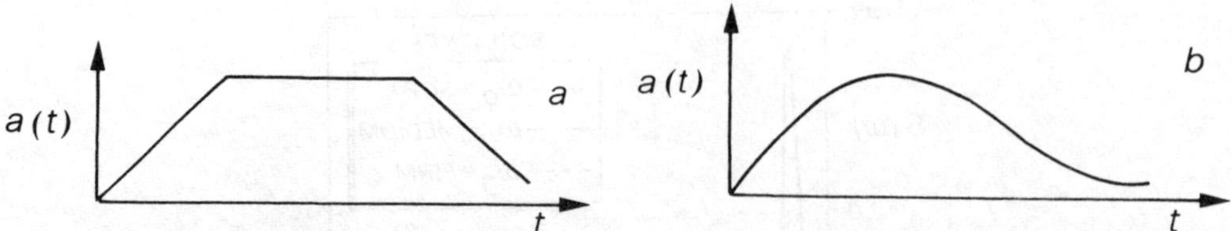

Fig. 5a) - Deterministic function: Fig. 5b) - Deterministic function:
    piecewise linear function   exponential function

It may be recognized that the mean number of zero crossings is identical for both $Q(t)$ and $\ddot{Z}_g(t)$, that means that the frequency content of the uniformly modulated model does not changes in time.

Recorded earthquake ground motions usually exhibit nonstationarity in both frequency content and intensity. This is because high frequency components of seismic waves travel faster and arrive at the recording station earlier than the low frequency components. In order to retain this property a lot of research works have been made by several authors (see e.g. [8] ÷ [13]). An evolutionary model is usually given in the Priestley's [14] form, that is

$$\ddot{Z}_g(t) = \int_{-\infty}^{\infty} M(t,\omega)e^{i\omega t}\,dZ(\omega) \tag{2.11}$$

in which $M(t, \omega)$ is a deterministic modulating function (generally complex valued) and $dZ(\omega)$ is an orthogonal increment process, that is

$$E[dZ(\omega_j)dZ^*(\omega_k)] = \delta_{jk}\,S(\omega_j)d\omega_j \tag{2.12}$$

the star means complex conjugate and δ_{jk} is the Kronecker delta, $S(\omega)$ is the power spectral density of the stationary part of the process. In passing it can be seen that by selecting $M(t, \omega) = 1$ one obtain the stationary model discussed in Section 2.1, while by selecting $M(t, \omega) = a(t)$ one obtains the uniformly modulated process defined in equation (2.9). There are several models for modeling evolutionary power spectrum however the simplest one, from a mathematical point of view, is the model proposed in [14]. It mainly consists in assuming that the filter parameters ζ_g and ω_g in equation (2.2a), are time dependent

$$\ddot{Z}_g + 2\zeta_g(t)\omega_g(t)\dot{Z}_g + \omega_g^2(t)Z_g = \ddot{W}(t) \tag{2.13}$$

the input on the superimposed structure, according to equation (2.3), will be given in the form

$$a(t)\,(2\,\xi_g(t)\,\omega_g(t)\dot{Z}_g + \omega_g^2(t)Z_g) \tag{2.14}$$

from these equations it can be easily recognized that the evolutionary model outlined above is a modification of the Kanai-Tajimi model, $a(t)$ in equation (2.14) mainly controls the intensity, while $\omega_g(t)$ mainly controls the frequency variation in time.

In Fig. 6, the variation of zero crossing rate of a displacements $Z_g(t)$ of El Centro earthquake is plotted versus time. In this figure the dashed line corresponds to the mooth zero crossing rate function given by [14]

$$v(t) = v_0 + v_1(e^{-b_1 t} - e^{-b_2 t}) \tag{2.15}$$

where

$$v_0 = 3/sec\,; \quad v_1 = 19.01/sec\,; \quad b_1 = 0.0625/sec\,; \quad b_2 = 0.15\,sec^{-1} \tag{2.16}$$

once the zero crossing rate (with positive negative slopes) is identified for a given real record the estimated approximation parameters $\omega_g(t)$ in equation (2.13) is simply given as

$$\omega_g(t) \cong \pi\,v(t) \tag{2.17}$$

the other relevant parameters for defining the evolutionary El Centro 1940 earthquake $\zeta_g = 0.42$, $a(t) = 9.44\,t^3\,exp(-1.174\,t) + 3.723(0 \le t \le 29.880)$.

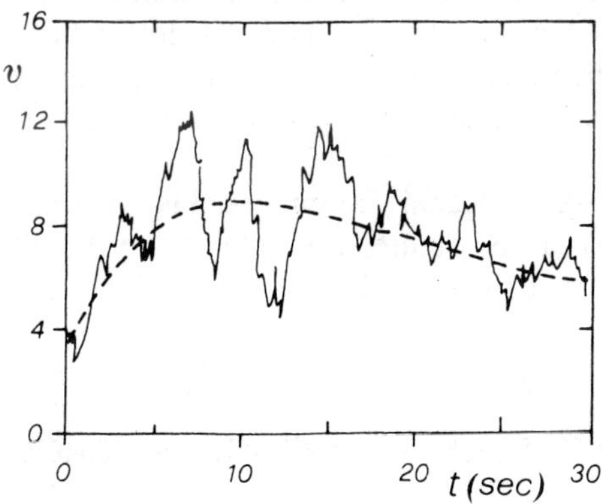

Fig. 6 - Variation of zero crossing rate with rim and smooth zero
crossing rate (El Centro 1940 earthquake from [14])

Typical records exhibit smoothed variation in time. This suggest that the Priestley form of $Z_g(t)$ and $\dot{Z}_g(t)$ outputs of equation (13) can be approximated in the form

$$\frac{d^k}{dt^k}Z_g(t) = \int_{-\infty}^{\infty} \frac{(i\omega)^k \, exp\,(i\,\omega\,t)}{\omega_g^2(t) - \omega^2 + 2\,i\,\zeta_g(t)\,\omega_g(t)\,\omega}\, dZ(\omega)\,; \quad k = 0.1 \tag{2.18}$$

where $dZ(\omega)$ is the orthogonal increment process of the original white noise source

$$E[dZ(\omega_j)dZ^*(\omega_k)] = \delta_{jk}\, S_0\, d\omega_j \tag{2.19}$$

S_0 being the intensity of the white noise at the bedrock.

2.3 Non-normal seismic acceleration model

Here another relevant model of seismic acceleration is examined using the mathematical concepts of random pulse train, that is a filtered Poisson white noise already described in [16]. Most earthquakes are caused by slips in a fault zone in the ground. Assuming that slippings occur in independent short spurts, almost impulsively [17], the acceleration at they can be represented in the form of Poisson white noise

$$\ddot{W}(t) = \sum_{k=1}^{N(t)} Y_k\, \delta(t - T_k) \tag{2.20}$$

In this case the magnitudes Y_k are assumed to be zero mean independent accelerations but having identical distribution. The Poisson white noise (2.20) is characterized by the correlation functions given as product of Dirac's deltas [16]. It is a non-stationary stochastic process under either one of two assumptions: the average impulse occurrence rate λ is time dependent; or λ is time-dependent, or $\lambda = const$ and Y_k depends on the occurrence time T_k.

The equation of motion for the Kanai-Tajimi model is given as in Eq. (2.2), the response of the filter equation (2.2b) can be obtained by integrating the differential equation of the filter. The statistical moments can be easily evaluated using the tools already described in [16].

3. RESPONSE OF SDOF SYSTEMS TO EARTHQUAKE GROUND MOTION

The response of linear system under seismic acceleration of the support can be conducted in different ways. In the case of stationary model and response the analysis can be performed in frequency domain. At this end if $S(\omega)$ is the PSD of the input either given in Eq. (2.4) or in Eq. (2.6) by means of the simple relationship

$$S_X(\omega) = |H(\omega)|^2 S(\omega) \tag{3.1}$$

$H(\omega)$ being the transfer function of the system given as

$$H(\omega) = \frac{1}{\omega_0^2 - \omega^2 + 2i\xi\omega_0\omega} \tag{3.2}$$

then the variance in terms of displacement of the response X is given in the form

$$\sigma_X^2 \equiv E[X^2] = \int_{-\infty}^{\infty} |H(\omega)|^2 S(\omega)\,d\omega$$

$$\sigma_{\dot{X}}^2 = [\dot{X}^2] = \int_{-\infty}^{\infty} |H(\omega)|^2 \omega^2 S(\omega)\,d\omega \tag{3.3a,b}$$

In the case of non stationary models and response the most convenient way is to perform the analysis in time domain using the proper tools of stochastic differential calculus. As an example for the Tajimi-Kanai model Eq. (2.2) are transformed into first order differential equations in the form

$$\dot{U} = D U + V \ddot{W} \tag{3.4}$$

where U is the vector of state variables $U^T [X\ \dot{X}\ Z_g\ \dot{Z}_g]$, the matrix D and the vector V are given as

$$D = \begin{bmatrix} 0 & 1 & 0 & 0 \\ -\omega_0^2 & -2\xi\omega_0 & \omega_g^2 & 2\xi_g\omega_g \\ 0 & 0 & 0 & 1 \\ 0 & 0 & -\omega_g^2 & -2\xi_g\omega_g \end{bmatrix} ; \quad V = -\begin{bmatrix} 0 \\ 0 \\ 0 \\ 1 \end{bmatrix} \tag{3.5}$$

all state variables are zero mean processes, the input $\ddot{W}(t)$ is normal the also the response is a normal process. It follows that, the complete characterization of the response U can be obtained by the second order statistics. In order to do this the second order moments of the state variables have to be obtained by using the Ito differential rule. Since the input is an external white noise, the Ito type differential equation associated to Eq. (3.4) is simply given

$$dU = D U\,dt + V\,dB \tag{3.6}$$

where $dB(t)$ is the Wiener process $(dB/dt = \ddot{W}, E[dB(t_1)\,dB(t_2)] = \pi S_0 \delta(t_1 - t_2)dt_1\,dt_2)$

The Ito differential rule is then given as

$$d(U_i\,U_j) = U_i\,dU_j + U_j\,dU_i + dU_i\,dU_j \qquad (3.7)$$

$U_i(t)$ and $U_j(t)$ being the *i-th* and *j-th* components of the vector U. Another hand

$$dU_i = \sum_{k=1}^{4} D_{ik}\,U_k\,dt + V_i\,dB \;;\quad dU_j = \sum_{r=1}^{4} D_{jr}\,U_r\,dt + V_j\,dB \qquad (3.8)$$

by inserting Eq. (3.8) in (3.7) one obtains

$$d(U_i\,U_j) = \sum_{r=1}^{4} D_{jr}\,U_i\,U_r\,dt + \sum_{k=1}^{4} D_{ik}\,U_k\,U_j\,dt + U_i\,V_j\,dB +$$

$$+ U_j\,V_i\,dB + V_i\,V_j\,(dB)^2 \qquad (3.9)$$

making stochastic average, dividing by dt and using the non anticipating property the second order moment equations of the response is written as

$$\dot{E}[U_i\,U_j] = \sum_{r=1}^{4} D_{jr}\,E[U_i\,U_j] + \sum_{k=1}^{4} D_{ik}\,E[U_j\,U_k] + V_i\,V_j\,\pi S_0 \qquad (3.10)$$

Equation (3.10) represent a set of sexteen deterministic differential equation in terms of moments of second order. In the case of stationary solution $\dot{E}[U_i\,U_j] = 0$ and equation (3.10) becomes a set of linear algebraic equations.

In the case of non-stationary model of seismic ground motion given in Eq. (2.14) the previously described procedure holds, the only difference is that the matrix D becomes time dependent and than the response evaluated by Eq. (3.10) is always non stationary.

4. MULTIDEGREE OF FREEDOM LINEAR SYSTEMS

The equation of motion of an *n-DOF* structural linear system excited by earthquake ground motion can be cast in the form

$$M\ddot{X} + C\dot{X} + K X = -M V \ddot{Z}_g \qquad (4.1)$$

in which M, C, K are the mass, damping and stiffness matrices of order *nxn*, $X(t)$ is an n-vector of displacement, respectively, V is a n-vector whose components are either one or zero depending upon the support motion is coincident with the corresponding direction of the structural displacement or not.

Introducting the state vector $Z^T(t) = [X^T(t)\ \dot{X}^T(t)]$, equation (4.1) becomes

$$\dot{Z} = D_s Z + G_s \ddot{Z}_g \tag{4.2}$$

where D_s and G_s are given as follows

$$D_s = \begin{bmatrix} 0 & I \\ -M^{-1}K & -M^{-1}C \end{bmatrix}; \quad G_s = \begin{bmatrix} 0 \\ V \end{bmatrix} \tag{4.3}$$

if $\ddot{Z}_g \equiv \ddot{W}(t)$ is a normal white noise input, $\ddot{W}(t)$ is characterized by a correlation function

$$E[\ddot{W}(t_1)\,\ddot{W}(t_2)] = q(t_1)\,\delta(t_1 - t_2) \tag{4.4}$$

where $q(t)$ is the strength of the white noise. If q does not depends on t, then the input is probabilistically stationary. Using Ito rule previously described, the covariance matrix $E[Z(t)\,Z^T(t)] = \Sigma_Z$ can be obtained from the Lyapunov form equation

$$\Sigma_Z = D_s\,\Sigma_Z + \Sigma_Z\,D_s^T + G_s\,G_s^T\,q(t) \tag{4.5}$$

Once the covariance matrix is evaluated by solving Eq. (4.5), the correlation matrix (statistic of second order) $R_Z(t_1,t_2) = E[Z(t_1)\,Z^T(t_2)]$ is given as

$$\begin{aligned}
R_Z(t_1,t_2) &= \Theta_s(t_1 - t_2)\Sigma_Z(t_1), \quad \text{for} \quad t_1 \le t_2 \\
R_Z(t_1,t_2) &= \Theta_s(t_2 - t_1)\Sigma_Z(t_2), \quad \text{for} \quad t_2 \le t_1
\end{aligned} \tag{4.6}$$

where $\Theta_s(t)$ is the so called *transition matrix*. Exact expression of the transition matrix can be found in Muscolino [18] for classically and non classically damped systems.

As shown in [19], by employing Ito stochastic differential rule, equation (4.5) can be rewritten in the form

$$\dot{\sigma}_Z = D_{2,s}\,\sigma_Z + G_{2,s}\,q(t) \tag{4.7}$$

where

$$\begin{aligned}
\sigma_Z &= Vec\,(\Sigma_Z) = E[Z \otimes Z] \\
G_{2,s} &= G_s^{[2]} = G_s \otimes G_s
\end{aligned} \tag{4.8}$$

In equation (4.8) the symbol $\otimes$ means Kronecker product (see Appendix A of [16]), the exponent in square brackets means Kronecker Power $(G_s^{[2]} = G_s \otimes G_s)$ and $Vec\ (\cdot)$ is a column vector obtained by putting each column of the matrix in parenthesis below each other. The matrix $D_{2,s}$ is related to the matrix D_s by the following relationship

$$D_{2,s} = D_s \otimes I + I \otimes D_s = D_s \oplus D_s \tag{4.9}$$

where the symbol $\oplus$ means Kronecker sum.

Equation (4.9) appears to be more illuminative than equation (4.5). Additionally for the equation (4.7), due to the Lyapunov form of $D_{2,s}$ some important eigenproperties relating the differential equation (4.2) and equation (4.8) can be derived. In particular let Ψ_s be the eigenmatrix of D_s, that is Ψ_s satisfies the equation

$$D_s\ \Psi_s = \Psi_s\ \Gamma_s \tag{4.10}$$

where Γ_s is a diagonal matrix listing the (complex) eigenvalues of D_s. It can be shown the eigenvectors of $D_{2,s}$ are contaned in the matrix $\Psi^{[2]}$ and the eigenvalues of $D_{2,s}$ have the Lyapunov form

$$\Gamma_{2,s} = \Gamma_s \otimes I + I \otimes \Gamma_s = \Gamma_s \oplus \Gamma_s \tag{4.11}$$

Finally, the transition matrix of equation (4.7) is given in the form

$$\Theta_{2,s}(t) = \Theta_s(t) \otimes \Theta_s(t) \tag{4.12}$$

then the solution of the differential equation (4.7) can be written in the form

$$\sigma_Z(t) = \Theta_s^{[2]}(t - t_0)\ \sigma_Z(t_0) + \left[\int_{t_0}^{t} \Theta_s^{[2]}(t - \tau)\ q(\tau)d\tau\right] G_{2,s} \tag{4.13}$$

$\sigma_Z(t_0)$ being the covariance vector at the time instant t_0. If $q(t)$ is constant, that is the white noise is stationary, and the system operate from $t = -\infty$, then the response vector attains a stationary state, and the covariance vector simply write

$$\sigma_Z = -D_{2,s}^{-1}\ G_{2,s}\ q \tag{4.14}$$

If the earthquake ground motion is considered as a filtered white noise, that is $\ddot{Z}_g(t)$ is the output of a linear system excited by a normal white noise, one can add to the equation of motion the filter equations how has been shown in the previous sections. In this case the enlarged systems can be written in the form

$$\dot{U} = DU + G\ddot{W} \tag{4.15}$$

where $U^T = [Z^T \ Z_g \ \dot{Z}_g]$, and the matrix D and the vector G are given in the form

$$D = \begin{bmatrix} D_s & D_{sf} \\ 0 & D_f \end{bmatrix}; \quad G = \begin{bmatrix} 0 \\ -1 \end{bmatrix} \tag{4.16}$$

where, according to the Tajimi-Kanai model the matrices D_f and D_{sf} are given as

$$D_f = \begin{bmatrix} 0 & 1 \\ -\omega_g^2 & -2\xi_g\omega_g \end{bmatrix}; \quad D_{sf} = -V \begin{bmatrix} 2\xi_g\omega_g & \omega_g^2 \end{bmatrix} \tag{4.17}$$

By indicating with Ψ_s and Λ_s the eigenproblem associated to D_s, with Ψ_f and Λ_f the eigenproblem associated to D_f, that is

$$D_s \Psi_s = \Psi_s \Lambda_s; \quad D_f \Psi_f = \Psi_f \Lambda_f \tag{4.18}$$

due to the particular form of D_f, the eigematrices of D are given as

$$\Psi = \begin{bmatrix} \Psi_s & N_{sf}\Psi_f \\ 0 & \Psi_f \end{bmatrix}; \quad \Lambda = \begin{bmatrix} \Lambda_s & 0 \\ 0 & \Lambda_f \end{bmatrix} \tag{4.19}$$

where the matrix N_{sf} in Eq. (4.19) can be evaluated in the form [20]

$$Vec\,(N_{sf}) = -\left(D_s \otimes I_{2m} - I_{2n} \otimes D_f^T\right)^{-1} Vec\,(D_{sf}) \tag{4.20}$$

m being the number of degree of freedom (DOF) of soil (two, in the case of Kanai-Tajimi model) and n the number of DOF of the structural systems and I_r is the identity matrix of order rxr.

The transition matrix $\Theta(t)$ related to the matrix D defined in Eq. (4.16) can be obtained in the form

$$\Theta(t) = \begin{bmatrix} \Theta_s(t) & \Theta_{sf}(t) \\ 0 & \Theta_f(t) \end{bmatrix} \tag{4.21}$$

where $\Theta_s(t)$ and $\Theta_f(t)$ are the fundamental matrices of the structure and of the soil, respectively, and $\Theta_{sf}(t)$ is given in the form

$$\boldsymbol{\Theta}_{sf}(t) = N_{sf}\,\boldsymbol{\Theta}_f(t) - \boldsymbol{\Theta}_s(t)\,N_{sf} \tag{4.22}$$

Once the fundamental matrix is evaluated the second order moment of the system subjected to the white process can be obtained in the form

$$\dot{\boldsymbol{\sigma}}_U = D_2\,\boldsymbol{\sigma}_U + G^{[2]}q(t) \tag{4.23}$$

where D_2 is given in the form

$$D_2 = D \otimes I_{2(m+n)} + I_{2(m+n)} \otimes D \tag{4.24}$$

By solving equation (4.23) all second order moments can be evaluated.

An overview of the methods of linear systems under normal or non-normal white noise processes can be found in [21].

5. NONLINEAR SYSTEMS UNDER EARTHQUAKE GROUND MOTION

In the nonlinear case two several methods can be adopted here only the Moment Equation (ME) approach and the Pseudo Force Method (PF) are discussed. The former consists in writing the differential equation of the nonlinear system under the white noise or the filtered white noise and by truncating the hierarchy by some closure scheme. The PF consists in expanding the solution in an asymptotic form. In fact in the nonlinear case, especially for MDOF systems and or non stationary input finding the solution of the FPK equation is quite impossible.

5.1 ME approach

Let the equation of motion be given in the form

$$M\ddot{X} + C\dot{X} + K X + f(X) = -M V \ddot{Z}_g \tag{5.1}$$

where $f(X)$ is a deterministic vector of the state variable X which represents the nonlinearities in the system, $\ddot{Z}_g(t)$ is the ground acceleration modeled as a filtered normal or non normal white noise.

The equation of motion can be now written in Ito standard form as follows

$$dU = D U dt + \boldsymbol{\psi}(U)dt + G dB(t) \tag{5.2}$$

where U, D and G have been already defined in equation (4.15) and (4.16) while $\boldsymbol{\psi}(U)$ is the vector of nonlinearities defined as

$$\psi^T(U) = \begin{bmatrix} 0^T & f^T(X) & 0^T \end{bmatrix} \tag{5.3}$$

The moment equation of order k for the system (5.2) can be obtained by putting in the Ito rule $\phi(U) = \phi(U_1^r \; U_2^s \; ... \; U_{2n+2}^p)$, $r+s+...+p=k$ dividing by dt and making stochastic average one get the moment equations has been discussed in [16].

Because the moment equations of a fixed order, say s contains moments of higher order than s the system of equation constitutes an infinite hierarchy, then it cannot exactly solved. A truncation procedure is then necessary in order to make the system equation finite.

5.2 Statistical linearization

The simplest closure scheme is the so-called Gaussian closure, it consists in assuming that the probability distribution of U is a normal vector process. It has been demonstrated that this closure scheme coincides with the well known statistical linearization. In the follows without loss of generality we confine ourselves to the case in which $f(X) = -f(X)$, in this case the output X is a zero mean non normal process. The statistical linearization can be brevely summarized as follows instead of Eq. (5.1) one work with the linearized system

$$M\ddot{X} + C\dot{X} + K_e X = -MV\ddot{Z}_g \tag{5.4}$$

where K_e is a matrix whose elements are such that the mean square difference between eq. (5.1) and (5.4) is the minimum in some statistical sense. The most common way is finding the unkown parameters in K_e in a such a way that the mean square difference between $KX + f(X) - K_e X$ is minimum, that is

$$E[(KX + f(X) - K_e X)^T (KX + f(X) - K_e X)] = min \; K_e \tag{5.5}$$

it leads to

$$K_e = K + \left\{ E[f(X)X^T] + E[Xf^T(X)] \right\} \Sigma_X^{-1} \tag{5.6}$$

now the various elements of the matrices $E[f(X)X^T]$ and its transpose are evaluated by considering that the response vector X is a normal process having zero mean and covariance matrix Σ_X, in this case the matrix $E[f(X)X^T]$ can be easily evaluate especially in the case in which $f(X)$ is expressed in polynomial form by taking into account that the moments of any order of normal processes can be evaluated simply as a non linear combination of the moments up to the second order. The matrix Σ_X in Eq. (5.6) is already unknown and then an iteration procedure is necessary in order to find K_e. In the first step one assume $K_e = K$, the linear system (5.4) can easily solved in terms of moments of second order, then one with these values of Σ_X evaluate the new value of K_e by equation (5.6), from the linear equation (5.4). A new attempt of Σ_X is then evaluated and a new matrix K_e is evaluated, the procedure ends when the matrix Σ_X remain almost unchanged between two subsequent iterations.

Equation (5.6) can be rewritten in the more expressive form

$$K_e = K + K' \tag{5.7}$$

where

$$K' = E\left[\nabla_X^T \otimes f(X) \right] \tag{5.8}$$

∇_X being the gradient vector ($\nabla_X^T = [\partial/\partial X_1 \dots \partial/\partial X_n]$), then the various elements of the matrix K' are given as [23]

$$K'_{ij} = E[\partial f_i(X) / \partial X_j] \tag{5.9}$$

Further hint in the statistical linearization can be found in [24].

5.3 Non linear system handled by pseudo force method

The statistical linearization approach is simple but effects of the non-normality of the response process $X(t)$ is not retained. In order to take into account this effect some approximate methods can be applied. Here the pseudo force method will be discussed. For clarity sake's the scalar case is discussed, but extension to multidegree of freedom and filtered normal or non normal white noise is strigthforeward.

Let the equation of motion be given in the form

$$\dot{X} + aX - f(X) = \ddot{W}(t) \tag{5.10}$$

$\ddot{W}(t)$ being a white noise and $f(X)$ any non linear function of X.

The pseudo force principle solve the nonlinear system equation (5.10) as a cascade of linear differential equation in the form

$$\dot{X}_0 + aX_0 = \ddot{W}(t)$$
$$\dot{X}_1 + aX_1 = f(X_0) + \ddot{W}(t)$$
$$\vdots$$
$$\dot{X}_n + aX_n = f(X_{n-1}) + \ddot{W}(t) \tag{5.11}$$

As $n \to \infty$, $X_n \to X(t)$. If $f(X)$ is given in a polynomial form $f(X) = \sum\limits_{k=1}^{N} b_k X^k$ Eq. (5.11) can be rewritten in the form

$$\dot{X}_0 + a X_0 = \ddot{W}(t)$$

$$\dot{X}_1 + a X_1 = \sum_{k=1}^{N} b_k X_0^k + \ddot{W}(t)$$

$$\vdots$$

$$\dot{X}_n + a X_n = \sum_{k=1}^{N} b_k X_{n-1}^k + \ddot{W}(t)$$

(5.12)

by means of suitable transformations the nonlinear system equation (5.12) can be reconducted to a quasi-linear one, as in fact by truncating for sake of simplicity the equations (5.12) up to $n = 1$ and putting

$$X_0^N = Z_{N+1}, \ X_0^{N-1} = Z_N, \dots, X_0^2 = Z_3, \ X_0 = Z, \ X_1 = Z_2 \tag{5.13}$$

$$\dot{Z}_1 + a Z_1 = W(t)$$

$$\dot{Z}_2 + a Z_2 = W(t) + \sum_{k=1}^{N} b_k Z_{k+2}$$

(5.14)

Now by differentiating equation (5.13) with respect to the time one get [25]

$$\dot{Z}_{N+1} = N \dot{X}_0 X_0^{N-1}; \dot{Z}_N = (N-1) \dot{X}_0 X_0^{N-2}, \dots, \dot{Z}_3 = 2 X_0 \dot{X}_0 \tag{5.15}$$

By inserting $\dot{X}_0 = (-a X_0 + W(t))$ in Eq. (5.15) one obtains

$$\dot{Z}_{N+1} = -N a Z_{N+1} + N \ddot{W} Z_N$$

$$\vdots$$

$$\dot{Z}_3 = -2 a Z_4 + 2 \ddot{W} \ddot{Z}_1$$

(5.16)

and hence Eq. (5.14) and (5.16) can be rewritten in the standard form

$$\dot{Z} = D Z + (R Z + \mu) \ddot{W} \tag{5.17}$$

where D, R and μ are given as

$$
D = \begin{bmatrix} -a & 0 & 0 & 0 & \cdots & 0 \\ 0 & -a & b_1 & b_2 & \cdots & b_N \\ 0 & 0 & -2a & 0 & \cdots & 0 \\ \vdots & \vdots & \vdots & \vdots & \cdots & \vdots \\ 0 & 0 & 0 & 0 & \cdots & -Na \end{bmatrix} ; \quad R = \begin{bmatrix} 0 & 0 & \cdots & 0 \\ 0 & 0 & \cdots & 0 \\ 2 & 0 & \cdots & 0 \\ 0 & 3 & \cdots & 0 \\ \vdots & \vdots & \cdots & \vdots \\ 0 & 0 & \cdots & 0 \end{bmatrix} ; \quad \mu = \begin{bmatrix} 1 \\ 1 \\ 0 \\ \vdots \\ 0 \end{bmatrix} \qquad (5.18)
$$

Equation (5.17) is a quasi-linear system and the response in terms of state variables Z is not normal. If $\ddot{W}(t)$ is a white noise (normal or non normal) the moment equations can be derived in a closed form solution since they do not constitute an infinite hierarchy how has been discussed in [16].

Extension to multidegree of freedom systems as well as higher order truncations can be easily obtained. An improuvement of the method herein presented can be made by substituting as the first equation (5.11) $\dot{X}_0 + a_e X_0 = \ddot{W}$, where a_e is obtained by the statistical linearization [26], in this way the first Eq. (5.11) give the solution of the statistical linearization, while non normality of the response are accounted by the quasi-linearization method.

REFERENCES

1. Tajimi, H.: A statistical method of determining the maximum response of a building structure during an earthquake, Proc. of 2nd World Conference on Earthquake Engineering, Tokyo, Japan 1960.

2. Ruiz, P. and Penzien, J.: Probabilistic study of the behaviour of structures during earthquake, Report No. EERC 69.03, Earthquake Engineering Research Center, University of California, Berkeley, California, USA, 1989.

3. Shinozuka, M., Zhang, R. and Deodatis, G.: Sinesquare modification to Kanai-Tajimi earthquake ground motion spectrum, ICOSSAR 93, Insbruck, Austria 1993.

4. Bolotin, V.V.: Statistical theory of the aseismic design of structures, Proc. of 2nd World Conference on Earthquake Engineering, Tokyo, Japan 1960.

5. Bogdanoff, J.E., Goldberg J.L. and Sharpe, D.R.: The response of simple nonlinear structure to a random disturbance of earthquake type, Bull. Seismological Society of America, 54, 263-276, 1964.

6. Amin, M. and Ang, A.H.S.: A nonstationary stochastic model for strong motion earthquakes, Structural Research Series, No. 306, University of Illinois, Urbana, IL., USA, 1966.

7. Jennings, P.C., Housner, G.W. and Tsai, N.C.: Simulated earthquake ground motions, Earthquake Engineering Research Laboratory, California Institute of Technology, Pasadena, CA., USA, 1968.

8. Iyengar, R.N. and Iyengar, K.T.: A nonstationary random process model for earthquake accelerations, Bulletin Seismological Society of America, 59, 1969, 1163-1188.

9. Lin, S.C.: Evolutionary spectra and nonstationary processes, Bulletin Seismological Society of America, 30(3), 1970, 891-900.

10. Saragoni, G.R. and Hart, G.C.: Simulation of artificial earthquakes, Earthquake Engineering and Structureal Dynamics, 2(3), 1974, 249-267.

11. Kameda, H.: Evolutionary spectra of seismogram by multifilter, Journal Engineering Mechanics Division, ASCE, 113(8), 1987, 787-801.

12. Deodatis, G., Shinozuka, M.: Auto-regressive model for nonstationary stochastic processes, Journal of Engineering Mechanics, ASCE, Vol. 114, No. 11, 1988, 1195-2012.

13. Shinozuka, M. and Deodatis, G.: Stochastic process models for earthquake ground motion, Probabilistic Engineering Mechanics, Vol. 3, No. 3, 1988, 114-125.

14. Fan, F.G. and Ahmadi, G.: Nonstationary Kanai-Tajimi models for El Centro 1940 and Mexico City 1985 earthquakes, Probabilistic Engineering Mecha-nics, Vol. 5, No. 4, 1990, 171-181.

15. Priestley, M.B.: Evolutionary spectra and nonstationary processes, Journal of Royal Statistical Society, Ser. B, 28(2), 1965, 204-230.

16. Di Paola, M.: Stochastic differential calculus, non linear systems under Poisson white noise, CISM, Udine 1998 (companion paper).

17. Lin, Y.K. and Yong, Y.: Evolutionary Kanai-Tajimi earthquake models, Journal Engineering Mechanics Division, ASCE, 113(8), 1987, 1119-1137.

18. Muscolino, G.: Non stationary pre-envelope covariances of non-classically damped systems, Journal of Sound and Vibration, 149, 1991, 107-123.

19. Di Paola, M.: Moments of nonlinear systems: Probabilistic methods in civil engineering, Proc. of Fifth ASCE Special Conference, Blacksburg, 1988, 285-288.

20. Falsone, G., Muscolino, G. and Ricciardi, G.: Combined dynamic response of primary and mulitply connected cascaded secodary subsystems, Earthquake Engineering and Structural Dynamics, 20, 1991, 749-767.

21. Di Paola, M. and Elishakoff, I.: Non-stationary response of linear systems under stochastic Gaussian and non-Gaussian excitation: a Brief overview of recent results, Chaos, Solitons & Fractals, Vol. 7, No. 7, 1996, 961-971.

22. Elishakoff, I.: Probabilistic methods in the theory of structures, Wiley, New York, 1985.

23. Atalik, T.S. and Utku: Stochastic linearization of multidegree of freedom non linear systems, Earthquake Engineering and Structural Dynamics, 4, 1976, 441-420.

24. Roberts, J.B. and Spanos, P.D.: Random vibration and statistical linearization, John Wiley & Sons, New York, 1990.

25. Di Paola, M.: Linear systems excited by polynomials of filtered Poisson pulses, Journal of Applied Mechanics, ASME, Vol. 64, 1997, 712-717.

26. Benfratello, S.: Pseudo-force method for a stochastic analysis of non linear systems, Probabilistic Engineering Mechanics, 11, 1996, 113-123.

CHAPTER 5

ARE PROBABILISTIC AND ANTI-OPTIMIZATION APPROACHES COMPATIBLE?

I. Elishakoff
Florida Atlantic University, Boca Raton, FL, USA

WHAT MAY GO WRONG WITH PROBABILISTIC METHODS?

Isaac Elishakoff
Department of Mechanical Engineering
Florida Atlantic University
Boca Raton, FL 33431-0991

ABSTRACT

This study is directed to a single objective: to illustrate the possible error associated with the effect of a small perturbation in the probability density on the structural reliability. This perturbation is associated with interpretation of the experimental data, which lies as a basis of the probabilistic model involved. Moreover, the small perturbation in the probability density is still associated with the same probabilistic moments possessed by an unperturbed density. It appears that the analysis of a possible error must become a part of a meaningful probabilistic analysis.

INTRODUCTION

Apparently the first study in which probabilistic methods were applied to structures is the book by Mayer (1926). Since then probabilistic methods passed the age of adolescence (Cornell, 1981).

Usually, in probabilistic reliability studies the assumptions on the probabilistic densities are invariably made. The knowledge of the probabilistic densities of involved quantities is central. Once these are known, the probabilistic properties of the output quantities can be determined. As Wentzel (1980) stresses, probabilistic methods are

> *"frequently regarded as a kind of 'magic wand' which produces information out of a void. This is a fallacy; the theory of probability only enables information to be transformed, and conclusions on inaccessible phenomena to be drawn from data on observable ones."*

Probabilistic methods must result in a central probabilistic quantity-reliability of the structure, namely, the probability that the structure will perform its mission satisfactorily. Modern society rightfully should expect extremely high reliabilities, and, consequently, extremely small probabilities of failure.

The deterministic mechanic's theories represent a cornerstone of the probabilistic mechanics. First the phenomenon should be understood qualitatively; then it should be understood sufficiently well quantitatively. Then the relationship of the output quantities in their intricate dependence of the input quantities is developed either as a set of equations, algebraic expressions, or numerical codes of various complexities. At this stage the parameters which must be treated as uncertain variables or functions are identified. The deterministic relation or the numerical code then serve as a transfer function which allows to determine the probabilistic characteristics of the output and the reliability of the structure.

Since we are looking for extremely high reliabilities, it is immediately understood, that the deterministic relations or numerical codes must be of supreme accuracy, in order to avoid "garbage in - garbage out" situation. But each deterministic relation is obtained based on simplifying assumptions which themselves have some specified degree of accuracy. Legitimate questions arise: can one use less accurate tools to determine the desired, extremely accurate probabilities of failure? What happens with the reliability of the structure whose design is based on the increasingly accurate theories? Do these invalidate previous designs?

Not less central is an input data. In most cases the accurate data is missing. Probabilistic analysts maintain that their theoretical analyses, that are presently devoid of the experimental inputs, still are of value; they maintain that once the data will become available, these can be introduced into the 'ready-steady' theoretical analyses.

But these analyses have taken the lives of their own: in order to get analytical results the researchers resort to various assumptions which allow for seemingly elegant solutions: the assumptions of normal distribution, or time-wise stationarity of the random process, or space-wise homogeneity of the random field, or of ergodicity of the involved process, or Markov property are almost invariably found in the exponentially growing literature. Often, however, these assumptions have no bearing to reality, and as an English biologist Thomas Henry Huxley maintains, what often occurs, when these theories are confronted with real-world situations, is "the great tragedy of science – the slaying of a beautiful hypothesis by an ugly fact." When confronted with ordinary life situations – these assumptions are failing and most of the theories are, therefore, inapplicable.

Thus the questions "What can go wrong with probabilistic methods? To use or not to use them?" becomes extremely relevant. Their various aspects have been addressed in monographs by Ben-Haim and Elishakoff (1990) and Ben-Haim (1996), and in some papers, see, e.g. by Elishakoff and Hasofer (1996). The necessity of the major revision in our understanding of the role of probabilistic methods is stressed in the monograph by Ben-Haim and Elishakoff (1990) and in the articles by Kalman (1994) and Elishakoff (1995).

In this article we elucidate inherent difficulties arising in the structural reliability due to imperfect information on the data in conjunction with the ever-increasing desire to allow only extremely small probabilities of failure.

Probabilistic Interpretation of Safety Factors

First of all we will address the following cardinal question:
Why may one need probabilistic methods? Once we elucidate this question, we will be better equipped to answer the Hamletian questions posed above. The reply to this question is best given in terms of design of a structure. In deterministic design process, one requires

$$\sigma \leq \sigma_y \tag{1}$$

where σ is an actual stress, whereas σ_y is an yield stress. But then one introduces uncertainty considerations into the picture by stating that we may know the loads precisely, or there may be an imprecision in measuring geometric parameters of the cross-sectional area, or we may have built an imperfect mechanical model to describe the behavior of the structure. Thus, the required safety factor k_{req} is introduced, and Eq. (1) is replaced by

$$\sigma \leq \frac{\sigma_y}{k_{req}} \tag{2}$$

Once the structure is designed one can introduce the actual safety factor

$$k_{act} = \frac{\sigma_y}{\sigma_{max}} \tag{3}$$

where σ_{max} is the maximum actual stress occurring in the structure. Thus the design requirement can be formulated as follows

$$k_{act} \geq k_{req} \tag{4}$$

In other words, the actual safety factor should not be less than the required one.
    Can one quantify the actual safety factor of the uncertainty is not hidden, but directly introduced into the scene? Let us attempt to answer this question. Let the force P acting on a tension-compression element with cross-sectional area have a Weibull distribution with the following probability distribution [9]

$$F_P(p) = 1 - \exp\left[-\left(\frac{p - p_0}{w - p_0}\right)^k\right], \ (k > 0, \quad w > p_0, \quad p \geq p_0) \tag{5}$$

For $p < P_0$, $F_P(p) \equiv 0$. We are interested in the reliability of the structure, ie. the probability that the structure will perform its intended mission satisfactorily. Such a performance is identified with holding relationship (1) true. Thus, reliability becomes:

$$R = \text{Prob}\left(\sum \leq \sigma_y\right) = \text{Prob}\left(P/b \leq \sigma_y\right)$$
$$= \text{Prob}\left(P \leq \sigma_y\, b\right) = F_P\left(\sigma_y\, b\right) \tag{6}$$

Thus

$$R = 1 - \exp\left[-\left(\frac{\sigma_{yb} - P_o}{w - P_o}\right)^k\right] \tag{7}$$

How to define the safety factor in the context of probabilistic design? One natural way is to relate it to some characteristic load, say the average load. The average load equals

$$E(P) = p_0 + (w - p_0)\,\Gamma\,(1 + 1/k) \tag{8}$$

where $\Gamma(x)$ is the Gamma function. Variance of the load is

$$V(P) = (w - p_0)\left[\Gamma(1 + 2/k) - \Gamma^2(1 + 1/k)\right] \tag{9}$$

Central safety factor is defined as the ratio of yield stress over the average stress $E(P)/b$:

$$k_c = \frac{\sigma_y b}{E(P)} = \frac{\sigma_y b}{p_o + (w - p_o)\,\Gamma(1 + 1/k)} \tag{10}$$

Let us design the structure probabilistically. Probabilistic design requires that the reliability be not less than a codified value r:

$$R \geq r \tag{11}$$

Thus, in view of Eq. (7) we get

$$1 - \exp\left[-\left(\frac{\sigma_y b - p_0}{w - p_o}\right)^k\right] \geq r \tag{12}$$

The design value of the cross-sectional area b_{design} is found from an equality $R = r$, and becomes:

$$b_{\text{design}} = \frac{P_o + (w - P_o)\,[\ell n\,1/(1 - r)]^{1/k}}{\sigma_y} \qquad (13)$$

Substitution of this expression Eq. (10) allows us to explicitly write the central safety factor in terms of the codified required reliability r:

$$k_c = \frac{P_o + (w - p_o)\,[\ell n\,1/(1 - r)]^{1/k}}{P_o + (w - p_o)\,\Gamma\,(1 + 1/k)} \qquad (14)$$

Consider as an example, the following set of parameters:

$$w = 3p_o,\ k = 4 \qquad (15)$$

Then

$$k_c = \frac{1 + 2\ell n\,[1/(1 - r)]^{1/k}}{1 + 2\Gamma\,(1.25)} \qquad (16)$$

For $r = 0.9$, $k_c = 1.2314$; for $r = 0.99$, $k_c = 1.3971$; for $r = 0.999$, $k_c = 1.5082$; for $r = 0.9999$, $k_c = 1.5942$; for $r = 0.99999$, $k_c = 1.6653$; for $r = 0.999999$, $k_c = 1.7263$, etc. As we see, probabilistic model allows associate the required reliability directly with the safety factor. We learn several lessons from this simplest example:

(1) Concept of the safety factor, which is so often criticized by practitioners and researchers alike, is, in actuality, a powerful concept, which could be given a probabilistic interpretation.

(2) Within probabilistic interpretation, safety factor is not a mysterious quantity determined by the will of the designer, for whom this may a "personal factor of safety." If one can quantify, say through legislation, the required reliability, one can then also quantify the celebrated safety factors.

HOW RELIABLE ARE RELIABILITY CALCULATIONS?

Let us investigate now how a small error may affect the reliability calculations. Let us assume that instead of parameters w, p_0 and k we have measured other values, w_1, p_1 and k_1, respectively. The analysis, however was performed for values w, p_0 and k. We may ask ourselves: What is the actual reliability R_{act}, or actual probability of failure $P_{f,act}$?

$$P_{f,act} = 1 - R_{act} \qquad (17)$$

Actual reliability is given by Eq. (7) with substituted actual values of the parameters

$$R_{act} = 1 - \exp\left[-\left(\frac{\sigma_y b - p_1}{w_1 - p_1}\right)^{k_1}\right] \qquad (18)$$

actual probability of failure being

$$P_{f,act} = \exp\left[-\left(\frac{\sigma_y b - p_1}{w_1 - p_1}\right)^{k_1}\right] \qquad (19)$$

Design of the structure has been performed using the values w, p_0 and k, respectively. The appropriate value of the cross-sectional area is given by Eq. (13). To calculate the actual probability of failure, corresponding to the design value, b_{design}, we substitute the expression b_{design} into Eq. (19) to result in

$$P_{f,act} = \exp\left[-\left(\frac{\sigma_y b_{design} - p_1}{w_1 - p_1}\right)^{k_1}\right] \qquad (20)$$

or

$$P_{f,act} = \exp\left[-\left(\frac{p_0 - p_1 + (w - p_0)\,[\ell n\,1/(1-r)]^{1/k}}{w_1 - p_1}\right)^{k_1}\right] \qquad (21)$$

Let us investigate some particular cases. In the simplest case $p_1 = p_0$, $w = w_1$, but $k_1 \neq k$. Then

$$P_{f,act} = \exp\left[-\left(\ell n\,\frac{1}{1-r}\right)^{\frac{k_1}{k}}\right] \qquad (22)$$

Since r is the required reliability, $1 - r$ is recognized in

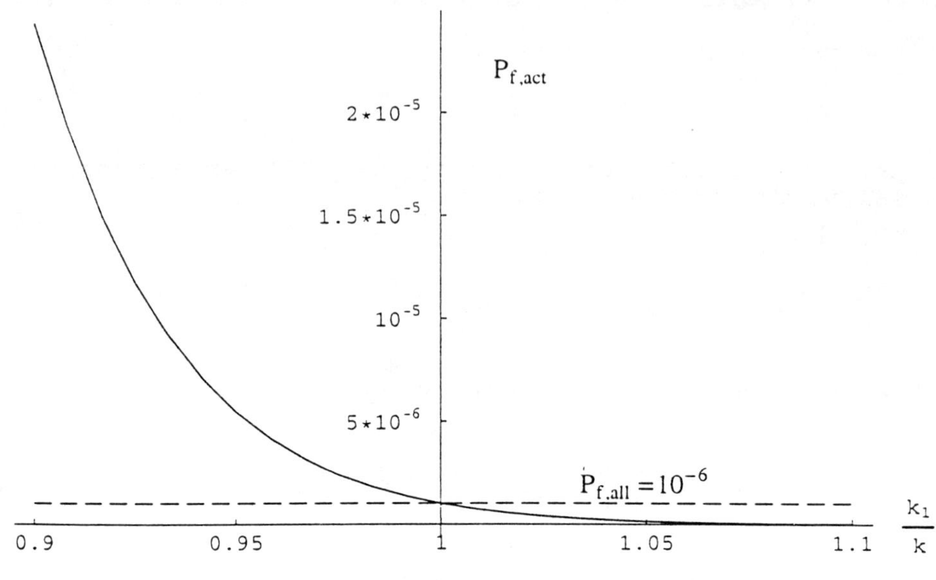

Fig. 1

Actual probability of failure as a function of the ratio k_1/k : for $k_1 = k$ the actual probability of failure coincides with the required one; for $k_1/k>1$, the actual probability of failure is less than the required one, whereas for $k_1/k<1$, the actual probability of failure may well exceed the allowed value, resulting in a detrimental state.

Eq. 22 as the allowed probability of failure $P_{f,all}$. Thus, Eq. (22) can be rewritten as:

$$P_{f,act} = \exp\left[-\left(\ell n\ \frac{1}{P_{f,all}} \right)^{k_1/k} \right] \qquad (23)$$

Let $P_{f,all} = 10^{-6}$. Then

$$P_{f,act} = \exp\left[-13.81551056^{k_1/k} \right] \qquad (24)$$

This can be viewed as a function of the ratio k_1/k. This function is depicted in Fig. 1. As is seen, when

$$\frac{k_1}{k} = 1,\ P_{f,act} = P_{f,all} = 10^{-6}$$

Yet, if $k_1 \neq k$, the actual probability of failure may differ from the allowed one. It is remarkable that there could be a serendipitous situation, i.e. when an error in measurement of k_1 may be of a "favorable" nature: for $k_1/k>1$, the actual probability of failure is less than the allowable one. Yet, when $k_1/k<1$, the effect of a small error in evaluating k may be detrimental. If $k_1/k = 0.95$, the actual probability of failure is about five times larger than the allowed one. If $k_1/k = 0.93$, the actual probability of failure is approximately 10 times larger than the one which was permitted! We conclude that the probability of failure is too sensitive a parameter to allow for an imprecise input characteristics. We conclude that the accurate determination of the input's probabilistic characteristics must become an integral part of the rigorous probabilistic analysis. Paradoxically, probabilistic engineers advocate a detachment from experiments, since they neither try to validate their assumptions via experiments, nor introduce the study of the errors in their analyses. Yet, mathematicians do not advocate such a detachment; as Richard Bellman (1962) mentions

"We postulate then that one of the basic responsibilities of the mathematician is to examine the structure of the problems of society and to provide mathematical formulations which are easily susceptible to numerical solution in terms of the current technology. This means not only an examination of the

*computational aspects, but also of the experimental
aspects. What information is required for which
formulation, what sensing devices are available, what
accuracy do they possess, what should be measured when,
and so on?"*

Engineers dealing with uncertainty are well advised to
follow these recommendations by a mathematician.

EFFECT OF A SMALL DEVIATION IN THE PROBABILITY DENSITY

We consider a fundamental problem in structural
reliability, namely a bar with a cross-sectional area b and
subjected to a load N, that is a random variable with the
probability density f_P (p), where p is a possible value the
load may take on. The following assumption is made for
facilitating a simple analysis: The bar's material is
perfectly elastic in compression, but has a yield stress in
tension σ_y. We also assume that the bar cannot lose its
stability, or have any other form of failure. The
reliability of the structure is determined as the probability
that the stress $\Sigma = P/b$ does not exceed the value of the
yield stress:

$$R = \text{Prob} \left(\sum \leq \sigma_y \right) \tag{25}$$

In other words, let us visualize the situation in which the
analyst has assumed the probability density of the load as
that of the log-normal variable (Elishakoff, 1983):

$$f_P(p) = \begin{cases} \dfrac{1}{p\sigma_P \sqrt{2\pi}} \exp\left[\dfrac{-(\ell np - a)^2}{2\sigma_P^2} \right], & \text{for } p > 0 \\ 0, & \text{otherwise} \end{cases} \tag{26}$$

The parameters a and σ_F characterize the probability
density, namely, the mean value $E(P)$ and the variance $V(P)$
are expressed as:

$$E(P) = \exp\left(a + \frac{1}{2} \sigma_P^2 \right)$$

$$V(P) = \exp\left(2a + \sigma_P^2 \right) [\exp(\sigma_P^2) - 1] \tag{27}$$

We find the reliability of the structure:

$$R = \text{Prob } (P \le \sigma_y b) = \text{Prob } (\ell nP \le \ell n \, \sigma_y b)$$

$$= \frac{1}{2} + \text{erf} \left(\frac{\ell n \, \sigma_y \, b - a}{\sigma_P} \right) \tag{28}$$

$$\text{erf}(x) = \frac{1}{\sqrt{2\pi}} \int_0^x \exp(-t^2/2) \, dt$$

since ℓnP is a normal variable with mean "a", and variance σ^2.

Let us visualize now that the actual probability density slightly differs from the true one, and reads, for $p>o$:

$$f_P^{(\epsilon)}(p) = \frac{C}{p \, \sigma \sqrt{2\pi}} \exp\left[- \frac{\ell np - a^2}{2\sigma_P^2} \right] \quad , C = \{1 + \epsilon \sin[2\pi \, (\ell np - a)] \tag{29}$$

where ϵ is a constant, belonging to an interval $[-1,1]$. For $p<o$, the probability density vanishes. It can be shown that $f_P^{(\epsilon)}(p)$ is indeed a probability density. It is non-negave and satisfies the equality

$$\int_{-\infty}^{\infty} f_P^{(\epsilon)}(p) \, dp = 1 \tag{30}$$

In order to prove this property, we have to demonstrate that

$$\int_{-\infty}^{\infty} f_P(p) \, \sin[2\pi(\ell np - a)] \, dp = 0 \tag{31}$$

To do this we make a substitution $\ell np = u$. Thus, we should calculate the integral

$$I = \int_{-\infty}^{\infty} \frac{1}{\sigma\sqrt{2\pi}} \exp\left[- \frac{(t-a)^2}{2\sigma^2} \right] \sin[2\pi(t-a)] \, dt \tag{32}$$

Further substitution $t-a = u$ leads to

$$I = \int_{-\infty}^{\infty} \frac{1}{\sigma\sqrt{2\pi}} \exp\left(-\frac{u^2}{2\sigma^2}\right) \sin(2\pi u)\, du \tag{33}$$

which vanishes since the integral is an odd function. Thus, the function defined in Eq. (29) represents a probability density of some random variable, denoted $P^{(\varepsilon)}$. Obviously, if $\varepsilon=0$, $f^{(\varepsilon)}$ and $P^{(\varepsilon)}$ are, respectively, f and P defined in the beginning of this section. Stoyanov (1987) demonstrates that for any $k = 1,2,\ldots$, we have

$$E\left(P^{(\varepsilon)}\right)^k = E\left(P^k\right) \tag{34}$$

i.e. the "perturbed" random variable $P^{(\varepsilon)}$ has the same moments as those of a "unperturbed" random variable P. Let us calculate now the true reliability associated with the random variable $P^{(\varepsilon)}$:

$$R = \mathrm{Prob}\,(P \le \sigma_y b) = \int_0^{\sigma_y b} f_P^{(\varepsilon)}(p)\, dp$$

$$= \int_0^{\sigma_y b} \frac{1}{p\,\sigma\sqrt{2\pi}} \exp\left[-\frac{(\ell np - a)^2}{2\sigma^2}\right] \times \tag{35}$$

$$\times \{1 + \epsilon \sin[2\pi(\ell np - a)]\}\, dp$$

or

$$R = \int_0^{\sigma_y b} \frac{1}{p\,\sigma\sqrt{2\pi}} \exp\left[-\frac{(\ell np - a)^2}{2\sigma^2}\right] \times \tag{36}$$

$$\times \{1 + \epsilon \sin[2\pi(\ell np - a)]\}\, dp$$

Introducing again the new variable of integration $\ell np = t$, we reduce reliability to:

$$R = \int_{-\infty}^{\ell n\,\sigma_y b} \frac{1}{\sigma\sqrt{2\pi}} \exp\left[-\frac{(t-a)^2}{2\sigma^2}\right] \{1 + \epsilon \sin[2\pi(t-a)]\}\, dt \tag{37}$$

with $t - a = u$, we get

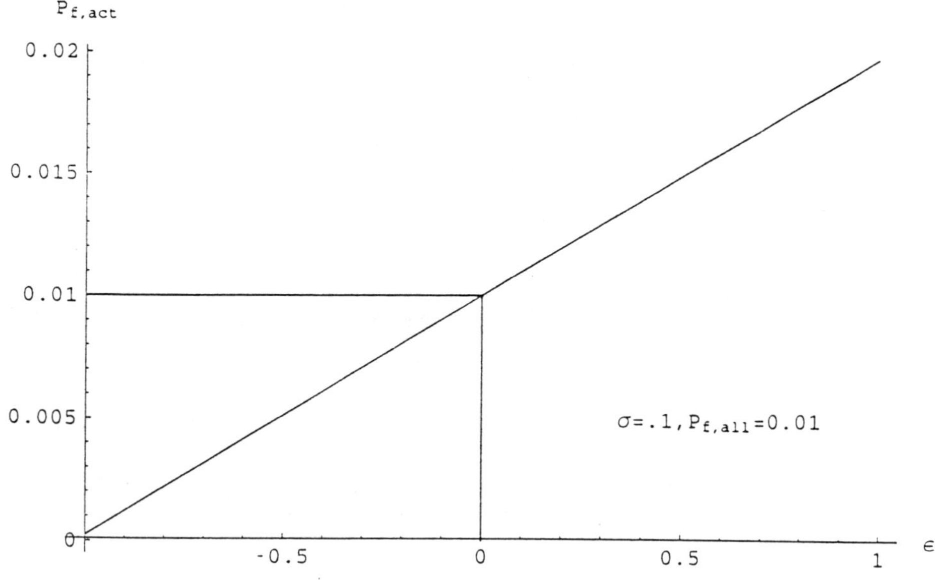

Fig. 2

Actual probability of failure as a function of the control parameter ϵ; the allowed probability of failure is set at 0.01.

$$R = \int_{-\infty}^{\ln\sigma_y b - a} \frac{1}{\sigma\sqrt{2\pi}} \, \exp\left(-\frac{u^2}{2\sigma^2}\right) \{1 + \epsilon \sin[2\pi \, (t-a)]\} \, dt \tag{38}$$

$$= \frac{1}{2} + \mathrm{erf}\left(\frac{\ln\sigma_y b - a}{\sigma}\right) + \frac{\epsilon}{\sigma\sqrt{2\pi}} \int_{-\infty}^{\ln\sigma_y b - a} \exp\left(-\frac{u^2}{2\sigma^2}\right) \sin(2\pi u) \, du$$

Evaluation of this integral yields

$$R = \frac{1}{2} + \mathrm{erf}\left(\frac{\ln\sigma_y b - a}{\sigma}\right) + \frac{\epsilon D}{\sigma\sqrt{2\pi}} \left\{ - e^{2\pi^2\sigma^2} \sqrt{\frac{\pi}{2}} \right. \tag{39}$$

where

$$D = \left(\mathrm{erfi}\left[\frac{2\pi\sigma^2 - i(a + \ln[\sigma_y b])}{\sqrt{2}\sigma}\right] + \mathrm{erfi}\left[\frac{2\pi\sigma^2 + i(-a + \ln[\sigma_y b])}{\sqrt{2}\sigma}\right] \right) \tag{40}$$

where erfi(z) is an imaginary error function, defined as (Wolfram, 1996)

$$\mathrm{erfi}(z) = \frac{\mathrm{erf}(iz)}{i} \tag{41}$$

where $i = \sqrt{-1}$. Let us visualize that the reliability calculations have been performed so that the probability density of the load was postulated to equal $f_P^{(0)}(p)$. Let required reliability be r, so that the design was performed using expression (28). Figures 2-4 are associated with $r = 0.99$, $r = 0.999$ and $r = 0.9999$, respectively. The figures depict the actual situation in which the parameter ε describing perturbation from reality does not vanish identically. Thus, actual probability of failure $P_{f,act}$ varies with the control parameter ε. Naturally, when $\varepsilon=0$, the actual probability of failure takes a value $1-r$, or 0.01, 0.001 and 0.0001, respectively, in Figures 2-4. An integrand in the integral (36) is an oscillatory function. Therefore, depending on the value of σ, the value of the integral is either positive or negative. Hence different behaviors are exhibited in Figs. 2-4. Figures 2-4 demonstrate that one can make "good" errors, i.e. deviation of the model from the reality may turn to be on the safe side. Indeed, if in Fig. 2, the control parameter ε tends to -1 from above, the actual probability of failure tends to zero. Likewise, in Fig. 3,

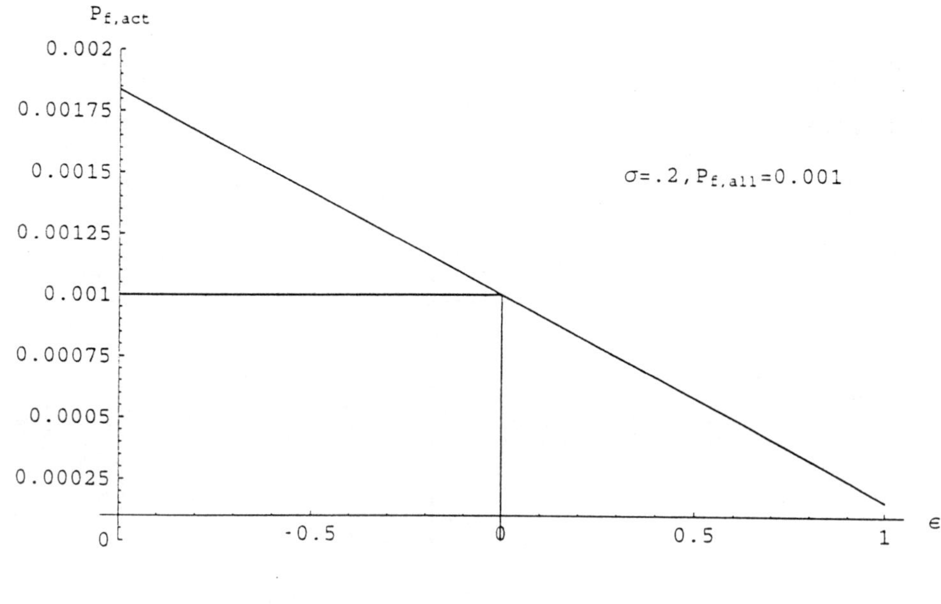

Fig. 3

Actual probability of failure as a function of the control parameter ϵ; the allowed probability of failure is fixed at 0.001.

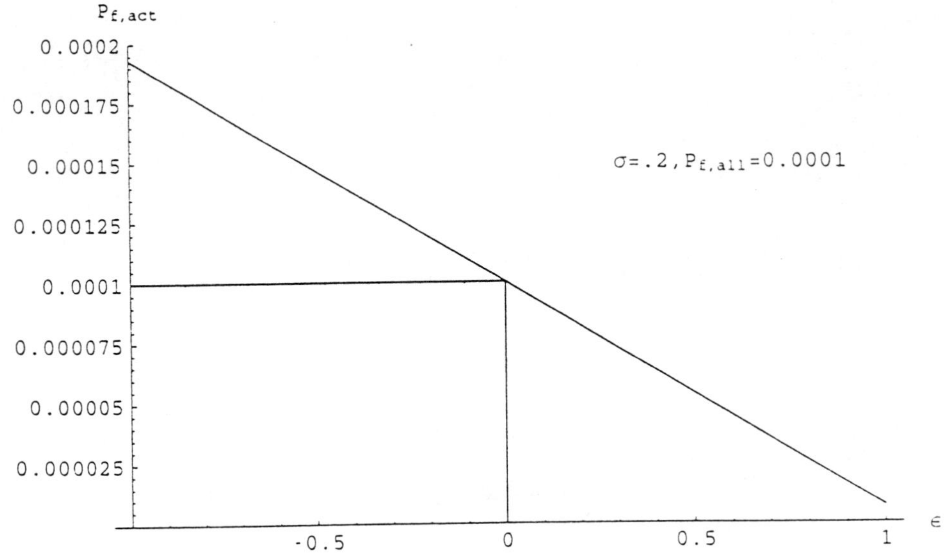

Fig. 4

Actual probability of failure as a function of the
control parameter ϵ; the allowed probability of failure
is fixed at 0.0001.

associated with $\sigma = 0.2$ and $P_{f,all} = 10^{-3}$, if ε reaches unity from below, the actual reliability too will reach unity. Analogously, in Fig. 4, with $\sigma = 0.2$ and $P_{f,all} = 10^{-4}$, with ε approaching unity, from below, the actual probability of failure will approach zero.

It is prudent, however, to look what can go wrong, when the theoretical model differs from the reality. As Fig. 2 demonstrates, when ε tends to unity, the actual probability of failure becomes twice as much as the allowable one. This implies, through the use of the frequency interpretation of the probability notion, that the number of unsuccessful performances may be double of the allowed quantity. Nearly analogous situations are portrayed in Fig. 3 and 4 when ε approaches -1.

It must be stressed that bad things happen to good probabilistic analyses for different parameters of ε: sometimes the unfavorable situation occurs when ε approaches unity (Fig. 2), whereas in other circumstances the detrimental effect takes place in the vicinity of $\varepsilon=-1$.

CONCLUSION

The leitmotif of this paper is as follows: Beware of uncritical use of the probabilistic methods when extremely reliable performance of the structure is called for! A colleague of the present writer, in his lecture on the safety of the nuclear power plants conjectured that the probability of failure had to be of order 10^{-42}. As this study shows the accuracy of the prediction in the absurdly small probabilities of failure may turn to be illusory!

Our conclusions are in agreement with the statement made by Ekeland (1993): "Even in circumstances in which there are no material interests at stake the probabilities calculated by competent people in good faith may turn out to be systematically underrated." This paper quantifies Ekeland's qualitative statement in terms of simple examples. We still fall short of the far-going recommendation made by Kalman (1994): "Define randomness without probability." Perhaps, a less antagonistic statement would read "Define uncertainty without probability." Still Kalman's idea must become a focus of inquiring thoughts of both probabilistic and non-probabilistic analysts of uncertainty. In this study we have demonstrated the existence of possibly dangerous "underground stones" for those who want to "swim" in the sea of randomness and probability.

ACKNOWLEDGEMENT

Author greatfully appreciates partial support of the National Science Foundation through the SGER grant (Dr. K.P. Chong, Program Director). Help provided by Nicola Impollonia and Ruijia Mu in carrying numerical calculations is appreciated.

REFERENCES

Bellman R., The Roles of the Mathematician in Applied Mechanics, U.S. National Congress in Applied Mechanics, ASME Press, New York, pp. 195-204, 1962.

Ben-Haim Y., *Robust Reliability in the Mechanical Systems*, Springer Verlag, Berlin, 1996.

Ben-Haim Y., Convex Models of Uncertainty: Applications and Implications, *Erkenntnis: An International Journal of Analytic Philosophy*, Vol. 41, 139-156, 1994.

Ben-Haim Y. and Elishakoff I., *Convex Models of Uncertainty in Applied Mechanics*, Elsevier, Amsterdam, 1990.

Cornell C.A., Structural Safety : Some Historical Evidence That Is a Healthy Adolescent, *Structural Safety and Reliability*, (T. Moan and M. Shinozuka, eds.), Elsevier, Amsterdam, pp. 19-20, 1981.

Drenick R.F., On a Class of Non-Robust Problems in Stochastic Dynamics, *Stochastic Problems in Dynamics*, (B.L. Clarkson, ed.), Pitman, pp. 237-255, 1977.

Elishakoff I., *Probabilistic Methods in the Theory of Structures*, Wiley, New York, 1983.

Elishakoff I., Essay on Uncertainties in Elastic and Viscoelastic Structures: from A.M. Frendenthal's Criticisms to Modern Convex Modeling, *Computers and Structures*, Vol. 56, 871-895, 1995.

Elishakoff I. and Hasofer A.M., Detrimental or Serendipitous Effect of Human Error on Reliability of Structures, *Computer Methods in Applied Mechanics and Engineering*, Vol. 129, 1-7, 1996.

Elishakoff I. and Nordstrand T., Probabilistic Analysis of Uncertain Eccentricities in a Model Structure, Proceedings of Sixth International Conference on Applications of Statistics and Probability in Civil Engineering (L. Esteva and S.E. Suiz, eds.), Vol. 1, 250-256, Mexico City, 1991.

Ekeland I., *The Broken Dice and* Other Mathematical *Tales of Chance*, The University of Chicago Press, p. 142, 1993.

Freudenthal A.M., Introductory Remarks, *Proceedings, International Conference on Structural Safety and Reliability*, (A.M. Freudenthan, ed), Pergamen Press, Oxford, pp. 5-6, 1972.

Freudenthal A.M., Fatigue Sensitivity and Reliability of Mechanical Systems, Especially Aircraft Structures, WADD Technical Report 61-53, Wright Patterson, AFB, Ohio 1961.

Freudenthal A.M., Safety and Probability of Structural Failure, *Transactions ASCE*, Vol. 121, 1337-1375, 1956.

Kalman R.E., Randomness Re-examined, *Modeling, Identification and Control*, Vol. 15, 141-151, 1994.

Mayer M., *Die Sicherheit des Bauwerke und Ihre Berechnung nach Grenzkräften anstatt nach Zulassigen Spannungen*, Springer Verlag, Berlin, 1926 (in German).

Stoyanov J., *Counterexamples in Probability*, Wiley, Chichester, pp. 89-91, 1987.

Weibull W., A Statistical Theory of the Strength of Materials, Proceedings of Royal Swedish Institute for Engineering Research, Stockholm, No. 151, 1939.

Wentzel E.C., *Operations Research: Problems, Principles, Methodology*, Nauka, Moscow, 1980 (in Russian).

Wolfram S., *The Mathematica Book*, Third Edition, Cambridge University Press, New York, p. 745, 1996.

ARE PROBABILISTIC AND ANTIOPTIMIZATION METHODS

INTERRELATED?

Isaac Elishakoff
Department of Mechanical Engineering
Florida Atlantic University
Boca Raton, FL 33431-0991

ABSTRACT

This paper is devoted to a single objective: to find, if any, an interrelation between seemingly totally different analyses of uncertainty: the classical probabilistic method and the antioptimization method. The former is based on the concept of reliability of the structure, whereas the latter utilizes the idea of the maximum, least favorable response, or minimum, least favorable buckling load. Antioptimization method is an antidote of the optimization method, where the structure is designed in such a manner as to make for the minimum, most favorable response. Extremely simple examples are chosen in order the reader to follow them without unnecessary complications. These examples are analytically evaluated in order to elucidate an interrelation between the probabilistic and anti-optimization methods.

INTRODUCTION

Recent decades witnessed an explosive interest in probabilistic methods in the various fields of engineering. Some researchers even talk about "probabilistic revolution" (Morgan et al, 1987).   Others are extremely skeptical (Kalman, 1994).   Some researchers propagate alternative methods of uncertainty analyses. For example, fuzzy sets gained a wide popularity, especially due to items of popular and mass consumption, (camcorders, rice cooking machines, refrigerators, various control systems etc.), developed, as their researchers maintain, utilizing fuzzy logic, in Japan and elsewhere.

Yet this method is also criticized quite energetically. The common feature of these two methods is presence of the measure.   Namely, probabilistic approach necessitates a knowledge of the probability density function. If X is a random variable, then the probability density $f_X(x)$ defines a probability $P(a,b)$ that random variable X will take on values in an interval (a,b); this probability is found as the integral

$$P(a,b) = \int_a^b f_X(x)\,dx \qquad (1)$$

Fuzzy sets based methods involve the notion of the membership function which represents the degree of belonging to a specific set.

As Kunzevich and Lychak (1992) maintain,

"The assumption of stochastic nature of uncertainty cannot be accepted at least in two cases : (a) when the volume of a priori experimental data on the nature of the uncertain factors is so small that it does not allow the conclusion on the availability of stable statistic characteristics; (b) when it is known a priori that the uncertainty basically cannot be considered to be produced by some probabilistic mechanisms."

Kalman (1984) stresses, that:

"it would be a great untruth to claim that ... the whole uncertainty arises by virtue of the mechanisms of statistical choice. The nature does not confirm to the rules of traditional probability."

Analogously, according to Israeli philosopher Leibovitch (1992):

"The nature did not attend the course of the theory of probability."

The idea of fuzzy sets apparently was originated by Black (1937) but was energetically developed, apparently independently of Black, by Zadeh (1965). As Kosko (1993) puts it "Black saw that everything is to some degree *A* and to some degree not *A*. Everything is red and not-red, large and not-large, smooth or non-smooth." Zadeh's (1965) considerations were based upon his principle of incompatibility:

"As the complexity of a system increases, our ability to make precise and yet significant statements about its behavior diminishes until a threshold is reached beyond which precision and significance (or relevance) become almost mutually exclusive characteristics."

Additional idea, propagated by Zadeh is the imprecision of the human reasoning itself:

"Almost all human reasoning is approximate in nature. It is the approximate rather than exact reasoning that accounts for the remarkable human ability to recognize distorted speech, decipher sloppy handwriting and, more generally, make rational decisions in an environment of uncertainty and imprecision."

By now there are well over 5,000 publications in the fuzzy-sets-based analyses. But it did not escape criticism as sharp as the one enjoyed by the probabilistic methods. Kosko (1993) quotes from R. E. Kalman:

"Fuzzification is a kind of scientific permissiveness. It tends to result in socially appealing slogans unaccompanied by the discipline of hard scientific work and patient observation."

According to W. Kahan:

"Fuzzy theory is wrong, wrong, and pernicious. What we need is more logical thinking, not less. The danger of fuzzy logic is that it will encourage the sort of imprecise thinking that has brought us so much trouble. Fuzzy logic is the cocaine of science."

Moreover, W. Kahan maintains: "I can not think of any problem that could not be solved better by ordinary logic." Kosko (p.3, 1993) also cites M. Tribus:

> *"Fuzziness probability in disguise. I can design a controller with probability that could do the same thing that you could do with fuzzy logic."*

Finally, Cox (1995) stresses:

> *"Fuzzy systems are not a panacea for the problems encountered in the real world of information and process modeling. They have limitations and representational constraints that make them difficult to use them in some situations. Further, some developers, unfamiliar with fuzzy logic, believe this technology is a way of handling missing data or a way of calculating confidences from arbitrary noisy or missing data."*

As is clearly detected, this criticism is highly correlated with Wentzel's (1980) criticisms of probabilistic methods. She mentions that the probabilistic methods are:

> *"Frequently regarded as a kind of magic wand which produces information out of a void. This is a fallacy; the theory of probability only enables information to be transformed, and conclusions on in accessible phenomena to be drawn from data on observable ones."*

The above quotations are not reproduced here in order to agree with both of the quarrelling groups and then proceed to the third, alternative treatment of the uncertainty. Above thoughts are recapitulated here in order to stress that neither the concept of probability nor the notion of fuzziness, are universally accepted tools in the uncertainty analysis.

The third group of researchers notice the similarity between the probability methods and those based on the concept of fuzziness. These are both defined on the measure of belonging to the set of variation of the uncertain variable. Often a rigorous definition of such a measure is either cumbersome or unfeasible, especially in the absence of the data. If we do not know the data, we are still not allowed to treat them as random or fuzzy. In these circumstances the ideas of guaranteed estimates of the system's response, irrespective of the measure of belonging, appear to be extremely attractive. The first book on this topic was published by Schweppe (1973). He discussed the unknown-but bounded uncertainty modeled by set-theoretic models. Schweppe (p.2, 1973) defined such scaler functions $x(t)$ that

(a) $x(t)$ is bounded for all t, $0 \le t \le T$; for example, $|x(t)| < 1$.

(b) $x(t)$ must satisfy some integral constraints, for example

$$\int_0^T x^2(t)\,dt \le 1 \qquad (2)$$

(c) the spectrum of $x(t)$ is considered as constrained

$$|X(\omega)|^2 < \epsilon, \text{all } \omega > \omega_0, \omega < -\omega_0$$

$$X(\omega) = \int_0^T x(t)\,e^{-i\omega t}dt \qquad (3)$$

Schweppe (1973) pioneered an ellipsoidal approach in which uncertain variables belong to ellipsoidal sets

$$(x - m)^T \Omega^{-1} (x - m) \le 1 \qquad (4)$$

where Ω is a positive-definite $n \times n$ matrix, m = center, x and m are n - dimensional vectors. It was noted that all these inequalities (Ben-Haim and Elishakoff, 1990) qualify $x(t)$ as set of convex functions. Therefore, the appropriate modeling was dubbed as a *convex modeling of uncertainty*.

Yet, it was realized by the present writer that uncertain sets may also be non-convex. In these circumstances one should utilize the nonlinear programming methods for numerical evaluation of the least favorable responses (Li, et al, 1996). There, it appeared, that the method must convey an information not necessarily on the nature of the underlying sets (convex or non-convex), but, rather, the physical idea behind it. The idea is somewhat opposite to the classical optimization, where an engineer tries to design the structure which performs in the best way, depending on the definition of the performance criterion. Yet, when uncertainty is involved, with uncertainty specified as belonging to a set, one is looking for worst scenario under given constraints. Thus, the term *antioptimization* was coined by the present writer (1990). The combined optimization and anti-optimization were studied by Elishakoff et al (1994b).

It is natural to attempt to contrast various uncertainty analyses on identical problems.

This paper was motivated, inter alia, by a question posed by one of the organizers of the Lambrecht Conference on Uncertainty Modeling held in 1996; after present writer made a statement that the probabilistic and convex modeling techniques must be extensively contrasted (as it was done by Elishakoff et al, 1994a), the following question was posed: "Why do we have to compare totally different techniques?" In terminology of Shakespeare, "comparisons are odious," but in all probability the above quotation does not apply to different scientific methods. Indeed, if we would not compare various techniques dealing with the same problem how would we conclude which of the techniques is more effective, or at least more computation-friendly? We can establish some meaningful inner connection between different methods through comparison only. This thought guided present writer in this study, which attempts to extensively contrast the performance of probabilistic and antioptimization methods.

A BAR WITH RANDOM MODULUS OF ELASTICITY

In order to contrast probabilistic method with that of anti-optimization, let us construct simplest cases that will allow us not to overlook a forest when observing the trees. We will start with a one-dimensional example. Consider a bar of length L, modulus of elasticity E and moment of inertia I. The bar is subjected to an axial force P. The bar is simply supported at its ends. Let the material of the bar be perfectly elastic. The classical buckling load of such a bar is found via classical Euler formula

$$P_{c\ell} = \frac{\pi^2 EI}{L} \tag{5}$$

Let now the modulus of elasticity be a random variable with known probability density which is a truncated normal

$$f_E(e) = \frac{A}{\sigma_1\sqrt{2\pi}}\exp\left[-\frac{(e - e_0)}{2\sigma_1^2}\right], \text{ for } e_L \le e \le e_U \tag{6}$$

where e_L = lower value the modulus of elasticity can take, e_U = upper value. Outside interval $[e_L, e_U]$ probability density vanishes. Normalization requirement yields

$$A_1 = \left[erf\left(\frac{e_U - e_0}{\sigma_1}\right) - erf\left(\frac{e_L - e_0}{\sigma_1}\right) \right]^{-1} \tag{7}$$

Average value of the modulus of elasticity reads

$$M(E) = e_0 + B\sigma_1 \tag{8}$$

$$B_1 = \frac{1}{A_1}\left\{ \exp\left[-\frac{(e_L - e_0)^2}{2\sigma_1^2}\right] - \exp\left[-\frac{(e_U - e_0)^2}{2\sigma_1^2}\right]\right\} \tag{9}$$

whereas variance equals

$$\frac{V(E)}{\sigma^2} = 1 - B^2 - A_1\left\{\frac{(e_U - e_0)}{\sigma_1\sqrt{2\pi}}\exp\left[-\frac{(e_U - e_0)^2}{2\sigma_1^2}\right]\right.$$

$$\left. -\frac{(e_L - e_0)}{\sigma_1\sqrt{2\pi}}\exp\left[-\frac{(e_L - e_0)^2}{2\sigma_1^2}\right]\right\} \tag{10}$$

Note that if $e_L \to -\infty$ and $e_U \to +\infty$ then $A \to 1$ and E becomes a normally distributed random variable with $M(E) = e_0$ and $V(E) = \sigma_1^2$. In Eq. (8) M denotes the mathematical expectation.

We are interested in calculating the reliability of the structure.

$$R = Prob\left(P_{c\ell} > \lambda\right) \tag{11}$$

i.e. probability that the buckling load will exceed some preselected load level. We have

$$R = 1 - Prob\left(P_{c\ell} \leq \lambda\right)$$

$$= 1 - Prob\left[\frac{\pi^2 EI}{L^2} \leq \lambda\right] \tag{12}$$

$$= 1 - Prob\left[E \leq \frac{\lambda L^2}{\pi^2 I}\right] = 1 - F_E\left(\frac{\lambda L^2}{\pi^2 I}\right)$$

where F_E is the probability distribution function of the elastic modulus. The latter function equals

$$F_E(e) = A_1\left[erf\left(\frac{e - e_0}{\sigma_1}\right) - erf\left(\frac{e_L - e_1}{\sigma_1}\right)\right] \tag{13}$$

This function vanishes for $e \leq e_L$, and equals unity for $e > e_U$. Consider the case when

$$e_L < \frac{\lambda L^2}{\pi^2 I} \, e_U \tag{14}$$

Then

$$R = 1 - A_1\left\{erf\left[\frac{1}{\sigma_1}\left[\frac{\lambda L^2}{\pi^2 I} - e_0\right]\right] - erf\left[\frac{1}{\sigma_1}(e_L - e_0)\right]\right\} \tag{15}$$

Let us now design the bar, so that the reliability is not less than the codified reliability r:

$$R \geq r \tag{16}$$

This, we have to find the allowable value $\left(L^2/I\right)_{design}$ such that

$$r = 1 - A_1\left\{erf\left[\frac{1}{\sigma_1}\left[\frac{\lambda L^2}{\pi^2 I} - e_0\right]\right] - erf\left[\frac{1}{\sigma_1}(e_L - e_0)\right]\right\} \tag{17}$$

This leads to

$$\left[\frac{L^2}{I}\right]_{design} = \frac{\pi^2 e_0}{\lambda} + \frac{\pi^2 \sigma_1}{\lambda} erf^{-1}\left\{erf\left[\frac{1}{\sigma_1}(e_U - e_0)\right]\frac{r}{A_1}\right\} \tag{18}$$

The following question arises: Does the geometric characteristic L^2/I have a limit when we demand increasingly high reliability tending to unity?

The answer to this question can be directly deduced from Eq. (15). When r tends to unity the expression

$$A_1\left\{erf\left[\frac{1}{\sigma_1}\left[\frac{\lambda L^2}{\pi^2 I} - e_0\right]\right] - erf\left[\frac{1}{\sigma_1}(e_L - e_0)\right]\right\} \to 0 \tag{19}$$

Since A_1 is a non-zero constant, the expression in parenthesis tends to zero, or, alternatively

$$erf\left[\frac{1}{\sigma_1}\left[\frac{\lambda L^2}{\pi^2 I} - e_0\right]\right] \rightarrow erf\left[\frac{1}{\sigma_1}\left(e_L - e_0\right)\right] \qquad (20)$$

or

$$\frac{\lambda L^2}{\pi^2 I} \rightarrow e_L \qquad (21)$$

and for absurdly small probabilities of failure, the geometric parameter $\left(L^2/I\right)_{design}$ is found as

$$\left[\frac{L^2}{I}\right]_{design} = \frac{\pi^2 e_L}{\lambda} \qquad (22)$$

As is clearly seen, when the probability of failure tends to zero, the analysis should be performed in according to the lowest value the elastic modulus can take. We will now illustrate the opposite case, in which the design will be based on the upper value taken by the random variable.

Fig. 1 depicts variation of $\left(L^2/I\right)_{design}$ for a bar with following characteristics $e_L = 60°$ GPa , $e_L = 60°$ GPa ,

$e_0 = 70$ GPa, $\sigma = $ GPa, $\lambda = 400 kN$. Then $\left(L^2/I\right)_{design}$ when r tends to unity, reaches the value 1.48, as demonstrated in Fig. 1.

A BAR WITH RANDOM LOADING

Consider now the case when the applied load is a random variable, with truncated normal density

$$f_p(p) = \frac{A_2}{\sigma\sqrt{2\pi}}\exp\left[-\frac{(p - p_0)}{2\sigma_2^2}\right], \ p_L \leq p \leq p_U \qquad (23)$$

and zero outside the interval (p_L, p_U). The mathematical expectation $M(P)$ and the variance $V(P)$ are defined by formulas analogous to Eqs. (8)-(10) in which one should replace e_0 by p_0, e_L by p_L, e_U by p_U, σ_1 by σ_2, A_1 by A_2, B_1 by B_2. The yield stress in tension is a deterministic quantity. The material is perfectly elastic in compression. In the new circumstances the reliability reads

$$R = Prob\left(P \leq \sigma_y b\right) \qquad (24)$$

where b is the cross sectional area. Reliability reads

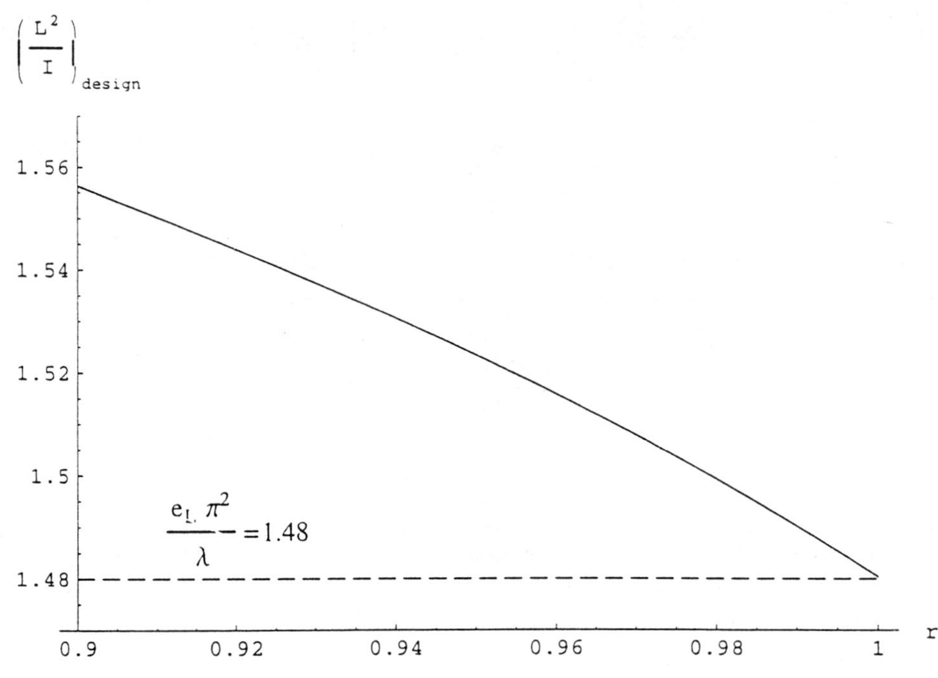

Fig. 1

Variation of the design value of L^2/I versus required reliability r for a bar under compressive load; when r tends to unity, the design value of L^2/I reaches 1.48, that can be found by anti-optimization, interval analysis.

$$R = A_2 \left[erf \left(\frac{\sigma_y b - p_0}{\sigma^2} \right) - erf \left(\frac{p_L - p_0}{\sigma_2} \right) \right] \qquad (25)$$

It equals zero, if $\sigma_y b < p_L$, and equals unity if $\sigma_y b > p_U$. Consider the intermediate case when

$$p_L < \frac{\sigma_{yb} - p_0}{\sigma_2} < p_U \qquad (26)$$

The design value b_{design} is found from the requirement $R = r$:

$$b_{design} = \frac{p_0}{\sigma_y} + \sigma_2 erf^{-1} \left\{ \frac{r}{A_2} + erf \left(\frac{p_L - p_0}{\sigma_2} \right) \right\} \qquad (27)$$

Again, the following question arises: What is the limit value for b_{design} when reliability tends to unity? The reply to this question is immediately obtained from the Eq. 21. Then the right-hand side of Eq. (22) tends to unity

$$A_2 \left[erf \left(\frac{\sigma_y b - p_0}{\sigma_2} \right) - erf \left(\frac{p_L - p_0}{\sigma_2} \right) \right] \to 1 \qquad (28)$$

or, taking into account the definition of A_2,

$$erf \left(\frac{\sigma_y b - p_0}{\sigma_2} \right) - erf \left(\frac{p_L - p_0}{\sigma_2} \right) \to \frac{1}{A_2} =$$

$$= erf \left(\frac{p_L - p_0}{\sigma_2} \right) - erf \left(\frac{p_L - p_0}{\sigma_2} \right) \qquad (29)$$

Comparison of the left and right sides shows that when the reliability tends to unity,

$$\sigma_y b_{design} \to p_L \qquad (30)$$

or, the design value $b_{design} = b_{1-}$ tends to p_L/σ_y, or the value determined by the upper value of the load. Fig.2 portrays the variation of b_{design} versus codified reliability. When $r \to 1-$, the design value of b approaches $p_L/\sigma_y = 0.03$.

COMPARISON WITH ANTIOPTIMIZATION

The following comments appear here in order. Results reported in both examples can be obtained extremely simply

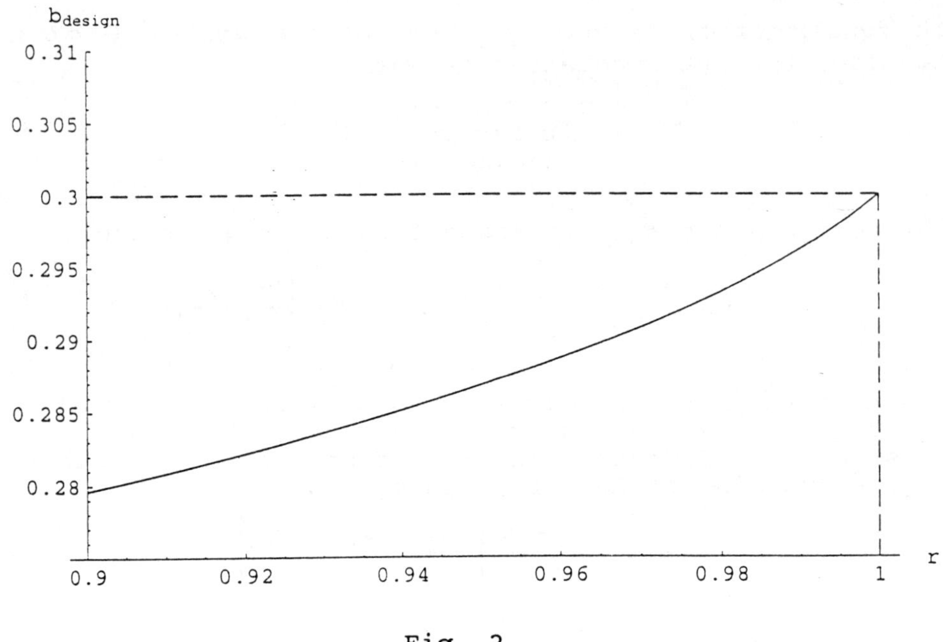

Fig. 2

Variation of the design value of the cross-sectional area
versus required reliability; when r tends to unity, the
design value tends to 0.3, obtainable via anti-
optimization, interval analysis.

without resorting to probabilistic methods.   Since the
problem in each case is one-dimensional, we will utilize
interval analysis to describe the uncertain variables.   Let
us consider modulus of elasticity in the first example as an
interval variable

$$E = \left[e_L, e_U\right] \tag{31}$$

If we want the buckling load to be not less than a
preselected value λ, i.e.

$$\frac{\pi^2 EI}{L^2} \geq \lambda \tag{32}$$

Then, naturally, the geometric parameter L^2/I should be
chosen so that

$$\frac{L^2}{I} \leq \frac{\pi^2 E}{\lambda} \tag{33}$$

Since E is an interval variable so is the quantity $\pi^2 E/\lambda$

$$\frac{\pi^2 E}{\lambda} = \left[\frac{\pi^2 e_L}{\lambda}, \frac{\pi^2 e_U}{\lambda}\right] \tag{34}$$

Thus, Eq. 29 is equivalent to

$$\frac{L^2}{I} \leq \frac{\pi^2 e_L}{\lambda} \tag{35}$$

and the design value of $\left(L^2/I\right)_{design}$ is determined my the minimum
value of the modulus of elasticity e_L.   This coincides with
the reliability analysis.

Analogously, the results of the second example can be
directly obtained by utilizing the interval analysis.   The
strength requirement reads

$$\frac{P}{b} \leq \sigma_y \tag{36}$$

We treat load as an interval variable

$$P = \left[p_L, p_U\right] \tag{37}$$

Then the actual stress $\Sigma = P/b$ is also an interval variable

$$\frac{P}{b} = \left[\frac{p_L}{b}, \frac{p_U}{b}\right] \tag{38}$$

The strength requirement becomes

$$\left[\frac{p_L}{b}, \frac{p_U}{b}\right] \le \sigma_y \tag{39}$$

This leads to a conclusion that

$$\frac{p_U}{b} \le \sigma_y \tag{40}$$

Minimum acceptable value of the cross-sectional area becomes $b_{design} = p_U/\sigma_y$, as is the result of the probabilistic analysis.

As Ang and Tang (1984 p. 388) note

"In engineering, the information may often have to be expressed in terms of the lower and upper limits of the variable. For example, when judgement is necessary it is often convenient and perhaps more realistic to express judgmental information in the form of a range of possibilities. Given the range of possible values of a random variable and the underlying uncertainty may be evaluated by prescribing a suitable distribution within the range."

We will show now that irrespective of the probability distribution that one may "prescribe" to a random variable in a finite range of its variation, the high-reliability design will tend to the antioptimization result. Indeed, let probability density of the load be $f_p(p)$ in the range $[p_L, p_U]$ and be zero outside this interval. Reliability reads $R = F_P(\sigma_y b)$ where $F_p(p)$ is the probability distribution of P. Let $p_L \le \sigma_y b \le p_U$. The value $\sigma_y b$ equals $F_p^{-1}(R)$. For the required reliability level r, $\sigma_y b = F_p^{-1}(r)$. It is easy to observe, that $F_p^{-1}(r)$ tends to p_U when the required reliability approaches unity from below. Thus, for highly reliable structures, probabilistic design tends to one based on the antioptimization, namely to $b = p_U/\sigma_y$.

RELIABILITY IN A TWO-DIMENSIONAL CASE

Consider now the case with two random variables

involved.   Let us again be given an uniform bar with the cross-sectional area b subjected to an axial load P.   We treat now the yield stress also as a random variable Σ_y. That joint probability density of the load and the yield stress is given as

$$f_{p\Sigma_y}(p,\sigma_y) = \frac{1}{4mn} \tag{41}$$

in the region $\alpha-m \leq p \leq \alpha+m, \beta-n \leq \sigma_y \leq \beta+n$, and zero outside this region.   Reliability reads

$$R = Prob(P \leq \Sigma_y b) \tag{42}$$

In order to obtain the reliability the joint probability density $f_{p\Sigma_y}$ must be integrated in the region specified in the inequality (39).   Depending on the value of the cross-sectional area b, these could be four different cases.   We will concern ourselves with the most important case leading to highest reliability range.   It is easier to determine structural reliability from geometric consideration than by the direct integration.   The line $p = b\sigma_y$ (see Fig. 3) intersects the uncertainty region in points A_1 and A_2. Coordinates of the point A_1 are $\alpha - m, b(\alpha - m)$, respectively. Coordinates of the point A_2 are $(\beta + n)/b, \beta + n$, respectively. The reliability equals the integral of the probability density over the hatched area, or, alternatively, it equals unity minus the integral over the unhatched area.   The latter is the area of the unhatched triangle.   Thus

$$R = 1 - \frac{1}{2}DA_1 \times DA_2$$

$$= 1 - \frac{1}{2}\left[\beta + n - b(\alpha - m)\right]\left[\frac{\beta+n}{b} - (\beta-n)\right] \tag{43}$$

Design value b_{design} will be found by solving an equation $R = r$ leading to

$$b_{design} = \alpha\beta + \beta m - \alpha n + 3mn - 4mnr - \frac{2\sqrt{2}}{(n-\beta)^2}\,\omega_1 \tag{44}$$

$$\omega_1\left[mn(1-r)(\alpha\beta+\beta m-\alpha n+mn-2mnr)\right]^{\frac{1}{2}}$$

Variation of the reliability R with the value of the cross-sectional area b is depicted in Fig.4.   When required reliability increases and tends to unity from below, the

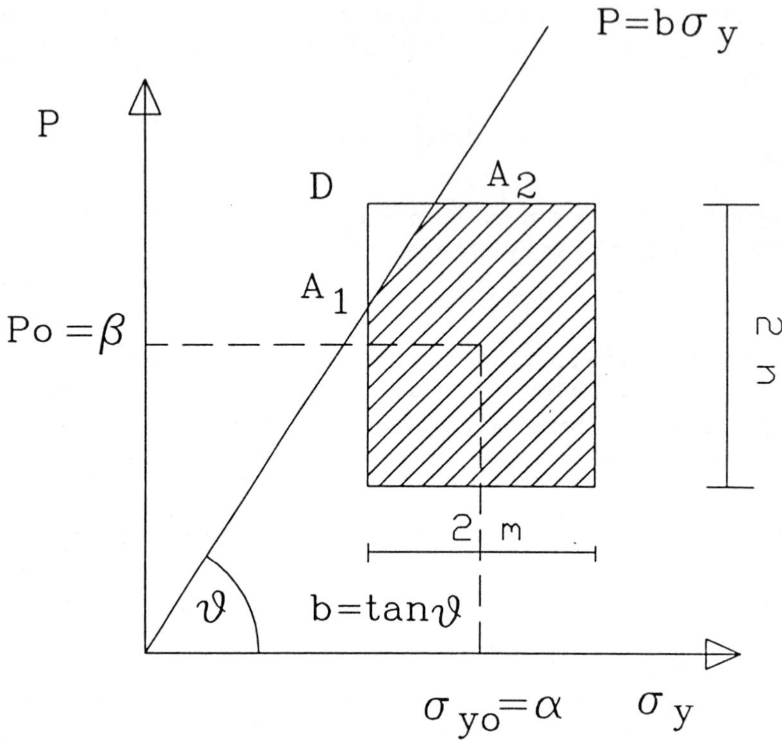

Fig. 3

Reliability of a bar possessing the random yield stress, under random axial loading; the reliability equals the hatched area; when required reliability tends to unity the limit value of the design cross-sectional area is given by Eq. (45).

limit value of design cross-sectional area tends to

$$b_{design} = \frac{\alpha + m}{\beta - n} \qquad (45)$$

Following numerical values were utilized: $\alpha = 75,000 \; kgf/cm^2$, $\beta = 50kgf$, $m = 12,500kgf/cm^2$, $n = 10kgf$. The design value b_{design} tends to 2185 cm^2 when r approaches unity from below.

Let us contrast this result with those obtained from the anti-optimization analysis. Since both the load and the yield stress are interval variables,

$$P = \left[\alpha - m, \alpha + m \right], \; \Sigma_y = \left[\beta - n, \beta + n \right] \qquad (46)$$

so is the cross-sectional area

$$b = \frac{P}{\Sigma_y} = \frac{[\alpha - m, \alpha + m]}{[\beta - n, \beta + n]} \qquad (47)$$

According to the definition of the ratio of two interval variables (see, e.g.,Moore, 1979), b is also an interval variable

$$b = [b_L, b_U] = \left[\frac{\alpha - m}{\beta + n}, \frac{\alpha + m}{\beta - n} \right] \qquad (48)$$

As we see, the probabilistic approach yields the same design value, as per Eq. 42 as the interval analysis, since design value clearly must be the upper bound the cross-section areas may take on. The anti-optimization result clearly corresponds to unity reliability, i.e. the line $p = b\sigma_y$, passing so that the probability density must be integrated over the entire area of variation of the uncertain variables Fig. 5).

INTERRELATION BETWEEN PROBABILITY AND CONVEXITY

In order to find the interrelation between the probabilistic methods and convex modeling of uncertainty consider the same bar subjected to a random load P, with bar's material having a random yield stress Σ_y. Let a joint probability density be given by

$$f_{p\Sigma_y}(p\sigma_y) = \begin{cases} \dfrac{1}{2\pi c^2} & ,\text{for } \left[\dfrac{p - p_0}{b_0}\right]^2 + (\sigma_y - \sigma_{y0})^2 < c^2 \\ 0, & \text{otherwise} \end{cases} \qquad (49)$$

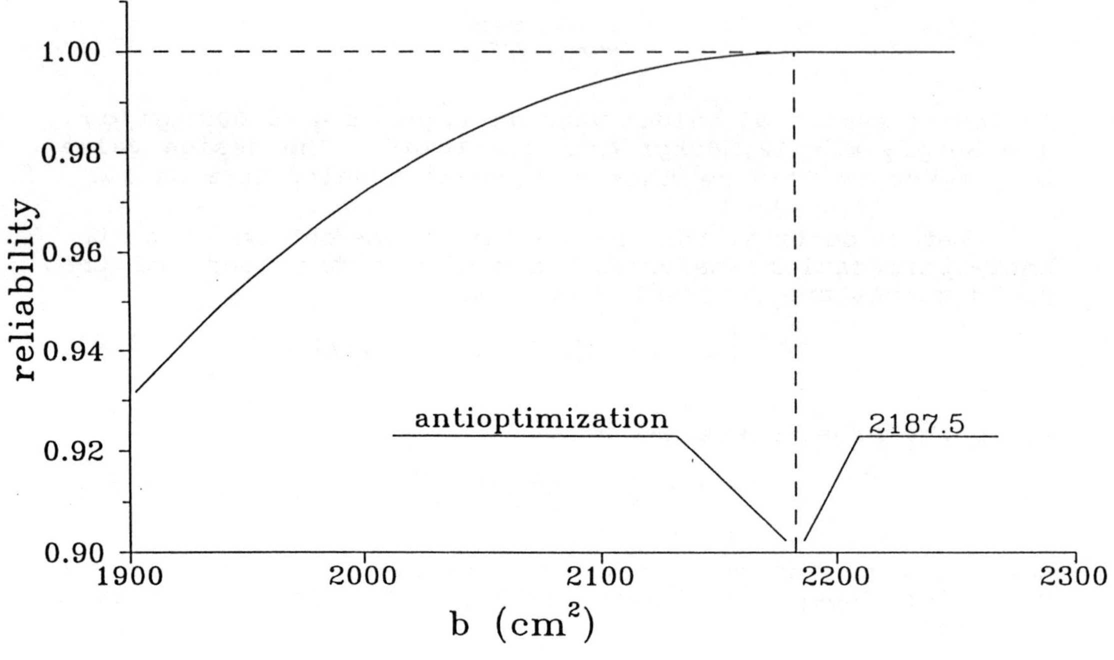

Fig. 4

Variation of reliability versus the cross-sectional area; the anti-optimization result a compatible with the reliability approach.

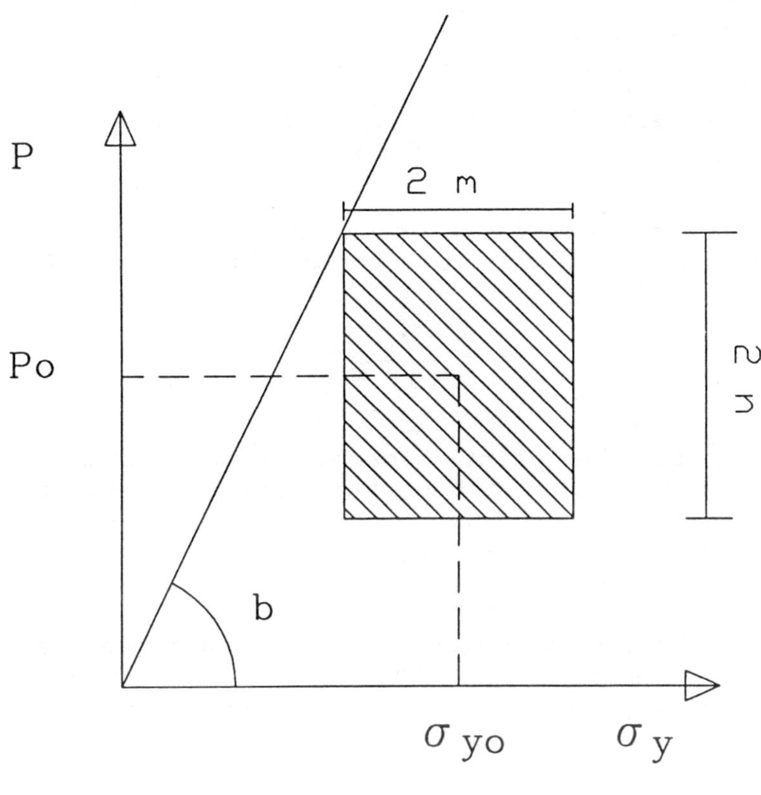

Fig. 5

The anti-optimization result corresponds to unity reliability, i.e. the line $p = b\sigma_y$, passing so that the probability density must be integrated over the entire area of variation of uncertain variables.

In other words, let P and Σ_y have an uniform distribution in a circle (Elishakoff, 1983, p. 233), p_0 and σ_{y0} are constants; b_0 is a constant having a dimension of the cross-section area such that the terms $(p - p_0)/b_0$ and $\sigma_y - \sigma_{y0}$ have the same dimension. Reliability of the bar equals (Fig. 6)

$$R = Prob(P \le \sigma_y b) = \int\!\!\!\int_{p \le \sigma_y b} f_{p\Sigma_y}(p, \sigma_y)\, dp\, d\sigma_y \tag{50}$$

There are two different cases which should be distinguished when performing the integration in Eq. 50. Denote $b = b_1$ a value such that

$$b_1 = \frac{p_0}{\sigma_{y0} - c} \tag{51}$$

corresponding to the point closest to the axis of the ordinates. If the tangent k of the line $p/b_0 = k\sigma_y$ is less than b_1 then reliability calculation includes an integration in the hatched area. We are however interested in a design of a highly reliable structure, with nearly unity or unity reliability. Assume therefore that $k > b_1$. Reliability then is easier to calculate as unity minus the unhatched area over the line whose tangent is greater than b_1 (Fig.6). The abscissas σ_1 and σ_2 of the points of intersection A_1 and A_2 with the circle of the line $p = \sigma_y b$ are found by solving an equation

$$\left[\frac{\sigma_y b - p_0}{b_0}\right]^2 + (\sigma_y - \sigma_{y0})^2 = c^2 \tag{52}$$

Thus, σ_1 and σ_2 are given by

$$\sigma_{1,2} = \left[\sigma_{y0} + \frac{p_0 b}{b_0^2}\right] \pm \left[\sigma_{y0}^2 + \frac{p_0^2 b^2}{b_0^4} + \frac{2\sigma_{y0} p_0 b}{b_0^2} \right. $$
$$\left. + \left[1 + \frac{b^2}{b_0^2}\right]\left[\frac{p_0}{b^2} - \sigma_{y0}^2 - c^2\right]\right]^{1/2} \tag{53}$$

Reliability in the vicinity near unity becomes

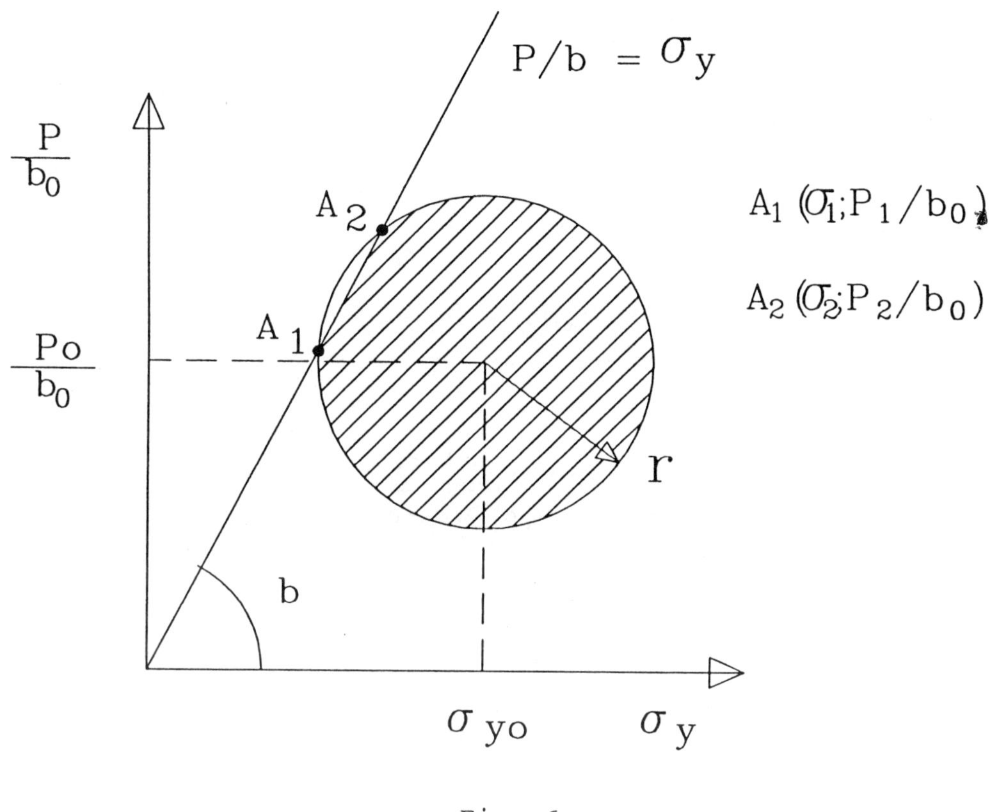

Fig. 6

Illustration of the interrelation between probabilistic
and anti-optimization, convex modeling; the load and the
yield stress have jointly uniform density in a circle;
reliability equals to the hatched area.

$$R = 1 - \frac{1}{\pi c^2} \int_{\sigma_1}^{\sigma_2} d\sigma_y \int_{b\sigma_y/b_0}^{\frac{p_0}{\sigma_0} + \sqrt{c^2 - (\sigma_y - \sigma_{y0})^2}} dp \tag{54}$$

Final result reads

$$R = 1 - \frac{1}{\pi c^2} \left\{ \frac{b(\sigma_1^2 - \sigma_2^2) + 2(\sigma_2 \sigma_1)p_0}{2b_0} \right.$$

$$- \frac{1}{2}(\sigma_1 - \sigma_{y0})\left[c^2 - (\sigma_1 - \sigma_{y0})^2\right]^{1/2}$$

$$+ \frac{1}{2}(\sigma_2 - \sigma_{y0})\left[c^2 - (\sigma_2 - \sigma_{y0})^2\right]^{1/2} \tag{55}$$

$$+ \frac{c^2}{2}\tan^{-1}\left[\frac{\sigma_{y0} - \sigma_1}{\sqrt{c^2 - (\sigma_1 - \sigma_{y0})^2}}\right]$$

$$\left. + \frac{c^2}{2}\tan^{-1}\left[\frac{\sigma_{y0} - \sigma_2}{\sqrt{c^2 - (\sigma_2 - \sigma_{y0})^2}}\right] \right\}$$

Let us perform design calculations to calculate b_{design} such that reliability is at least r. We fix the numerical data at

$$p_0 = 75000 \ kg/cm^2$$

$$\sigma_{y0} = 50 \ kg/cm^2$$

$$b_0 = p_0/\sigma_{y0} = 1500 \ cm^2 \tag{56}$$

$$c = 10 \ kg/cm^2$$

Then

$$\sigma_{1,2} = \frac{1}{(225 \times 10^4 + b^2)}\left[1125 \times 10^5 + 75 \times 10^3 \pm 1,500 \times \right.$$

$$\left. \times (54 \times 10^8 + 75 \times 10^5 b - 2400 b^2)^{1/2}\right] \tag{57}$$

The reliability in accordance to formula (55) must be calculated if the following conditions is met:

$$b > b_1 = \frac{p_0}{\sigma_{y0} - c} = \frac{7500}{50 - 10} = 1875 \ cm^2 \tag{58}$$

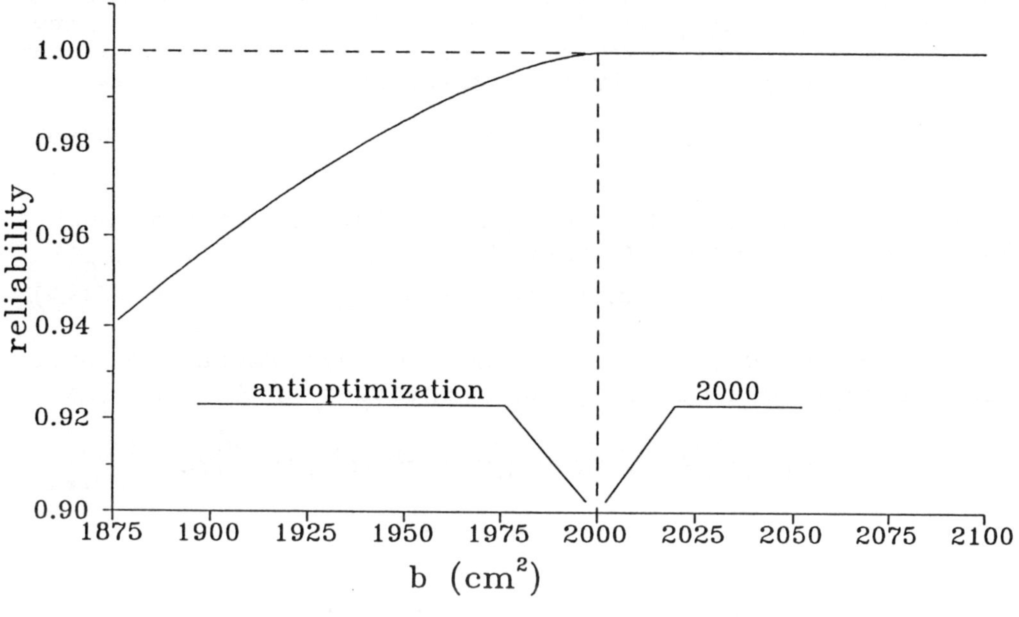

Fig. 7

Reliability versus design value of the cross-sectional area; when required reliability is unity design cross-sectional area equals 2,000 cm^2, obtainable by the anti-optimization analysis.

The value $b=b_1$ corresponds to reliability $R = 0.940496$. The dependence of reliability versus the cross-sectional are b is portrayed in Fig.7. This dependence allows one to determine the design value b_{design} so that $R = r$. When r tends to unity, the cross sectional area reaches 2,000 cm^2. This value can also be found graphically as a value at which the line

$$\frac{p}{p_0} = k\sigma_y \tag{59}$$

represents a tangent to the circle (Fig.8). To find such a value of b^*, we solve an equation

$$\sigma_1(b^*) = \sigma_2(b^*) \tag{60}$$

in other words, the points of intersection A_1 and A_2 degenerate into a single joint. Solution of Eq. (60) is

$$b^* = \left\{ 2p_0\sigma_{y0} + \left[4p_0^2\sigma_{y0}^2 - 4(p_0^2 - b_0^2 c^2)(\sigma_{y0}^2 - c^2) \right]^{1/2} \right\} / \omega_2 \tag{61}$$

$$\omega_2 = 2\left(\sigma_{y0}^2 - c^2 \right)$$

For the data utilized above b equals 2,000 cm^2, as expected.

BOUNDEDNESS VERSUS UNBOUNDEDNESS OF RANDOM VARIABLES

As we have seen in several one-dimensional and two-dimensional examples, for the random variables which take on values in bounded regions, like intervals, rectangular or circular areas, the probabilistic design with high required reliability tends, in the limit, to the results obtainable directly by the antioptimization analysis.

Natural question arises: What happens when involved random variables are not bounded? In which interrelation are probabilistic and antioptimization analyses in these circumstances?

It is immediately understood, as noted by Professor Raphael T. Haftka (1995), that for random variables whose values are not bounded, requirement of extremely high reliability may lead to a design that is more conservative than any design obtained by antioptimization. This implies that the probabilistic design may be more conservative, or less conservative than the antioptimization analysis, depending on the required reliability.

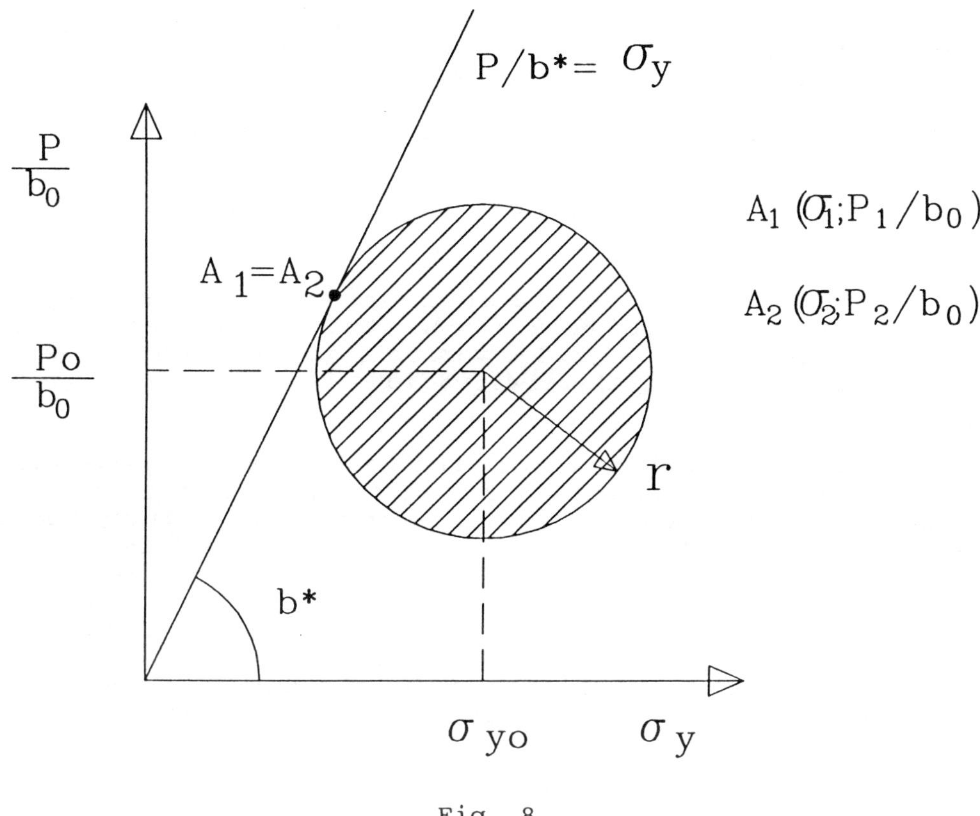

Fig. 8

Line $p/b^* = \sigma_y$ corresponds, on one hand, to unity reliability, and, on the other hand, to the anti-optimization solution.

Thus, the antioptimization analysis, including the convex modeling of uncertainty, does not have a monopoly over robust design. This interesting observation is never stressed (at least until now) by the researchers developing non-probabilistic approaches of dealing with uncertainty.

This discussion naturally leads to new questions. As a proverb maintains to every answer you can find a new question.

Indeed an additional question arises: The observations of the random variable always constitute finite values. How can one establish if a random variable is unbounded? How can one demonstrate that the observations belong to a truncated random variable?

It appears that these are the most intriguing questions in the problems of uncertainty analyses in the context of the present study. Indeed, as is known normal distribution assumption is widely utilized in probabilistic design. A popular maxim maintains that theoreticians think that normal distribution arises in experimental settings, whereas experimentalists trust that it is stemming from theoretical analyses. How to evaluate if a random process follows a normal distribution? The reply to this question appears to be given in a personal communication by Professor Stephen Crandall (1993):

"A good way to visualize the rapid thinning out of the tails of the normal distribution is to consider a stationary normal process with zero mean. Consider the fraction of time such a process spends above various thresholds. For example, it is an easy exercise to show the process spends:
16% of its time above the one-sigma level,
2.3% of its time above the two-sigma level,
0.13% of its time above the three-sigma level.
The fraction of time above the threshold decreases dramatically as the threshold is raised further.
My favorite example is for a threshold level slightly higher than ten-sigma. The fraction of the time the precess spends above this level is represented by the fraction below:

$$\frac{one\ microsecond}{Age\ of\ the\ earth\ (5\ billion\ years)}$$

This shows that one ever live long enough to verify experimentally that any given random process is truly normal."

In light of this remark, one must be extremely doubtful when one states that the experimental evidence has supported the normality distribution. It appears that likewise, experimental proofs that other distributions extend to infinity must be doubtful.

Strongest objection the present writer has heard to the finiteness of the uncertain variables was about the wind velocity. One expert in probabilistic methods expressed an opinion, as well as quoted another expert, to the effect that the wind velocity can not be a bounded quantity. Yet, in the recent article by Simiu and Heckert (1996) authors mention *"It is a physical fact that extreme winds are bounded, and one would expect the probabilistic model to reflect this fact."* Analogous claim on boundedness of the wind velocity was expressed by Professor Arvid Naess in the lecture delivered at the Florida Atlantic University (1997).

Most researchers anyway are theoreticians. They are often are undle and/or disinterested in collecting experimental data, but even conceptually do not think in terms of availability of data to substantiate their assumptions. As English inventor and mathematicians Charles Babbage mentions, *"Errors using inadequate data are much less than those using no data at all."*

It is rightful then to wonder, if Nalimov (1981) was right when he remarked that *"Science models are rather a simulation of human consciousness than the reality of the universe."*

HISTORICAL COMMENTS ABOUT NON-PROBABILISTIC, NON-FUZZY-SETS-BASED APPROACHES

Present author had a long-term interest in various difficulties of probabilistic methods and especially effect of various simplifying assumptions (see e.g. Elishakoff 1977, 1983, 1986; Scheurkogel and Elishakoff, 1985; Scheurkogel et al, 1981). This constituted a fertile ground for tolerance to other, non-probabilistic ideas of uncertainty modeling. Present writer has been introduced to the non-probabilistic as well as non-fuzzy-sets based approaches by Professors Yakov Ben-Haim and Ezra Zeheb of the Technion-Israel Institute of Technology, in 1987 and in 1988, respectively. In addition, Professor Ben-Haim was looking for expanding applicability of the subject which he called *convex modeling*, which he utilized in the context of nuclear engineering (Ben-Haim, 1985) and fault identification (Ben-Haim , 1985; Ben-Haim and Elias, 1987). Our collaboration resulted in a joint monograph (1990) which presented, we hope, a first rigorous treatment of convex modeling to the various *applied mechanics* problems. Professor Zeheb's lectures concentrated on the celebrated Kharitonov theorem and its extensions (present article will not touch upon this theorem).

Later, the present writer has realized that he actually has attended a lecture 10 years prior to that, which possessed an analogous, non-probabilistic spirit, by Professor Rudolf Drenick of the Brooklyn Polytechnic Institute at the Conference in Southampton (1977). Even later it was realized that in its simplest setting, the convex modeling was utilized by the present writer in his book (1983, p. 109), as a worst case design.

Earliest paper traced by us in convex modeling of uncertainty was written by Bulgakov (1940, 1946). In Russia, there is a large body of references devoted to Bulgakov's problem, or, as it is known there, the problem of "accumulation the disturbances." The extensive works by Leitmann (1993) and his school can also be identified as studies in antioptimization.

The existence of the invertal analysis was pointed to the present writer by Professor Walter Wedig of the University of Karlsruhe, in 1990. This led to the idea (Elishakoff, 1991) that the interval analysis can be utilized in the uncertainty modeling. The papers by Qiu et al (1994) and Köylüğlu et al (19) followed. It must be noted that the convex modeling is a generalization of the interval analysis (Elishakoff, 1995). Recent reviews on convex modeling of uncertainty were prepared by Elishakoff (1995) and Ben-Haim (1997). Probabilistic and convex modeling were combined by Elishakoff and Colombi (1993) and Zhu and Elishakoff (1996). A monograph (Elishakoff et al, 1994) was dedicated to hybrid analysis in the space shuttle applications.

Earliest works in applications of interval analysis were written by Wiener (1914, 1921), whereas theoretical foundations are found first in the paper by Young (1931).

It will be no surprise to find that the idea of antioptimization in its various forms stems from the times immemorial. This anticipation is strengthened by the fact that its main idea resonates well with a universal wisdom: "Make the best out of the worst."

CONCLUSION

This study does not aim to make a bombastic statement like that of de Finneti (1974) "Probability does not exist". We can take the liberty of suggesting to consider two possible scenarios: either probability does not exist or it does. Consider first the former case: if probability does not exist then obviously antroptimization method is of utmost importance, and it must be utilized for design. Consider now the latter case: if probability exists, then for the structure to be acceptable it must possess high, near-unity reliability; for the case of bounded random variables, the probabilistic method tends to one furnished by the antioptimization. Thus, irrespective of existence or nonexistence of probability, antioptimization method must be incorporated.

Likewise, we do not state that "Fuzziness does not exist," (especially since this essay did not touch upon fuzzy-sets-based design) or more generally that "Probability and fuzziness do not exist." We make a conjecture that even if both probability and fuzziness exist the analyses based on them would tend to the antioptimization.

It is shown in this paper that the antioptimization methods, like interval analysis or convex analysis can be deduced from probabilistic considerations. Yet the antioptimization methods are much more simple both conceptually and numerically than the probabilistic analysis. We may conclude by quoting from Minsky (1987):

> *"If you understand something in only one way, then you do not really understand it at all...The secret of what anything means to us depends on how we have connected it to all the other things we know."*

We hope that we have shown design against uncertainty from two different perspectives.

ACKNOWLEDGMENT

Help provided by Dr. Piero Colajanni and Mr. Nicola Impollonia, of the University of Messina, Italy in numerical calculations is gratefully acknowledged. Financial support by the NASA Langley Research Center through grant NAG-1-2115 (Dr. J. H. Starnes, Jr.-Program Manager) is appreciated.

References

Barmish B.R., *New Tools for Robustness of Linear Systems*, Macmillan Publishing Company, New York, 1994.

Ben-Haim Y., *The Essay of Spatially Random Material*, Kluwer, Dortrecht, 1985.

Ben-Haim Y., *Robust Reliability in the Mechanical Sciences*, Springer-Verlag, Berlin, 1996.

Ben-Haim Y., Robust Reliability of Structures, *Advances in Applied Mechanics*, (J. W. Hutchinson and T. Y. Wu, eds.,) Academic Press, New York, pp. 1-41, 1997.

Ben-Haim Y. and Elias E., Indirect Measurement of Surface Temperature and Heat Flux: Optimal Design using Convexity Analysis, *International Journal of Heat and Mass Transfer*, Vol. 30, 1673-1683, 1987.

Ben-Haim Y. and Elishakoff I., *Convex Models of Uncertainty in Applied Mechanics*, Elsevier, Amsterdam, 1990.

Black M., Vagueness: An Exercise in Logical Analysis, *Philosophy of Science*, Vol. 4, 427-455, 1937.

Bulgakov B. V., Fehleranhaeufung bei Kreiselapparaten, *Ingenieur-Archiv*, Vol. 11, 461-469, 1940 (in German).

Bulgakov B. V., On the Accumulation of Disturbances in Linear Systems with Constant Coefficients, *Doklady Akadekiii Nauk SSR*, Vol. L. I. No. 5, 339-342, 1946 (in Russian).

Chernousko F. L., *State Estimation for Dynamic Systems*, CRC Press, Boca Raton, 1994.

Chmielewski M. A., Elliptically Symmetric Distributions: A Review and Bibliography, *International Statistical Review*, Vol. 49, 67-74, 1981.

Cox E.D., *Fuzzy Logic for Business and Industry*, Charles River Media, Inc., Rockland, MA, p. 44, 1955.

Crandall S. H., Private Communication, 1993.

De Finetti B., *Theory of Probability: A Critical Introductory Treatment*, Vol. 1, Wiley, London, 1974.

Drenick R.F., On the Class of Non-robust Problems in Stochastic Dynamics, *Stochastic Problems in Dynamics* (B.L. Clarkson, ed.), Pitman, London, pp. 237-255, 1977.

Elishakoff I., On the Role of Cross-Correlations in Random Vibrations of Shells, *Journal of Sound and Vibration*, Vol. 50, 239-252, 1977.

Elishakoff I., *Probabilistic Methods in the Theory of Structures*, Wiley, New York, 1983.

Elishakoff I., A Model Elucidating Significance of Cross-Correlations in Random Vibration Analysis, *Random Vibration-Status and Recent Developments* (I. Elishakoff and R. H. Lyon, eds.), pp. 101-112, Elsevier, Amsterdam, 1986.

Elishakoff, I., An Idea of the Uncertainty Triangle, *The Shock and Vibration Digest*, Vol. 22, No. 10, 1, 1990 (editorial).

Elishakoff I., Some Questions in Engineering Eigenvalue Problems in Natural Sciences, *Numerical Treatment of Eigenvalue Problems*, (J. Albrecht, L. Collatz, P. Hagerdorn, and W. Velte, eds.), Birkauser Publisher, Basel, pp. 71-107, 1991.

Elishakoff I., Essay on Uncertainties in Elastic and Viscoelastic Structures: from A. M. Freudenthal's Criticizms to Modern Convex Modeling, *Computers and Structures*, Vol. 56, 871-895, 1995a.

Elishakoff I., Convex Modeling – a Generalization of Interval Analysis for Nonprobabilistic Treatment of Uncertainty, *International Journal of Reliable Computing,* Supplement, 1995b.

Elishakoff I. and Colombi P., Combination of Probabilistic and Convex Models of Uncertainty when Scarce Knowledge is Present on Acoustic Excitation Parameters, *Computer Methods in Applied Mechanics and Engineering,* Vol. 104, 187-209, 1993.

Elishakoff I., Cai G. Q. and Starnes J. H., Jr., Non-Linear Buckling of a Column with Initial Imperfections via Stochastic and Non-Stochastic Convex Models, *International Journal of Non-Linear Mechanics*, Vol. 29, 71-82, 1994a.

Elishakoff I., Haftka R. T. and Fang J. J., Structural Design under Bounded Uncertainty – Optimization with Anti-Optimization, *Computers and Structures*, Vol. 53, 1401-1405, 1994b.

Elishakoff I., Lin Y.K. and Zhu L.P. *Probabilistic and Convex Modelling of Acoustically Excited Structures,* Elsevier, Amsterdam, 1994.

Grandori g., Paradigms and Falsification in Earthquake Engineering, *Meccanica*, Vol. 26, 17-21, 1991.

Haftka R. T., Private Communication, 1995.

Hirota K., Fuzzy Concept Very Clear in Japan, *Asahi Evening News*, Aug. 20, 1991.

Kalman R.E., Randomness Reexamined, *Modeling Identification and Control*, Vol. 15, 141-151, 1994.

Kalman R. E., Identification of Noisy Systems, *Uspekhi Matematicheskikh Nauk (Progress in Mathematical Sciences)*, Vol. 10, No. 4, 1984 (in Russian).
32. Kosko B., *Fuzzy Thinking*, Hyperion, New York, 1993.

Köylüoğlu H. U., Cakmak A. S. and Nielsen S. R. K. Applications of Interval Algebra to Deal with Pattern Loading and Structural Uncertainties, *Journal of Engineering Mechanics,* Vol., 121, 1995 (see also a discussion by Modaressi H. and authors closure Vol. 123, p 645, 1997).

Köylüoğlu H. U. and Elishakoff I., A Comparison of Stochastic and Interval Finite Elements Applied to Shear Frames with Uncertain Stiffness Properties, *Computers and Structures*, Vol. 67, 91-98, 1998.

Kunzevich V-M. and Lychak M., *Guaranteed Estimates Adaptation and Robustness in Control Systems*, Lecture Notes in Control and Information Science, Vol. 169, Springer Verlag, New York, 1992.

Leibovich Y., Philosophical Essay, *Makhshavot (Thoughts)*, 1992 (in Hebrew).

Leitmann G., On One Approach to the Control of Uncertain Systems, *Journal of Dynamics of Systems, Measurements and Control*, Vol. 115, 373-380, 1993.

Li Y. W., Elishakoff I., Starnes J. H., Jr. and Shinozuka M., Prediction of Natural Frequency and Buckling Load Variability due to Uncertainty in Material Properties by Convex Modeling, *Fields Institute Communication*, Vol. 9, 139-154, 1996.

McNeil D. and Freiberger P., *Fuzzy Logic*, Simon and Schuster, New York, 1993.

Minsky M., *The Society of Mind*, Simon and Schuster, New York, 1987.

Moore R., *Methods and Applications of Interval Analysis*, SIAM, Philadelphia, 1979.

Morgan M., Gigerenzer G. and Krüger L. (eds), *The Probabilistic Revolution*, MIT Press, Cambridge, 1987.

Naess A., Lectures delivered in the College of Engineering, Florida Atlantic University, Jan. 1997.

Nalimov V. V., *Faces of Science*, ISI Press, Philadelphia, p. XIII, 1981.

Natke H. G., Uncertainties in Mechanical System Modelling: Definitions, Models, Measures, Applications, *Uncertainty: Models and Measures,* Akademie Verlag, Berlin, pp. 44-67, 1997.

Qiu Z. P., Chen S. H. and Elishakoff I., Natural Frequencies of Structures with Uncertain but Non-Random Parameters, *Journal of Optimization Theory and Applications*, Vol. 86, 669-683, 1995.

Scheurkogel A. and Elishakoff I., On Ergodicity Assumption in an Applied Mechanics Problem, *Journal of Applied Mechanics,* Vol. 52, 133-136, 1985.

Scheurkogel A., Elishakoff I. and Kalker J., On the Error That Can be Induced by an Ergodicity Assumption, *Journal of Applied Mechanics*, Vol. 48, 654-656, 1981.

Schweppe E.C., *Uncertain Dynamic Systems*, Prentice Hall, Englewood Cliffs, 1973.

Simiu E. and Heckert N. A., Extreme Wind Distribution Tails: A "Peak Over Threshold" Approach, *Journal of Engineering Mechanics*, Vol. 122, 539-547, 1996.

Wentzel E.S., *Operations Research*, Mir Publishers, Moscow, p. 28, 1980 (in Russian).

Wiener N., A Contribution to the Theory of Relative Position, *Proceedings of the Cambridge Philosophical Society,* Vol. 17, 441-449, 1914.

Wiener N., A New Theory of Measurements: A Study in the Logic of Mathematics, *Proceedings of the London Mathematical Society*, Vol. 19, 181-205, 1921.

Zadeh L.A. Fuzzy Sets, *Information and Control*, Vol. 8, 338-353, 1965.

Zadeh L.A., *Fuzzy Sets and Applications: Selected Papers* (R.R. Yager, ed.), Wiley, New York, 1975.

Zhu L. P. and Elishakoff I., Hybrid Probabilistic and Convex Modeling of Excitation and Response of Periodic Structures, *Mathematical Problems in Engineering*, Vol. 2, 143-163, 1996.

HOW TO COMBINE PROBABILISTIC AND

ANTIOPTIMIZATION METHODS?

Isaac Elishakoff[1] **and Qiang Li**[2]
[1]Department of Mechanical Engineering
[2]Department of Mathematics

Florida Atlantic University,
Boca Raton, FL 33431-0991

ABSTRACT

Hybrid probabilistic-convex modelling is proposed in this study to deal with the situations when the loads applied to the structure are random, but their probabilistic parameters are uncertain. These parameters are assumed to be neither deterministic nor stochastic, rather, they are modeled as elements of a closed, convex set. Contrary to usual optimization studies where the best solutions are sought, here the problem is posed as that of anti-optimization, i.e. evaluating the worst possible probability of failure of a truss structure when the loads' mathematical expectations vary in an uncertainty ellipse. An additional illustration is provided for the fundamental case in the structural reliability: an element subjected to a random stress, with the material possessing with random strength. For the multidegree of freedom systems the necessity of automatic differentiation is indicated.

INTRODUCTION

Theory of reliability of structures is a subject which appears to be theoretically quite well developed (Cornell, 1981; Rosenblueth, 1991). Under its premise, the uncertain parameters are treated to be random variables or random functions. Then the appropriate reliability (the probability of successful performance of the structure) or unreliability (the probability of failure) can be efficiently calculated. For the present state of the structural reliability theory, readers may consult, for example, with studies by Shinozuka (1980), Thoft-Christensen and Baker (1982), Elishakoff (1983), Ang and Tang (1984), Madsen et al (1986), and other investigators. However, very often, sufficient information on the probabilistic characteristics is absent, and as a result, one may possess only an imprecise and/or limited data on the probabilistic characteristics. The natural question arises as to how to deal with this situation which is almost invariably encountered in an engineering practice. There are several alternative approaches to deal with this problem. Let us denote the probability of failure by P_f; as noted above it depends on probabilistic parameters of the loading, of strengths, of geometric characteristics, etc. These probabilistic parameters, which can be mathematical expectations, variances, covariances or higher-order moments, are denoted here as elements of vector $\boldsymbol{A}$. In addition, the probability of failure may depend on other, deterministic parameters, which are denoted here as elements of vector $\boldsymbol{b}$. Hence

$$P_f = P_f(\boldsymbol{A}, \boldsymbol{b}) \tag{1}$$

We assume that vector $\boldsymbol{A}$ is uncertain. One of the possible approaches to take this uncertainty into account is to consider vector $\boldsymbol{A}$ as a random vector. Indeed, if the joint probability distribution of components of the vector $\boldsymbol{A}$ is known, then we first calculate the conditional probability of failure $P_f(\boldsymbol{b} \mid \boldsymbol{A} = \boldsymbol{a})$, and then utilize the total probability theorem, to yield

$$P_f(\boldsymbol{b}) = \int \cdots \int P_f(\boldsymbol{b} \mid \boldsymbol{A} = \boldsymbol{a})\, f_A(\boldsymbol{a})\, \mathrm{d}a_1 \cdots \mathrm{d}a_n \tag{2}$$

where $f_A(a_1, \cdots, a_n)$ is the joint probabilistic density of the parameters $A_1, \cdots, A_n$, a_j is the possible value the random variable A_j can take. If the function $f_A(\boldsymbol{a})$ can be obtained from experiments or from theoretical reasoning, Eq. (2) could be utilized to evaluate the unconditional probability of failure of the structure.   However,   very   often   the probabilistic content of the vector A is unknown.

Indeed, consider a realistic situation when on one hand the vector A is uncertain, but on the other hand, insufficient information is available on A, to justify the above probabilistic framework. Assume that we possess only a scarce information on A, namely, that components of it belong to some bounded set (Ben-Haim and Elishakoff, 1990)

$$Z(G,\theta) = \{A: G(A) \le \theta^2\} \tag{3}$$

where G is some quadratic or other form, containing components of A, and θ is a positive constant. The proper choice of $G(A)$ and of θ is based on available limited information on A. In the study of Elishakoff and Colombi (1993), the special situation has been dealt with where some uncertain parameters were varying in an N-dimensional ellipsoidal set

$$G(A) = (A-A_0)^T W^{-1} (A-A_0) \tag{4}$$

where W is a positive definite matrix, describing the nature of the limited information available. A_0 is the nominal vector of uncertain parameters. Elishakoff and Colombi (1993) considered random vibrations of the space shuttle weather protection systems subjected to uncertain acoustic excitation. They have used an ellipsoidal model for uncertain parameters.

In this paper, the probability of failure of the structure under both the random and the set-valued uncertainties will be discussed.

BASIC EQUATIONS

We are interested in the maximum probability of failure that a structure may possess, i.e.

$$\tilde{P}_f(b) = \max_{A \in Z(W,\theta)} P_f(A,b) \tag{5}$$

Consider first the case when $P_f(A,b)$ is a linear function of uncertain vector A:

$$P_f(A,b) = A^T \varphi(b) \tag{6}$$

where $\varphi(b)$ are some deterministic functions. Once the specific form of $G(A)$ is given in Eq. (4), the set in Eq. (3) will be denoted hereinafter by $Z(W,\theta)$.

In these circumstances, since $Z(W,\theta)$ is a convex set, the maximum of P_f occurs on the set of extreme points of Z. Thus

$$P_{f,\max}(b) = \max_{A \in C(W,\theta)} P_f(A,b) \tag{7}$$

where $C(W,\theta)$ is the set of extreme points of the set $Z(W,\theta)$, and represents an ellipsoidal shell

$$C(W,\theta) = \{\, A : (A-A_0)^T W^{-1} (A-A_0) = \theta^2 \,\} \tag{8}$$

We wish to maximize $A^T\varphi$ subject to constraint $(A-A_0)^T W^{-1} (A-A_0) = \theta^2$. We utilize, following Ben-Haim and Elishakoff (1990), the method of Lagrange multipliers, and form the Lagrangean

$$L = A_0^T \varphi + D^T \varphi + \lambda\, (D^T W^{-1} D - \theta^2) \tag{9}$$

where the vector of deviation D from the nominal vector A_0 has been introduced

$$D = A - A_0 \tag{10}$$

A necessary condition for extremum is

$$\frac{\partial L}{\partial D} = 0 \tag{11}$$

yielding

$$\varphi + 2\lambda W^{-1} D = 0 \tag{12}$$

resulting in

$$D = -\frac{1}{2\lambda} W\varphi \tag{13}$$

Substituting this into the constraint

$$D^T W^{-1} D - \theta^2 = 0 \tag{14}$$

we obtain the expression for the Lagrange multiplier

$$\lambda = \pm \frac{\sqrt{\varphi^{\mathrm{T}} W^{\mathrm{T}} \varphi}}{2\,\theta} \qquad (15)$$

This yields the worst combination of deviations of uncertain coefficients from the nominal values

$$D = \mp \frac{\theta}{\sqrt{\varphi^{\mathrm{T}} W^{\mathrm{T}} \varphi}}\, W\varphi \qquad (16)$$

The worst probability of failure is obtained by substituting Eq. (16) into (6)

$$P_f = A^{\mathrm{T}}\varphi = A_0^{\mathrm{T}}\varphi + D^{\mathrm{T}}\varphi = A_0^{\mathrm{T}}\varphi + \theta\sqrt{\varphi^{\mathrm{T}} W^{\mathrm{T}} \varphi} \qquad (17)$$

The first term in Eq. (17) can be identified with the probability of failure associated with nominal values, A_0; thus

$$P_{f,\max} = P_f(A_0\,,b) + \theta\sqrt{\varphi^{\mathrm{T}} W^{\mathrm{T}} \varphi} \qquad (18)$$

In full analogy one can derive an expression for the best possible probability of failure

$$P_{f,\min} = P_f(A_0\,,b) - \theta\sqrt{\varphi^{\mathrm{T}} W^{\mathrm{T}} \varphi} \qquad (19)$$

Consider now the case when the function $P_f(A,b)$ is *nonlinear* in A. We can use then the linearization procedure in the vicinity of A_0:

$$P_f(A\,,b) \approx P_f(A_0\,,b) + \sum_{i=1}^{n} \frac{\partial P_f(A_0\,,b)}{\partial A_i}\,(A_i - A_{0,i}) \qquad (20)$$

For convenience we denote

$$\varphi^{\mathrm{T}} = \left(\frac{\partial P_f(A\,,b)}{\partial A_1}\,,\; \cdots\,,\; \frac{\partial P_f(A\,,b)}{\partial A_n} \right)\Bigg|_{A=A_0} \qquad (21)$$

The problem is then reduced to the one for the linear function $P_f(A,b)$ with attendant result given in Eq. (18). As is seen the sensitivity derivatives (Arora and Haug, 1979; Deif, 1981; Haug et al, 1986; Haftka and Adelman, 1989; Hou et al, 1990; Uber, 1990; Tortorelli, 1993; Zhang et al, 1993) appear explicitly in the equations. A truss example will be considered in the following section to illustrate the above analytical expressions for the probability of failure. It is worth mentioning that there are several papers dealing with the sensitivity in the reliability context. However the notion of *sensitivity* is not directly tied in these papers (see e.g., Karamchandani et al, 1988; Wu, 1993; Wu et al, 1993; Sorensen et al, 1993) with the *uncertainty* in parameters. In this studies authors compute the change in failure probability due to a change in parameter. However, the distribution parameters are not considered uncertain. In the study the concepts of sensitivity and uncertainty are *integrated* to yield a least favorable reliability estimate, needed for design purposes.

For large multidegree-of-freedom structures the analytical analysis becomes cumbersome; in the latter case, one should resort to the recently developed tool—automatic differentiation (Barthelemy et al, 1992; Bischof et al, 1994; Jankovic, 1994)

BAR UNDER TENSILE LOAD

Consider first a simplest possible result. Let a bar that is clamped at one end and free at the other be subjected to a random load P. The cross sectional area of the bar equals b. The bar's material in compression is perfectly elastic, but it exhibits an yield stress σ_y in tension. Let the load be normally distributed with mean p_0 and standard deviation σ_p. The reliability of the bar equals

$$R = Prob\left(\frac{P}{b} \leq \sigma_y\right) = F_P(\sigma_y b) = \frac{1}{2} + erf\left(\frac{\sigma_y b - P_0}{\sigma_p}\right) \qquad (22)$$

where F_P is the probability distribution function of the load, and $erf(x)$ is defined as

$$erf(x) = \frac{1}{\sqrt{2\pi}} \int_{-\infty}^{x} \exp\left(-\frac{t^2}{2}\right) dt \qquad (23)$$

if p_0 and σ_p are given, then the design of the bar proceeds by requiring that $R \geq r$ where r is the required codified reliability. The value of b corresponding to reliability r is designated as the minimum allowable cross sectional area b_{all}. It equals

$$b_{all} = \frac{P_0}{\sigma_y} + \sigma_p erf^{-1}\left(r - \frac{1}{2}\right) \qquad (24)$$

Often, however, the probabilistic characteristics p_0 and σ_p themselves exhibit variability. There are two possible avenues to deal with this situation. the first route is the purely probabilistic one. The reliability derived in Eq. (22) is considered as the conditional reliability $R(p_0,\sigma_p)$. The values p_0 and σ_p are considered random variables with specified joint probability density $f_{p_0\Sigma_p}(p_0,\sigma_p)$. Then the unconditional reliability equals

$$R = \iint R(p_0,\sigma_p)\, f_{p_0\Sigma_p}\,(p_0,\sigma_p)\, dp_0 d\sigma_p \qquad (25)$$

The integration extends over the domain of variation of p_0 and σ_p.

As Ang and Tang (1984) mention (p.406):"...the failure probability is calculated assuming (or conditional on) a given set mean values; since the mean values are subject to modeling errors, a range of calculated probabilities is possible. The resulting range, or error bound, of the calculated probability, therefore, represents the significance of the underlying modeling errors". Moreover Ang and Tang (1984, p. 407) consider the linear performance function $Y = C-D$, where C and D are statistically independent normal variables with mean values μ_C, μ_D and standard deviations σ_C, σ_D. Then they assume that μ_C and μ_D are also independent normal varieties with mean values $\bar{C},\bar{D}$ and standard deviations s_C, s_D. Since Y is a normal variable, the failure probability conditional on given values of μ_C and μ_D is

$$P_f = P\!\left(Y<0\,|\,\mu_C,\mu_D\right)$$
$$= \frac{1}{2} + erf\!\left(\frac{\mu}{\sigma}\right) \qquad (26)$$

where

$$\mu = \mu_D - \mu_C \;,\; \sigma = \sqrt{\sigma_C^2 + \sigma_D^2} \qquad (27)$$

Then Ang and Tang (1984) calculate the expected failure probability as

$$E(P_f) = \int_{-\infty}^{\infty} \left[\frac{1}{2} + erf\!\left(\frac{\mu}{\sigma}\right)\right] f_\mu(\mu)\, d_\mu \qquad (28)$$

with final result

$$E(P_f) = \frac{1}{2} + erf\left[\frac{\mu_{effective}}{\sigma_{effective}}\right] \tag{29}$$

where

$$\mu_{effective} = \overline{D} - \overline{C}, \sigma_{effective} = \sqrt{\sigma_C^2 + \sigma_D^2 + s_C^2 + s_D^2} \tag{30}$$

Probabilistic design then proceeds using eq. 29.

Often, however, the information for probabilistic treatment of the probabilistic characteristics themselves may be unavailable. In these circumstances, one may want to treat the unknown parameters as belonging to a set, that prescribes bounds of variation rather than its probability contents. Denote the vector A as

$$A^T = \begin{bmatrix} p_0 & \sigma_p \end{bmatrix} \tag{31}$$

Then one can visualize as vector A varying inside an ellipse $A^T W A \leq \alpha^2$ where W is a positive-definite matrix. Since the information is limited by Eq. (31), the following question arises: What quantity must be determined when the information about vector A is limited to a constraint (31)?. Naturally, we can find only extremal behavior of the reliability, namely we may ask for the maximum and the minimum values of reliability.

$$R_{\max} = \max R(p_0, \sigma_p), \text{ subjected to } A^T W A \leq \alpha^2 \tag{32}$$

$$R_{\min} = \min R(p_0, \sigma_p), \text{subjected to } A^T W A \leq \alpha^2 \tag{33}$$

Let the nominal values of p_0 and σ_p equal, respectively, $\overline{P}_0$ and $\overline{\sigma}_p$. We represent actual p_0 and σ_p as follows

$$p_0 = \overline{P}_0 + \delta p_0 \quad , \quad \sigma_p = \overline{\sigma}_p + \delta\sigma_p \tag{34}$$

Then, to the first order of δp_0 and $\delta\sigma_p$, we have

$$R\left(\overline{P}_p + \delta p_0, \ \overline{\sigma}_p + \delta\sigma_p\right) = R(\overline{P}_0, \overline{\sigma}_p) \tag{35}$$

$$+ \overline{R}_{,p_0} \delta p_0 + \overline{R}_{,\sigma_p} \delta\sigma_p \tag{36}$$

where $\overline{R}_{,p_0}$ and $\overline{R}_{,\sigma_p}$ read, respectively

$$\overline{R}_{,p_0} = \left.\frac{\partial R}{\partial p_0}\right|_{\substack{p = \overline{p}_0 \\ \sigma_p = \overline{\sigma}p}} \quad , \overline{R}_{,\sigma_0} = \left.\frac{\partial R}{\partial \sigma_0}\right|_{\substack{p = \overline{p}_0 \\ \sigma_p = \overline{\sigma}p}} \tag{37}$$

The limited information available about δp_0 and $\delta\sigma_p$ consists in their belonging to the following set (31),

$$\left[\frac{\delta p_0}{w_1}\right]^2 + \left[\frac{\delta\sigma_p}{w_2^2}\right]^2 \le \alpha^2 \tag{38}$$

The final results for the maximum and minimum reliability are:

$$R_{\max} = R(\overline{p}_0, \overline{\sigma}_p) + \alpha\sqrt{\left(w_1\overline{R}_{,p_0}\right)^2 + \left(w_2\overline{R}_{,\sigma_p}\right)^2}$$

$$R_{\min} = R(\overline{p}_0, \overline{\sigma}_p) - \alpha\sqrt{\left(w_1\overline{R}_{,p_0}\right)^2 + \left(w_2\overline{R}_{,\sigma_p}\right)^2} \tag{39}$$

For design then it is natural to utilize the minimum reliability reaching the level of r

$$R_{\min} \ge r \tag{40}$$

In Eqs (13) and (14)

$$\overline{R}_{,p_0} = -\frac{1}{\overline{\sigma}_p\sqrt{2\pi}} \exp\left[-\frac{\sigma_y b - \overline{p}_0}{\sigma_p}\right] \tag{41}$$

$$\overline{R}_{,\sigma_p} = -\frac{1}{\overline{\sigma}_p^2\sqrt{2\pi}} \exp\left[-\frac{\sigma_y b - \overline{p}_0}{\sigma_p}\right] \tag{42}$$

Alternatively, one may use the expression (24) and determine the maximum value b_{all} takes when p_0 and σ_p are varying in ellipse (38). This leads to

$$b_{\max} = \frac{\overline{p}_0}{\sigma_y} + \overline{\sigma}_p erf^{-1}\left(r - \frac{1}{2}\right)$$

$$+ \alpha\left\{\frac{w_1^2}{\sigma_y^2} + w_2^2\left[erf^{-1}\left(r - \frac{1}{2}\right)\right]^2\right\}^{\frac{1}{2}} \tag{43}$$

Then the minimum allowable cross-sectional area equals the maximum value found in Eq. (17):

$$b_{all,\,min} = b_{max} \tag{44}$$

PROBABILITY OF FAILURE OF A TRUSS

Consider a truss structure as shown in Fig. 1. The truss is subjected to loads P_1 applied at joint C, whereas the two other loads, each equal P_2 are applied at joints B and D. The analysis yields the following expressions for the axial forces in the members

$$F_1 = F_6 = \frac{1}{2\cos\eta\,(\tan\eta - \tan\epsilon)}\,(P_1 + 2P_2)$$

$$F_2 = F_5 = -\frac{1}{2\sin\epsilon\,(\tan\eta - \tan\epsilon)}\,(P_1\tan\eta + 2P_2\tan\epsilon)$$

$$F_3 = F_7 = -\frac{1}{2\cos\epsilon\,(\tan\eta - \tan\epsilon)}\,(P_1 + 2P_2) \tag{45}$$

$$F_4 = \frac{1}{2\,(\tan\eta - \tan\epsilon)}\left[\left(1 + \frac{\tan\eta}{\tan\epsilon}\right)P_1 + 4P_2\right]$$

For $0 < \epsilon < \eta < 90°$, it is obvious that $\cos\epsilon > \cos\eta$ and

$$|F_1| = |F_6| > |F_3| = |F_7| \tag{46}$$

Moreover, it can be shown that for $0 < \epsilon < \eta \leq 60°$

$$\begin{aligned} |F_4| &\geq |F_6| \geq |F_7| \\ |F_4| &\geq |F_5| \end{aligned} \tag{47}$$

This implies that for the latter case the maximum absolute value of the stress Σ always occurs in the fourth member

$$|\Sigma_{max}| = \frac{|F_4|}{c} \tag{48}$$

where c is the cross-sectional area of the members of the truss. We assume that the loads P_1 and P_2 form a bi-variable exponential distribution (Gumbel, 1960)

$$F_{P_1P_2}(p_1,p_2) = \left(1 - e^{-\frac{p_1}{\mu_1}}\right)\left(1 - e^{-\frac{p_2}{\mu_2}}\right)\left(1 + \delta\,e^{-\frac{p_1}{\mu_1}}e^{-\frac{p_2}{\mu_2}}\right) \tag{49}$$

where μ_1 is the mean value of P_1, whereas μ_2 is the mean value of P_2, $\delta = 4\rho$, where ρ is the coefficient of correlation

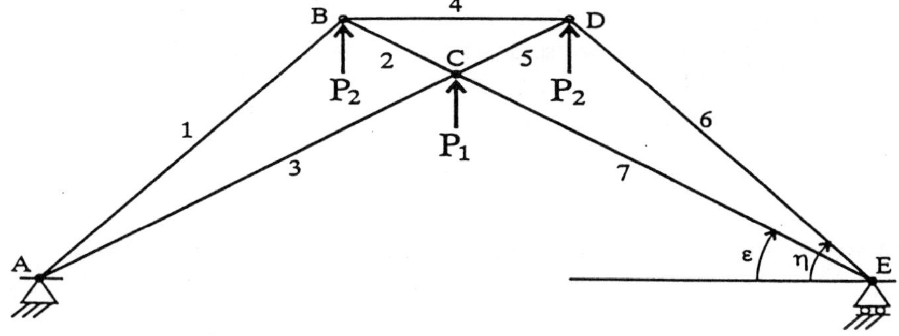

Fig. 1

A truss structure subjected to random loads P_1 and P_2 with uncertain values of mathematical expectations

between P_1 and P_2. We first assume that the yield stress σ_y and δ are deterministic quantities, as well as the values μ_1 and μ_2.

The reliability of the structure is calculated from

$$R = \text{Prob}\,(\,|\,\Sigma\,|\leq\sigma_Y) = \iint\limits_{|\Sigma|\leq\sigma_Y} f_{P_1P_2}(p_1,p_2)\,dp_1\,dp_2 \tag{50}$$

where $f_{P_1P_2}(p_1,p_2)$ is the joint probability density of loads. Since the P_1 and P_2 take on only positive values, Eq. (32) can be rewritten as follows

$$R = \int_0^{\sigma_Y/\beta} dp_2 \int_0^{(\sigma_Y-\beta p_2)/\alpha} f_{P_1P_2}(p_1,p_2)\,dp_1 \tag{51}$$

where the stress in the fourth bar was put in the form

$$\Sigma_4 = \alpha P_1 + \beta P_2 \tag{52}$$

where

$$\alpha = \frac{1+(\tan\epsilon)^{-1}\tan\eta}{2c(\tan\eta-\tan\epsilon)}\ ,\qquad \beta = \frac{2}{c(\tan\eta-\tan\epsilon)} \tag{53}$$

Final expression for reliability, obtained through utilization of computerized symbolic algebraic code *Mathematica* (Wolfram, 1991), reads, for particular combination of parameters $q_1\neq q_2$, $q_1\neq q_2/2$, $q_1\neq 2q_2$

$$R = 1 - \frac{1+\delta}{1-q_2/q_1}\left(e^{-\sigma_Y/q_1} - e^{-\sigma_Y/q_2}\right) - \delta\left[\frac{2}{2-q_2/q_1}e^{-\sigma_Y/q_1} + \frac{1}{1-2q_2/q_1}e^{-\sigma_Y/q_2}\right.$$
$$\left. - \frac{q_2/q_1}{(1-q_2/q_1)(1-2q_2/q_1)}e^{-2\sigma_Y/q_1} - \frac{q_2/q_1}{(1-q_2/q_2)(2-q_2/q_1)}e^{-2\sigma_Y/q_2}\right] \tag{54}$$

where

$$q_1 = \alpha\mu_1\ ,\qquad q_2 = \beta\mu_2 \tag{55}$$

For particular case $q_1 = q_2$, we obtain

$$R = 1 - [\,1 - 3\delta + (1+\delta)\,\frac{\sigma_Y}{q_1}\,]\,e^{-\sigma_Y/q_1} - (\,3 + \frac{\sigma_Y}{q_1}\,)\,\delta\,e^{-2\sigma_Y/q_1} \tag{56}$$

For $q_1 = 2q_2$ we have

$$R = 1 - 2\left(1 + \frac{\delta}{3}\right) e^{-\sigma_Y/q_1} + \left(1 + 2\delta\,\frac{\sigma_Y}{q_1}\right) e^{-2\sigma_Y/q_1} + \frac{2\delta}{3}\, e^{-4\sigma_Y/q_1} \qquad (57)$$

whereas for $q_1 = q_2/2$ we obtain

$$R = 1 - 2\left(1 + \frac{\delta}{3}\right) e^{-\sigma_Y/(2q_1)} + \left(1 + \delta\,\frac{\sigma_Y}{q_1}\right) e^{-\sigma_Y/q_1} + \frac{2\delta}{3}\, e^{-2\sigma_Y/q_1} \qquad (58)$$

The sensitivity derivatives $\partial R/\partial\mu_1$ and $\partial R/\partial\mu_2$ are obtained by differentiation. In the case $q_1\neq q_2$, $q_1\neq q_2/2$, $q_1\neq 2q_2$, they read

$$\frac{\partial R}{\partial\mu_1} = \alpha\left\{\left[\frac{(1+\delta)q_1}{(q_1-q_2)^2} - \frac{4\delta q_1}{(2q_1-q_2)^2} - \frac{(1+\delta)(1+\sigma_Y/q_1)}{q_1-q_2} + \frac{2\delta(1+\sigma_Y/q_1)}{2q_1-q_2}\right] e^{-\sigma_Y/q_1} \right.$$

$$+ \frac{\delta q_2(-q_1^3 + 2q_1 q_2^2 + 2q_1^2\sigma_Y - 6q_1 q_2\sigma_Y + 4q_2^2\sigma_Y)}{q_1(q_1-q_2)^2(q_1-2q_2)^2}\, e^{-2\sigma_Y/q_1}$$

$$\left. + \frac{q_2(-q_1^2 + \delta q_1^2 + 4q_1 q_2 - 4q_2^2 - 2\delta q_2^2)}{(q_1-q_2)^2(q_1-2q_2)^2}\, e^{-\sigma_Y/q_2} + \frac{\delta q_2(-2q_1^2 + q_2^2)}{(q_1-q_2)^2(2q_1-q_2)^2}\, e^{-2\sigma_Y/q_2}\right\} \qquad (59)$$

$$\frac{\partial R}{\partial\mu_2} = \beta\left\{ q_1\left[-\frac{1+\delta}{(q_1-q_2)^2} + \frac{2\delta}{(2q_1-q_2)^2}\right] e^{-\sigma_Y/q_1} + \delta q_1\left[-\frac{1}{(q_1-q_2)^2} + \frac{2}{(q_1-2q_2)^2}\right] e^{-2\sigma_Y/q_1} \right.$$

$$+ \left[\frac{(1+\delta)q_1}{(q_1-q_2)^2} - \frac{2\delta q_1}{(q_1-2q_2)^2} + \frac{\sigma_Y}{q_2^2}\left(\frac{q_2-\delta q_1}{q_1-q_2} - \frac{\delta q_1}{q_1-2q_2}\right)\right] e^{-\sigma_Y/q_2}$$

$$\left. + \frac{\delta q_1(2q_1^2 q_2 - q_2^3 + 4q_1^2\sigma_Y - 6q_1 q_2\sigma_Y + 2q_2^2\sigma_Y)}{q_2(q_1-q_2)^2(2q_1-q_2)^2}\, e^{-2\sigma_Y/q_2}\right\}$$

$$(60)$$

In case $q_1 = q_2$, we have

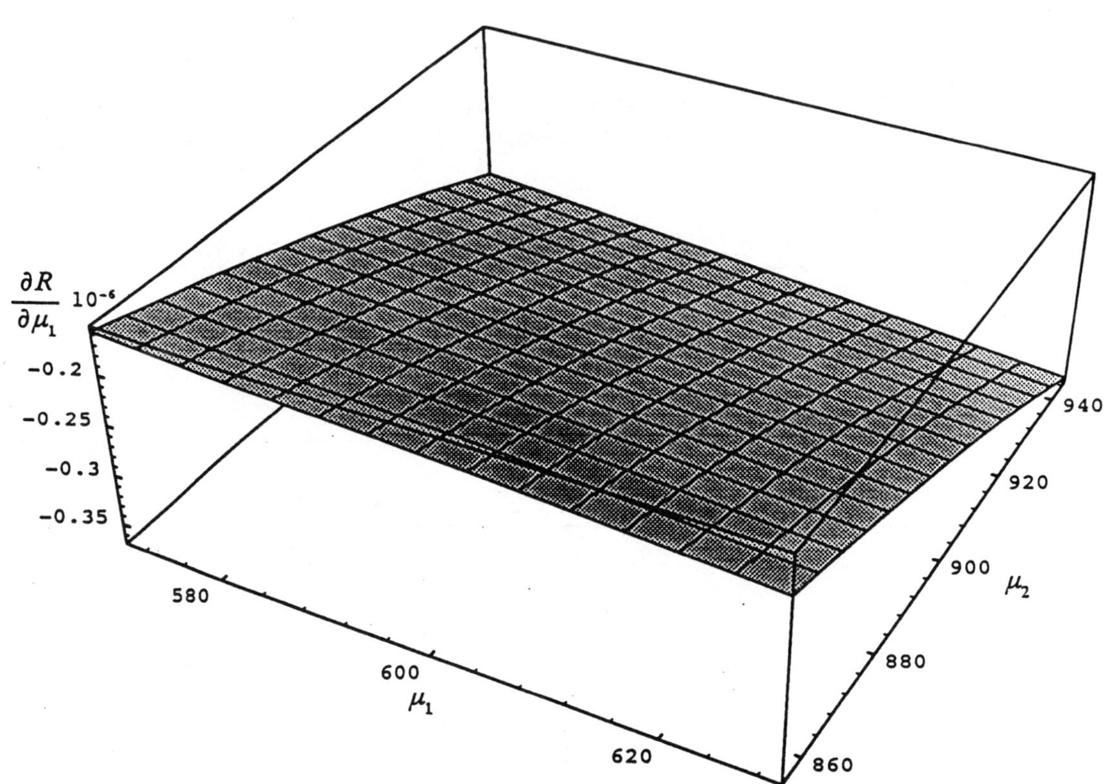

Fig. 2 Sensitivity derivative $\partial R/\partial \mu_1$

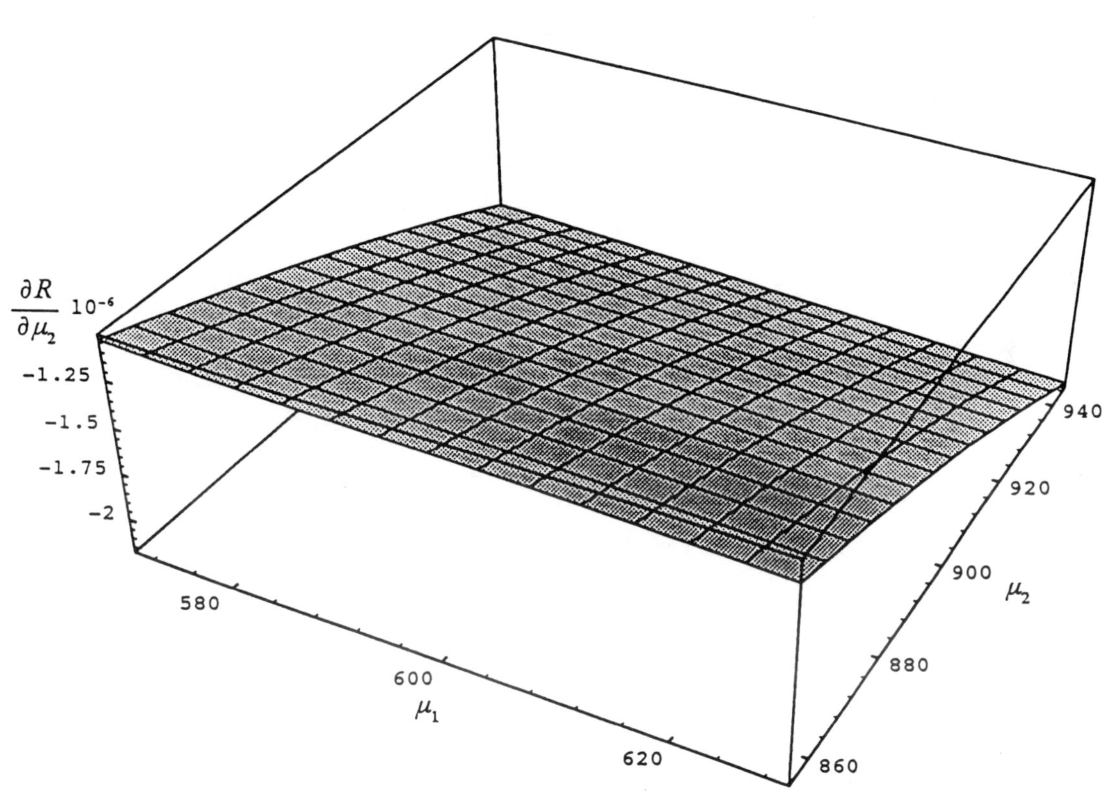

Fig. 3 Sensitivity derivative $\partial R/\partial \mu_2$

$$\frac{\partial R}{\partial \mu_1} = \frac{\alpha \sigma_Y}{q_1^2} \left\{ [\, 4\delta - (1+\delta)\, \frac{\sigma_Y}{q_1}\,]\, e^{-\sigma_Y/q_1} - (5+2\,\frac{\sigma_Y}{q_1})\, \delta\, e^{-2\sigma_Y/q_1} \right\} \tag{61}$$

$$\frac{\partial R}{\partial \mu_2} = \frac{\beta}{\alpha}\, \frac{\partial R}{\partial \mu_1} \tag{62}$$

whereas for $q_1 = 2q_2$, we obtain

$$\frac{\partial R}{\partial \mu_1} = \frac{2\alpha \sigma_Y}{q_1^2} \left[-(1+\frac{\delta}{3})\, e^{-\sigma_Y/q_1} + (1-\delta+2\delta\,\frac{\sigma_Y}{q_1})\, e^{-2\sigma_Y/q_1} + \frac{4\delta}{3}\, e^{-4\sigma_Y/q_1} \right] \tag{63}$$

$$\frac{\partial R}{\partial \mu_2} = \frac{2\beta}{\alpha}\, \frac{\partial R}{\partial \mu_1} \tag{64}$$

and for $q_1 = q_2/2$, we arrive at

$$\frac{\partial R}{\partial \mu_1} = \frac{\alpha \sigma_Y}{q_1^2} \left[-(1+\frac{\delta}{3})\, e^{-\sigma_Y/(2q_1)} + \delta\,\frac{\sigma_Y}{q_1}\, e^{-\sigma_Y/q_1} + \frac{4\delta}{3}\, e^{-2\sigma_Y/q_1} \right] \tag{65}$$

$$\frac{\partial R}{\partial \mu_2} = \frac{\beta}{2\alpha}\, \frac{\partial R}{\partial \mu_1} \tag{66}$$

Let us consider first the case when the parameters μ_1 and μ_2 are determinate. The data is set at $\epsilon=30°$, $\eta=45°$,

$$\alpha = \frac{3.232051}{c} \quad , \quad \beta = \frac{4.732051}{c} \tag{67}$$

For the data fixed at $\mu_1=600$ N, $\mu_2=900$ N, $\delta=0.2$, $\sigma_Y=2\times10^8$ N/m^2, $c=2\times10^{-4}$ m^2, the probability of failure is $P_f = 0.00016$. Assume now that the mathematical expectations μ_1 and μ_2 vary within the ellipse

$$\left[\frac{\mu_1-600}{a}\right]^2 + \left[\frac{\mu_2-900}{b}\right]^2 \leq 1 \tag{68}$$

The values of a and b are semiaxes of the uncertainty ellipse, and they are chosen in such a manner that the variations at most will correspond to ±5% variation of each parameter, i.e., $a=30$ N, $b=45$ N. The sensitivity derivatives $\partial R/\partial \mu_1$ and $\partial R/\partial \mu_2$ are depicted in Figs. 2 and 3, respectively. It is seen that the values of these derivatives are negative, indicating that the reliability decreases when the load increases, as expected. The surface plots also illustrate that the reliability is more sensitive to the values of load parameters in the region of larger loads. Calculation of the bounds of the probability of failure according to the formulas (18) and (19) yields

$$0.000093 \leq P_f \leq 0.00023 \qquad (69)$$

As is seen, the maximum probability of failure $P_f = 0.00023$ is 43% larger than the nominal value 0.00016. The former takes into account the uncertainty in the system parameters, whereas the latter does not. For $a=60$ N and $b=90$ N, corresponding to ±10% variation of parameters μ_1 and μ_2, the probability of failure variability region is [0.00002,0.00030], i.e, ±10% variation in parameters leads to 85% increase of the nominal probability of failure. It appears that for the robust engineering design one should utilize the notion of the maximum possible probability of failure rather than its ideal, nominal counterpart. It should be stressed that in the considered example the failure surface was linear. Additional studies appear to be needed for nonlinear and large scale structures with multiple degrees of freedom to ascertain the feasibility of the proposed study. The algorithms of numerical nonlinear optimization, along with modern automatic differentiation algorithms appear to prove useful (Barthelemy et al, 1992; Bischof et al, 1994; Jankovic, 1994).

CONCLUSION

Amongst uncertainty analysts nowadays there is a situation reminiscent of Babel Tower: almost total lack of communication. Stochasticians almost exclusively utilize probabilistic methods, analysts of fuzzy sets employ fuzzy logic, whereas investigators dealing with anti-optimization (i.e. convex modeling, interval analysis, nonlinear programming etc.) utilize models based on unknown-but-bounded, or, alternatively, uncertain-but-nonrandom quantities. It appear natural that "if all you have is a hammer, everything looks like a nail." (according to Cox 1995, p. 24, this statement is attributed, at various times, to Dionysis of Agapunta around 300 of Common Era, Abraham Maslow, and Lofti Zadeh). Indeed, researchers dealing with uncertainty modeling either randomize, fuzzify or convexity the uncertainty, depending on their preferences, without contrasting their methods. It appears that uncertainty analysts must explore other possibilities rather than to confine themselves solely to stochastic, fuzzy, or anti-optimization "hammers." As in many cases, the golden "middle road" could be found, or alternatively, one may get convinced that other approach(es) is (are) advantageous.

It appears that a continuous dialogue is needed between uncertainty analysts, on one hand, and program managers of federal research agencies to come up with integrative approaches, for meaningful incorporation of the notion of uncertainty into practice. As we have shown in this paper seemingly contradictory techniques of uncertainty modeling could be combined.

ACKNOWLEDGEMENTS

The financial support by NASA Langley Research Center through grant NAG -1-2115 is acknowledged (Dr. J. H. Starnes, Jr.- Program Manager). The help in numerical calculations provided by students A. Sne, and N. Impollonia and discussions with Professor M. Shinozuka are appreciated.

REFERENCES

Ang, A. H-S., and Tang, W. H., 1984, *Probability Concepts in Engineering Planning and Design*, Vol. 2, Wiley, New York.

Arora, J. S. and Haug, E. J. 1979, "Methods of Design Sensitivity Analysis in Structural Optimization", *AIAA Journal*, Vol. 17, pp. 970-974.

Bai E-W. and Andersland M.S., Stochastic and Worst Case System Identification Are not Necessarily Incompatible, *Automatica*, Vol. 30, 1491-1493, 1994.

Barthelemy, J. F. and Hall, L., 1992, "Automatic Differentiation as a Tool in Engineering Design", in *Fourth AIAA/USAF/NASA/OAI Symposium on Multidisciplinary Analysis and Optimization, A Collection of Papers,* Cleveland, OH, AIAA Paper 92-4743-CP, pp. 424-432.

Ben-Haim, Y., and Elishakoff, I., 1990, *Convex Models of Uncertainty in Applied Mechanics*, Elsevier Science Publishers, Elsevier, Amsterdam.

Bernard, J. E., Kwon, S. K. and Wilson, J. A., 1993, "Differentiation of Mass and Stiffness Matrices for High Order Sensitivity Calculations in Finite Element Based Equilibrium Problems", *Journal of Mechanical Design*, Vol. 115, pp. 829-832.

Bischof, C. H., Green, L. L., Haigler K. J. and Knauff, T. L. Jr., 1994, "Parallel Calculation of Sensitivity Derivatives for Aircraft Design using Automatic Differentiation", *5th AIAA/USAF/NASA/ISSMO Symposium on Multidisciplinary Analysis and Optimization*, AIAA Paper 94-4261, Panama City, FL, 1994.

Bjerager, P. and Krenk, S., 1987, "Sensitivity Measures in Structural Reliability Analysis", in *Reliability and Optimization of Structural Systems* (Thoft-Christensesn, P., ed.), Springer Verlag, Berlin, pp. 459-470.

Bjerager, P., and Krenk, S., 1989, "Parametric Sensitivity in First Order Reliability Analysis", *Journal of Engineering Mechanics*, Vol. 115, pp. 1577-1582.

Cornell, C. A., 1981, "Structural Safety: Some Historical Evidence That It Is a Healthy Adolescent," Keynote Lecture, *Proceedings of the Third International Conference on Structural Safety and Reliability*, T. Moan and M. Shinozuka, eds., Elsevier, Amsterdam, pp. 19-29.

Cox E.D., *Fuzzy Logic for Business and Industry*, Charles River Media, Inc., Rockland, Ma, p. 44, 1995.

Deif, A. S., 1981, "Sensitivity Analysis from the State Equations by Perturbation Techniques", *Applied Mathematical Modelling*, Vol. 5, pp. 405-408.

Downton, F., 1973, "The Estimation of $Pr(Y<X)$ in the Normal Case", *Technometrics*, Vol. 15, pp. 551-558.

Elishakoff, I., and Colombi, P., 1993, "Combination of Probabilistic and Convex Models of Uncertainty when Scarce Knowledge is Present on Acoustic Excitation Parameters," *Computer Methods in Applied Mechanics and Engineering*, Vol. 104, pp. 187-209.

Elishakoff, I., 1983, *Probabilistic Methods in the Theory of Structures*, Wiley-Interscience, New York.

Gumbel, E. J., 1960, "Bivariate Exponential Distribution," *American Statistical Association Journal*, pp. 698-707 (see Eq. 3.4).

Haftka, R. T. and Adelman, H. M., 1989, "Recent Developments in Structural Sensitivity Analysis, *Structural Optimization*, Vol. 1, pp. 137-151.

Haug, E. J., Choi, K. K. and Komkov, V., 1986, *Design Sensitivity Analysis of Structural Systems*, Academic Press, New York.

Haugen, E. B., 1980, *Probabilistic Mechanical Design*, John Wiley, New York.

Hou, J. W., Mei, C. and Xue, Y.X., 1990, "Design Sensitivity Analysis of Beams under Nonlinear Forced Vibrations", *AIAA Journal*, Vol. 28, pp. 1067-1068.

Jankovic, M.S., 1994, "Exact *n*th Derivatives of Eigenvalues and Eigenvectors", *Journal of Guidance, Control and Dynamics*, Vol. 17, pp. 136-144.

Karamchandani, A., Bjerager P. and Cornell C. A., 1988, "Methods to Estimate Parametric Sensitivity in Structural Reliability Analysis", *Probabilistic Methods in Civil Engineering*, Spanos, P.D., ed., ASCE Press, New York, pp. 56-89.

Karamchandani, A. and Cornell, C. A., 1990, "Sensitivity Estimation within First and Second Order Reliability Methods", *Journal of Structural Safety*, Vol. 7, pp. 115-123.

Kececioglu, D. and Lamarre, G., 1978, "Mechanical Reliability Confidence Limits", *Journal of Mechanical Design*, Vol. 100, pp. 607-612.

Lloyd, D. K., 1980, "Estimating the Life Cycle of Complex Modeled System", *Institute of Environmental Sciences Proceedings*, pp. 87-96.

Madsen, H. O., Krenk, S., and Lind, N. C., 1986, *Methods of Structural Safety*, Prentice-Hall Inc., Englewood Cliffs, New Jersey.

Reiser, B. and Guttman, I., 1984, "Statistical Inference for Reliability from Stress Strength Relationships: The Normal Case", *MRC Technical Summary Report No. 2695*, University of Wisconsin-Madison.

Rosenblueth, E., 1991, "Here and Henceforth," Keynote Lecture, *Proceedings of the Sixth International Conference on Applications of Statistics and Probability in Civil Engineering*, L. Esteva and S. E. Ruiz, eds., Mexico, Vol. 3, pp. 81-94.

Shinozuka, M., 1980, "Basic Analysis of Structural Safety," *Journal of Structural Engineering*, Vol. 109, pp. 721-740.

Sorensen, J. D. and Enevoldsen, I., 1993, "Sensitivity Weaknesses in Application of Some Statistical Distribution in First Order Reliability Methods" *Journal of Structural Safety*, Vol. 12, pp. 315-325.

Thoft-Christensen, P., and Baker, M. J., 1982, *Structural Reliability Theory and Its Applications*, Springer Verlag, Berlin.

Tortorelli, D. A., 1993, "Sensitivity Analysis for the Steady-State Response of Damped Linear Elastodynamic Systems Subject to Periodic Loads", *Journal of Mechanical Design*, Vol. 115, pp. 822-828.

Uber, J. G. and Brill, E. D. Jr., 1990, "Design Optimization with Sensitivity Constraints", *Engineering Optimization*, Vol. 16, pp. 15-28.

Wolfram, S., 1991, *Mathematica: a System for Doing Mathematics by Computer*, Addison-Wesley, Redwood City, California.

Wu, Y.-T. 1993, "Computational Methods for Efficient Structural Reliability and Reliability Sensitivity Analysis", *34th AIAA/ASME/ASCE/AHS/ASC Structures Structural Dynamics and Materials Conference, AIAA/ASME Adaptive Structures Forum*, AIAA-93-1626-CP, pp. 2817-2826.

Wu, Y.-T., Gureghian, A. B. , Codell, R. B. and Sagar, B., 1993, "Sensitivity and Uncertainty Analysis Applied to One-Dimensional Transport in a Layered Fractured Rock-Evaluation of the Limit State Approach", *Journal of Nuclear Technology*.

Zhang, Y. and Der Kiureghian, A., 1993, "Dynamic Response Sensitivity of Inelastic Structures", *Computer Methods in Applied Mechanics and Engineering*, Vol. 108, pp. 23-36.

HOW TO FIND THE RANGE OF EIGENVALUES DUE TO UNCERTAIN ELASTIC MODULUS AND MASS DENSITY?

Isaac Elishakoff[1], Dehe Duan[1],
Zhiping Qiu[2], and James H. Starnes, Jr.[3]

[1]Department of Mechanical Engineering
 Florida Atlantic University
 Boca Raton, FL 33431-0991

[2]Institute of Solid Mechanics
 Beijing University of Aeronautics and Astronautics
 Beijing 100083, P.R. China

[3]NASA Langley Research Center
 Hampton, VA 23665-5225

ABSTRACT

The theorem recently established by Qiu et al (1995) is exemplified by considering the free uncertain vibrations of strings and beams, with both spatially varying mass and stiffness distributions. In particular, interval finite element method is applied to this problem.

INTRODUCTION

The uncertainties in the parameters of the system are usually analyzed via stochastic modeling through the use of the concept of conditional probability density and the total probability formula. The first investigation of this kind was performed by Kozin (1961) over three decades ago. Since then numerous studies have been conducted utilizing Kozin's avenue of attack as well as other methods, amongst which the most popular method is the perturbation technique. The extensive review of the studies on this topic was given by Ibrahim (1987). More recent work includes, for example, investigations by Spencer and Elishakoff (1988), Khater and Grigoriu (1991), Iwan and Jensen (1992) and others. In all these works the probabilistic information in the uncertain parameters was assumed to be known. However, in most circumstances such a probabilistic information is unavailable. For example, the stochastic finite element method, utilized by Ramu and Ganesan (1993) requires knowledge of the auto-correlation function of the elastic modulus. However, as Shinozuka (1987) mentions, "...it is recognized that it is very difficult to estimate experimentally the auto-correlation function, or in the case of weak homogeneity, the spectral density function of the stochastic variation of material properties. In view of this, the upper bound approaches are particularly important, since the bounds derived to not require knowledge of the author-correlation function." In these cases, the probabilistic methods can not be used with full rigor, and alternative methods must be utilized. In a recent study by Elishakoff et al (1994) a convex model of uncertainty was developed to model uncertain elastic moduli of structures, following the general methodologies developed recent monographs (Ben Haim and Elishakoff, 1990; Elishakoff et al, 1994; Ben-Haim 1996). It was assumed that the stiffness characteristics fall into the four-dimensional ellipsoid or solid ball.

The analysis of eigenvalues of elastic bodies under partial information about the inner properties of structures was pioneered by Krein (1951). He dealt with the following differential equation

$$\frac{d^2u}{dx^2} + \lambda\rho(x)u = 0 \qquad (1)$$

with attendant boundary conditions

$$u(0) = u(\ell) = 0 \qquad (2)$$

The boundary value problem given in Eqs. (1)-(2) is valid for any non-negative integrable function $\rho(\lambda)$ $(0 \leq x \leq \ell)$, where ℓ is the length of the string. Krein (1951) dealt with determination of the maximum and minimum values of $\lambda_n(\rho)$ $(n = 1,2,..)$ when ρ denotes the class of all functions $\rho(\lambda)$ that satisfy the following conditions

$$\rho_L \leq \rho(x) \leq \rho_U, \quad \int_0^\ell \rho(x)\,dx = M \tag{3}$$

Krein (1951) has established that the maximum nth eigenvalue equals

$$\max\lambda_n(\rho) = \frac{\pi^2}{Md}n^2, (n = 1,2,...) \tag{4}$$

whereas the minimum eigenvalue is

$$\min\lambda_n(\rho) = \frac{4n^2}{Md}\chi\left(\frac{d}{L}\right) > \frac{4n^2}{M\ell}\frac{1}{1-(1-4/n^2)d/\ell} \tag{5}$$

where $d = M/\rho_U$, and the function $\chi(t)$ $(0 \leq t \leq 1)$ is defined as a minimal positive root of the equation

$$\sqrt{\chi}\tan\sqrt{\chi} = \frac{t}{1-t} \tag{6}$$

For the first natural frequency the inequality (5) was noted by Zhukovsky (1948). The reader may consult also with Courant and Hilber (1924). Expressions analogous to Krein's elegant expressions (4) and (5) for more complex structures, like beams, plates and shells have not been established yet. A general method appears to be needed to formulate the minimum and maximum values of eigenvalues due to uncertain inner properties. It appears that the interval analysis is a natural tool for problems with limited information on the stiffness and mass distributions in elastic bodies.

In this study the mathematical theory of interval analysis is applied to the uncertain vibrations of linear and nonlinear structures. It appears that the idea of the compatibility of interval analysis to the analysis of uncertainty was expressed apparently for the first time by the present writer (Elishakoff, 1991). Application of interval analysis for some engineering problems was facilitated by Qiu, et al (1994), Köylüoğlu et al (1994), Elishakoff and Duan (1994), Nakagiri and Suzuki (1996).

PROBLEM FORMULATION

As far as the interval computations is concerned, its applications in mathematics are ancient. Archimedes determines the bounds for π based on calculating perimeters of circumscribed and inscribed polygons. In this book "On the Measurements of the Circle" he determines π by its bounds $3^{10}/71 < \pi < 3\,^1/7$. To quote Archimedes, "The circumference of any circle equals three times the diameter plus an excess which is less than one seventh of the diameter but greater than 10/71 of it." In mechanics, often we do not determine the natural frequencies, but rather calculate an interval $\lambda_L \leq \lambda \leq \lambda_U$ where the lower bound λ_L is determined by Dunkerly or Temple or other methods, whereas the upper bound is computed by a single or multi-term Rayleigh-Ritz or Galerkin methods. In both of these circumstances no uncertainty is involved, but the interval result is obtained due to very nature of approximate methods utilized. The mathematical theory of interval analysis is developed by now rather well (see, e.g. monographs by Neumaier (1990) and Hansen (1992)). The question arises if the method will be applicable to incorporate scatter in material properties into the analysis. This idea was first entertained by the present writer in 1991.

Consider the eigenvalue problem described by the equation

$$Ku = \lambda Mu \tag{7}$$

subject to constraints representing the uncertainties in the stiffness and mass matrices

$$K_L \leq K \leq K_U \tag{8}$$

$$M_L \leq M \leq M_U \tag{9}$$

where $K_U = (k_{ij,U})$ and $K_L = (k_{ij,L})$ are, respectively, upper and lower bounds on the stiffness matrix $K = (k_{ij})$, which is uncertain but non-random. $M_U = (m_{ij,U})$ and $M_L = (m_{ij,L})$ are respectively upper and lower bounds on the mass matrix $M = (m_{ij})$, which is uncertain but non-random, $\lambda = \omega^2$ is a squared frequency of the uncertain but non-random matrix pair K and M (denoted $< K, M >$), u is an associated eigenvector.

We shall study a method computing the eigenvalues of Eq. (7) in which the elements k_{ij}, $(i, j = 1, 2, \ldots, n)$ and $M_{ij}, (i, j = 1, 2, \ldots, n)$ of matrices K and M are not known precisely.

By means of interval matrix notation, Eq. 8 and (9) can be written as

$$K \in K^I \quad , \quad M \in M^I \tag{10}$$

in which $K^I = [\underline{K}, \overline{K}]$ is a positive semi-definite interval matrix (Hansen, 1992) and $M^I = [M_L, M_U]$ is a positive definite interval matrix. In terms of Eq. (10), Eqs (7) – (9) can be simply written as

$$K^I u = \lambda M^I u \tag{11}$$

Eq. (11) is called a generalized interval eigenvalue problem. Because K^I and M^I are defined as interval matrices, the associated eigenvalues of K^I and M^I similarly constitute interval variables $\lambda^I = [\lambda_L, \lambda_U] = (\lambda_i^I)$.

The following theorem has been proved by Qiu et al (1991):

Theorem: If $K^I = [K_L, K_U] = [K^c - \Delta K, K^c + \Delta K]$ is a positive semi-definite interval matrix and $M^I = [M_L, M_U] = M^c - \Delta M, M^c + \Delta M]$ is positive definite interval matrix, ΔK and ΔM are also positive semi-definite real matrices, then eigenvalues λ_i, $i = 1, 2, \ldots, n$, of $K \in K^I$ and $M \in M^I$ range over the interval, i.e.

$$\lambda_i^I = \left[\lambda_{i,L}, \lambda_{i,U}\right], \quad i = 1, 2, \ldots, n , \tag{12}$$

where the lower bounds $\lambda_{i,U}$, satisfy

$$K_L u_{i,L} = \lambda_{i,L} M_U u_{i,L}, \quad i = 1, 2, \ldots, n , \tag{13}$$

the upper bounds $\lambda_{i,U}$ satisfy

$$K_U u_{i,U} = \lambda_{i,U} M_L u_{i,U} , \quad i = 1, 2, \ldots, n . \tag{14}$$

This theorem can be correlated with a simplest analysis of this kind, applied to a single-degree-of-freedom system. Indeed, if its mass is an interval variable $M = [m_L, m_U]$, and the stiffness is an interval variable $K = [k_L, k_U]$, then the squared natural frequency $\lambda = \omega^2$ is also an interval variable $\Lambda = [\lambda_L, \lambda_U]$, with $\lambda_L = K_L/m_U$ and $\lambda_U = K_U/m_L$.

Approximation of this kind is obtainable by utilizing a simple term Galerkin approximation. Consider for example a beam governed by the following differential equation:

$$\frac{d^2}{dx^2}\left[D\,\frac{d^2W}{dx^2}\right] - \rho A\omega^2 W = 0 \tag{15}$$

where $W(x)$ is the transverse displacement, $D = EI$ is the bending stiffness, ρ mass density, A = cross-sectional area, ω^2 = squared natural frequency. Let $\psi(x)$ satisfy all boundary conditions. Following Galerkin's formalism we get for the natural frequency

$$\omega^2 = \frac{\int_0^\ell (D\psi'')''\psi\,dx}{\int_0^\ell \rho A\psi^2\,dx} \tag{16}$$

Since $\psi(x)$ satisfies the boundary condition this integral can be rewritten as

$$\omega^2 = \frac{\int_0^\ell D(\psi'')^2\,dx}{\int_0^\ell \rho A\psi^2\,dx} \tag{17}$$

Since $(\psi'')^2$ and ψ^2 are non-negative functions, the minimum value of ω^2 is obtained when $D = D_L$ and $\rho = \rho_U$

$$\omega_L^2 = \frac{D_L}{\rho_U}\,\frac{\int_0^\ell (\psi'')^2\,dx}{\int_0^\ell A\psi^2\,dx} \tag{18}$$

Analogously, the maximum value of ω^2 is attained for $D = D_U$ and $\rho = \rho_L$

$$\omega_U^2 = \frac{D_U}{\rho_L}\,\frac{\int_0^\ell (\psi'')^2\,dx}{\int_0^\ell A\psi^2\,dx} \tag{19}$$

Now we will employ finite element technique to the string and beam vibrations to reduce the system to multi-degree-of-freedom systems and utilize the above formulated theorem of Qiu et al (1995).

APPLICATION TO UNCERTAIN STRING VIBRATION

Here we determine the frequencies of the string when its mass per unit length $m(x)$ and the extension force $T(x)$ are interval variables

$$m_L \leq m(x) \leq m_U \quad , \quad T_L \leq T(x) \leq T_U \tag{20}$$

The equation of motion reads

$$\frac{\partial}{\partial x}\left[T(x)\,\frac{\partial u}{\partial x}\right] = m(x)\,\frac{\partial^2 u}{\partial t^2} \tag{21}$$

We use the finite element method. The stiffness and mass matrices read:

$$k_{ij} = \int_0^L T(x)\,N_i'(x)\,N_j'(x)\,dx$$

$$m_{ij} = \int_0^L m(x)\,N_i(x)\,N_j(x)\,dx \tag{22}$$

where $N_i(x)$ are shape functions

$$N_1(x) = 1 - \xi \quad , \quad N_2(x) = \xi \quad , \quad \xi = x/L \tag{23}$$

The lower and upper elemental stiffness matrix elements can be written

$$K_{e,L} = \frac{1}{L}\begin{bmatrix} T_L & -T_U \\ -T_U & T_L \end{bmatrix}$$

$$K_{e,U} = \frac{1}{L}\begin{bmatrix} T_U & -T_L \\ -T_L & T_U \end{bmatrix} \tag{24}$$

The appropriate mass matrices read

$$M_{e,L} = \frac{m_L}{6}\begin{bmatrix} 2 & 1 \\ 1 & 2 \end{bmatrix} \quad , \quad M_{e,U} = \frac{m_U}{6}\begin{bmatrix} 2 & 1 \\ 1 & 2 \end{bmatrix} \tag{25}$$

Consider the string of length ℓ represented by N elements with span length $L = \ell/N$, $n \geq 2$. Considering boundary conditions $u(u = 0, t) = u(x = 1,t) = 0$ we get following global matrices, of dimension $(N - 1) \times (N - 1)$. The global stiffness matrix reads

$$
K_L = \frac{1}{L}
\begin{bmatrix}
2T_L & -T_U & 0 & 0 & \cdots & 0 & 0 \\
-T_U & 2T_L & -T & & & & \\
0 & -T_U & 2T_L & -T_U & \cdots & 0 & 0 \\
\vdots & & & & & & \\
0 & 0 & 0 & 0 & \cdots & 2T_L & -T_U \\
0 & 0 & 0 & 0 & \cdots & -T_U & 2T_L
\end{bmatrix}
\qquad (26)
$$

In order to obtain matrix K_U we must replace $T_L \to T_U$ and $T_U \to T_L$. The lower bound mass matrix reads

$$
M_L = \frac{m_L}{6}
\begin{bmatrix}
4 & 1 & 0 & 0 & \cdots & 0 & 0 \\
1 & 4 & 1 & 0 & \cdots & 0 & 0 \\
0 & 1 & 4 & 1 & \cdots & 0 & 0 \\
\vdots & & & & & \vdots & \\
0 & 0 & 0 & 0 & \cdots & 4 & 1 \\
0 & 0 & 0 & 0 & \cdots & 1 & 4
\end{bmatrix}
\qquad (27)
$$

To obtain M_U we must formally change m_L by m_U.

In order to obtain natural frequencies we must use two Rayleigh-Quotient type equations, specified by Eqs. (13) and (14), respectively. For example, for $n = 3$, we get the following equation for λ_L:

$$
\det\left\{ \frac{3}{\ell}
\begin{bmatrix}
2T_L & -T_U \\
-T_U & 2T_L
\end{bmatrix}
- \frac{m_U \ell}{18}
\begin{bmatrix}
4 & 1 \\
1 & 4
\end{bmatrix}
\lambda_L \right\} = 0
\qquad (28)
$$

With notation

$$
\alpha_L = \frac{54T_L}{m_U \ell^2} \quad , \quad
\alpha_U = \frac{54T_U}{m_L \ell^2} \quad , \quad
\beta_L = \frac{T_L}{T_U} \quad , \quad
\beta_U = \frac{T_U}{T_L}
\qquad (29)
$$

we get

$$
\det\left\{ \alpha_L
\begin{bmatrix}
2 & -\beta_U \\
-\beta_U & 2
\end{bmatrix}
-
\begin{bmatrix}
4 & 1 \\
1 & 4
\end{bmatrix}
\lambda_L \right\} = 0
\qquad (30)
$$

or

$$\lambda_{1,L} = \frac{\alpha_L(2 - \beta_U)}{5} = \frac{54}{5}\frac{T_L}{m_U\ell^2}(2 - \beta_U) \tag{31}$$

Analogously,

$$\lambda_{1,U} = \frac{\alpha_U(2 - \beta_L)}{5} = \frac{54}{5}\frac{T_U}{m_L\ell^2}(2 - \beta_L) \tag{32}$$

The middle point values are expressed as

$$T_C = \frac{1}{2}(T_L + T_U) \quad , \quad m_C = \frac{1}{2}(m_L + m_U) \tag{33}$$

If $T_L = 0.095T_C$, $T_U = 1.05T_C$, $m_L = 0.95m_C$, $m_U = 1.05m_C$, then $\beta_L = 1.10526$, $\beta_U = 0.90476$.   Hence

$$\lambda_{1,L} = 10.702\frac{T}{m\ell^2} \quad , \quad \lambda_{1,U} = 13.074\frac{T}{m\ell^2} \tag{34}$$

Which constitute the estimates of the lower and upper bounds of natural frequencies due to uncertainty in elastic modulus and material mass density.

APPLICATION TO UNCERTAIN BEAMS

For the beam the stiffness and mass matrices read

$$k_{ij} = \int_0^L D(x) N_i''(x) N_j''(x)\, dx$$

$$m_{ij} = \int_0^L \rho A(x) N_i(x) N_j(x)\, dx \tag{35}$$

where the shape functions read:

$$\begin{aligned}
N_1(\xi) &= 1 - 3\xi^2 + 2\xi^3 \\
N_2(\xi) &= L(\xi - 2\xi^2 + \xi^3) \\
N_3(\xi) &= 3\xi^2 - 2\xi^3 \\
N_4(\xi) &= L(-\xi^2 + \xi^3)
\end{aligned} \tag{36}$$

were $\xi = x/L$.   Derivation of lower and upper matrices K_L and K_U is more complex in this case.   Compute for example

$$k_{ij,L} = \int_0^L \min\left[D(x)N_i''(x)N_j''(x)\right]dx$$

$$k_{ij U} = \int_0^L \max\left[D(x)N_i''(x)N_j''(x)\right]dx \tag{37}$$

Let $i \neq j$; if $N_i''N_j'' > 0$ at a given x, then $\min[D(x)N_i''N_j''] = D_L N_i''N_j''$ and $\max[D(x)N_i''N_j''] = D_U N_i''N_j''$. However, if $N_i''N_j'' < 0$ then $\min[D(x)N_i''N_j''] = D_U N_i''N_j''$ whereas $\max[D(x)N_i''N_j''] = D_U N_i''N_j''$. For $i = j$, $N_i''N_i''$ is always greater than zero. For $i \neq j$ we must determine the regions where the product $N_i''N_j''$ changes its sign. For example

$$k_{21,L} = D_L \int_0^{1/2} N_1''N_2''d\xi + D_U \int_{1/2}^{2/3} N_1''N_2''d\xi$$

$$+ D_L \int_{2/3}^1 N_1''N_2''d\xi \tag{38}$$

The matrices K_L and K_U for uncertain beam were computed by Köylüoğlu et al (1994) in the static context:

$$K_L = \begin{bmatrix} 12D_L/L^3 & & & \\ & 4D_L/L^3 & & \\ & & 12D_L/L^3 & \\ & & & 4D_L/L^3 \end{bmatrix} \tag{39}$$

whereas the elemental matrix $\underline{M}$ is given below

$$M_L = \frac{AL}{420} \begin{bmatrix} 156\rho_L & 22\rho_L & 54\rho_L & -13\rho_U \\ 22\rho_L & 4\rho_L & 13\rho_L & -3\rho_U \\ 54\rho_L & 13\rho_L & 156\rho_L & -22\rho_U \\ -13\rho_U & -3\rho_U & -22\rho_U & 4\rho_L \end{bmatrix} \tag{40}$$

To obtain the upper elemental matrix M_U we must perform the formal substitution by replacing ρ_L by ρ_U, and ρ_U by ρ_L. For convenience we introduce mean matrices and deviations:

$$K_c = \frac{1}{2}(K_L + K_U) \quad , \quad \Delta K = \frac{1}{2}(K_U - K_L) \tag{41}$$

$$M_C = \frac{1}{2}(M_L + M_U) \quad , \quad \Delta M = \frac{1}{2}(M_U - M_L) \tag{42}$$

Hence

$$\Delta K = \frac{\Delta D}{L^3} \begin{bmatrix} 12 & \frac{55}{9} & 12 & \frac{55}{9} \\ \frac{55}{9} & 4 & \frac{55}{9} & \frac{22}{9} \\ 12 & \frac{55}{9} & 12 & \frac{55}{9} \\ \frac{55}{9} & \frac{22}{9} & \frac{55}{9} & 4 \end{bmatrix} \tag{43}$$

$$\Delta M = \frac{AL\Delta\rho}{420} \begin{bmatrix} 156 & 22 & 54 & 13 \\ 22 & 4 & 13 & 3 \\ 54 & 13 & 156 & -22 \\ 13 & 3 & -22 & 4 \end{bmatrix} \tag{44}$$

Then

$$K_U = K_C + \Delta K \quad , \quad K_L = K_C - \Delta K$$
$$M_U = M_C + \Delta M \quad , \quad M_L = M_C - \Delta M \tag{45}$$

Consider for illustration purposes a single element approximation for the damped free beam. The stiffness and mass matrices read:

$$K_C = \begin{bmatrix} 12 & -6 \\ -6 & 4 \end{bmatrix} \frac{D_C}{L^3} ,$$

$$\Delta K = \begin{bmatrix} 12 & 55/9 \\ 55/9 & 4 \end{bmatrix} \frac{\Delta D}{L^3}$$

$$M_C = \begin{bmatrix} 156 & -22 \\ -22 & 4 \end{bmatrix} \frac{\rho_C AL}{420} \tag{46}$$

$$\Delta M = \begin{bmatrix} 156 & 22 \\ 22 & 4 \end{bmatrix} \frac{\Delta\rho AL}{420}$$

The bounds of eigenvalues are found from

$$\det|K_L - \lambda_L M_U| = 0$$
$$\det|K_U - \lambda_U M_L| = 0 \tag{47}$$

Let

$$\alpha = \frac{\Delta D}{D_U} \ , \quad \beta = \frac{\Delta \rho}{\rho_U}$$

$$\zeta = \frac{\rho_U A L}{420} \ , \quad \eta = \frac{D_U}{L_3} \tag{48}$$

Then the equation for λ_U **becomes**

$$a_0 + a_1 \lambda_U + a_2 \lambda_U^2 = 0 \tag{49}$$

where

$$a_0 = 48(1 + \alpha)^2 \eta^2 - \left(\frac{55}{9}\alpha - 6\right)^2 \eta^2$$

$$a_1 = \left[-44(1 + \alpha)\left(\frac{55}{9}\alpha - 6\right) - 6\eta^2(1 - \alpha^2)\right]\zeta\eta \tag{50}$$

$$a_2 = \left[624(1 - \alpha)^2 - 484(1 + \alpha)^2\right]\zeta^2$$

Consider special cases. If $\alpha = \beta = 0$ **then**

$$a_0 = 12\eta^2 \ , \quad a_1 = -408\,\zeta\eta$$

$$a_2 = 140\,\zeta^2 \tag{51}$$

and

$$\lambda_{1,U} = 12.4803\,\frac{D}{\rho A} \tag{52}$$

For $\alpha = \beta = 10^{-3}$

$$a_0 = 12.1693\eta^2$$

$$a_1 = -408.00449\,\zeta\eta \tag{53}$$

$$a_2 = 137.7841\,\zeta^2$$

and

$$\lambda_{1,U} = 12.66 \frac{D_U}{\rho_U A \ell^4} \tag{54}$$

To determine $\lambda_{1,L}$ we should again solve Eq. (44) but with the following coefficients

$$a_0 = 11.8306773\eta^2 \quad , \quad a_1 = -407.9947085\zeta\eta$$
$$a_2 = 142.21614\zeta^2 \tag{55}$$

leading to

$$\lambda_{1,L} = 12.30 \frac{D_U}{\rho_U A \ell^4} \tag{56}$$

Further numerical results, including other boundary conditions and the dependence upon the number of elements one may consult with the study by Duan (1995).

CONCLUSION

If the limited information is available about the elastic moduli and the mass density, one can utilize the methods of non-probabilistic, interval analysis to determine the range of variations of the natural frequencies. Special efforts need to be directed towards development of tight bounds of the uncertain eigenvalues.

ACKNOWLEDGEMENT

The financial support by National Science Foundation through the SGER grant (Dr. K. P. Chong, Program Director) is gratefully appreciated.

REFERENCES

Ben-Haim Y., *Robust Reliability in the Mechanical Sciences*, Springer Verlag, Berlin, 1996.

Ben-Haim, Y. and Elishakoff, I., *Convex Models of Uncertainty in Applied Mechanics*, Elsevier Science Publishers, Amsterdam, The Netherlands, 1990.

Courant R. and Hilbert D, *Methoden der Mathematischen Physik*, Vol. 1, Springer, Berlin, 1924.

Köylüoğlu H. U., Cakmak A. S. and Nielsen S. R. K. Applications of Interval to Deal with Pattern Loading and Structural Uncertainties, *Journal of Engineering Mechanics,* Vol., 121, 1995 (see also a discussion by Modaressi H. and authors closure Vol. 123, p 645, 1997).

Krein M.G., About Some Problems on Maximum and Minimum for Characteristic Numbers and about Lyapunov Regions of Stability, *PMM-Prikladnaya Matematika i Mekhanika,* Vol. 15, 323-348, 1951 (in Russian).

Nakagiri S. and Suzuki K., Interval Estimation on Finite Element Sensitivity Analysis of Stiffness Equation, *Transactions of the Japan Society of Mechanical Engineers,* Vol. 62, No. 603, Series A, 2435-2439, 1996 (in Japanese).

Neumaier A. *Interval Methods for Systems of Equations,* Cambridge University Press, 1990.

Qiu Z. P., Chen S. H. and Elishakoff I., Natural Frequencies of Structures with Uncertain but Non-Random Parameters, *Journal of Optimization Theory and Applications,* Vol. 86, 669-683, 1995.

Ramu A. S. and Ganesan R., Parametric Stability of Stochastic Columns, *International Journal of Solids and Structures,* Vol. 30, 1339-1351, 1993.

Rao S.S. and Berke L., Analysis of Uncertain Structural Systems using Interval Analysis, *AIAA Journal,* Vol. 35, 727-735, 1997.

Shinozuka M., Structural Response Variability, *Journal of Engineering Mechanics,* Vol. 113, 825-842, 1987.

Young R. C., The Algebra of Many-Valued Quantities, *Math,* Vol. 104, 260-290, 1931.

Zhukovskii N.E., Conditions of Finiteness of Integrals of an Equation $d^2y/dx^2 + py = 0$, *Collection of Papers,* "ONTI" Publishers, Vol. 1, Moscow, pp.315-322, 1948 (in Russian).

Dong W. M., Chiang W. L. and Shah H. C., Fuzzy Information Processing in Seismic Hazard Analysis and Decision Making, *Soil Dynamics and Earthquake Engineering*, Vol. 6, 1987.

Duan D., Dynamic Response and Stability of Viscoelastic Structures by Interval Mathematics, *M. Sc. Thesis*, College of Engineering, Florida Atlantic University, Dec. 1994.

Elishakoff, I., *Probabilistic Methods in the Theory of Structures*, Wiley-Interscience, New York, 1983.

Elishakoff, I., Some Questions in Eigenvalue Problems in Engineering, *Numerical Treatment of Eigenvalue Problems,* (J. Albrecht, L. Collatz, P. Hagedorn and W. Velte, eds.), Birkhäuser Publishers, Basel, pp. 71-107, 1991.

Elishakoff I., Lin Y. K. and Zhu L. P., *Probabilistic and Convex Modelling for Acoustically Excited Structures*, Elsevier Science Publishers, Amsterdam, 1994.

Elishakoff I. Elisseeff, P. and Glegg, S., Convex Modeling of Material Uncertainty in Vibrations of a Viscoelastic Structure, *AIAA Journal*, Vol. 32, 843-849, 1994.

Elishakoff I., and Duan D., Application of the Mathematical Theory of Interval Analysis for Uncertain Vibrations, Proceedings, The 1994 National Conference on Noise Control Engineering (Cushieri J. M. et al, eds.), New York, pp. 519-524, 1994.

Gersch W., Mean Square Responses in Structural Systems, *Journal of the Acoustical Society of America*, Vol. 48, 403-413, 1970.

Guenther N. M., *Theory of Potentials and Its Applications to Basic Problems of Physics*, Fizmatgiz, Moscow, 1953 (in Russian).

Hansen E., *Global Optimization using Interval Analysis*, Marcel Dekker, New York, 1992.

Ibrahim R., Structural Dynamics with Parameter Uncertainties, *Applied Mechanics Reviews*, Vol. 40 (3), 309-328, 1987.

Jensen, H. and Iwan, W.D., Response of Systems with Uncertain Parameters to Stochastic Excitation, *Journal of Engineering Mechanics*, Vol. 118 (5), 1012-1025, 1992.

CHAPTER 6

**HOW TO DEVELOP AND UTILIZE NUMERICAL ALGORITHMS
WITH AUTOMATIC RESULT VERIFICATION?**

U. Kulisch
University of Karlsruhe, Karlsruhe, Germany

ABSTRACT

Advances in computer technology are now so profound that the arithmetic capability and repertoire of computers can and should be expanded. The quality of the elementary floating-point operations should be extended to the most frequent numerical data types or mathematical spaces of computation (vectors, matrices, complex numbers and intervals over these types). A VLSI coprocessor chip with integrated PCI-interface has been developed which provides these operations. The expanded capability is gained at modest hardware cost and does not implicate a performance penalty. Language and programming support (the -XSC languages) are available. There, the chip's functionality is directly coupled to the operator symbols for the corresponding data types. By operator overloading a long real arithmetic (array of reals) and long interval arithmetic as well as automatic differentiation arithmetic become part of the runtime system of the compiler. I. e. derivatives, Taylor-coefficients, gradients, Jacobian and Hessian matrices or enclosures of these are directly computed out of the expression by a simple type change of the operands. Techniques are now available so that with this expanded capability, the computer itself can be used to appraise the quality and the reliability of the computed results over a wide range of applications. Program packages for many standard problems of Numerical Analysis have been developed where the computer itself verifies the correctness of the computed result and proves existence and uniqueness of the solution within the computed bounds.

Many applications require that rigorous mathematics can be done with the computer using floating-point. As an example, this is essential in simulation runs (fusion reactor) or mathematical modeling where the user has to distinguish between computational artifacts and genuine reactions of the model. The model can only be developed systematically if errors entering into the computation can be essentially excluded. Automatic result verification re-integrates digital computing into real mathematics.

---

[1]This is a slightly changed version of the paper *Numerical Algorithms with Automatic Result Verification* originally published in *Lectures in Applied Mathematics*, vol. 32, 1996, pp. 471–502, edited by J. Renegar, M. Shub and St. Smale. Republished with the permission of the AMS.

1. INTRODUCTION

Floating-point arithmetic is the fast way to perform scientific engineering computations. It has been invented to avoid annoying scalings inherent in fixed-point computations. Floating-point arithmetic is now used for more than fifty years. Its quality has been improved over the years. The data format was extended from less to 64 bits and the IEEE arithmetic standard has finally taken the bugs out of it. IEEE arithmetic has been developed during the early eighties as a standard for microprocessors. Since that time the speed of microprocessors has been increased by a factor of more than 1000. IEEE arithmetic is now even provided by supercomputers, the speed of which is still faster by one or two magnitudes. This raises the question whether all this is still in balance.

Floating-point arithmetic has been used very successfully during the past. Very sophisticated and intelligent algorithms and libraries have been developed for particular problems. However, in general applications there are still inherent difficulties afflicted with floating-point arithmetic. The result of a floating-point computation can be correct, it can be inaccurate or even completely wrong. The computation itself as well as the computed data do not indicate which one of the three cases has occurred.

On the other hand computer technology has been improved tremendously during the last decades. In the early fifties computers were able to perform about 100 floating-point operations in a second. Today they are able to execute one hundred million floating-point operations per second on the chip level and the development will still go on. It can be expected that in fifteen years from now computers will again be 1000 times faster, i. e. they will double their speed in about every 18 months.

Emphasis in computing was traditionally on speed. The speed of an algorithm was often considered as being of higher priority than the accuracy and reliability of the computed result. This attitude should be reversed now. Advances in computer technology are now so profound that the mathematical and arithmetical capability and repertoire of computers can and should be expanded so that reliability and mathematical quality of computed results can obtain much higher priority.

There are many applications which require rigorous mathematics to be done with a computer using floating-point numbers. As an example, this is essential in simulation runs (fusion reactor) or mathematical modeling where the user has to distinguish between computational artifacts and genuine reactions of the model. The model can only be developed systematically if errors entered by the computation can be excluded essentially. There are many other applications, which can be solved more adequate and simple or can be solved at all, if computers with expanded arithmetic capability are available. We illustrate this by a few examples. To refrain from too many details, all examples are chosen particularly simple. More refined solution methods are available in all cases.

1.1 Non zero property of a function

We consider the question whether a given real function, defined by an arithmetic expression, has no zeros in a given interval X, Figure 1. This question cannot be answered with mathematical certainty if only floating-point arithmetic is available. All one can do, is to evaluate the function say at 1000 points in the interval X. If all computed function values are positive, it is very likely that the function does not have any zeros in X. However, this conclusion is certainly not reliable. Because of roundoff errors, a positive result could be computed for a negative function value. The function could also descend to negative values which was not detected due to the choice of the evaluation points, (see the lower part of Figure 1). The question can be answered much simpler, if interval arithmetic is available. A single interval evaluation of the function may suffice to solve the problem with complete mathematical certainty. We evaluate the function only once in interval arithmetic for the interval X. This delivers a superset of the range of all function values over X. If this superset does not contain zero, as in our example, the range, which is a subset, does not contain zero. As a consequence, the function does not have any zeros in the interval X.

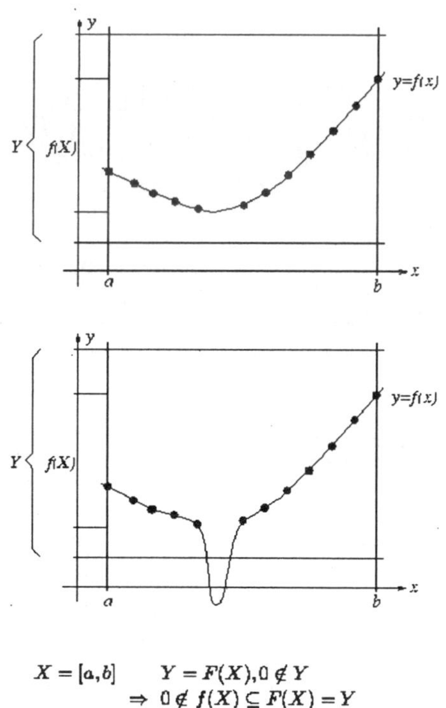

$$X = [a,b] \qquad Y = F(X), 0 \notin Y$$
$$\Rightarrow\ 0 \notin f(X) \subseteq F(X) = Y$$

Fig. 1, Non zero property of a function

This example contradicts already the rumor that is going around, that interval arithmetic is more costly than floating-point arithmetic. The interval evaluation of the function is only about twice as costly as a single floating-point evaluation – not counting the 1000 floating-point evaluations mentioned above.

1.2   Eigenfrequencies of a shaft or turbine

Figure 2. The heavy engine is working at high temperatures. When it is started the speed has to be increased slowly so that it warms up smoothly. While the speed is increased the eigenfrequencies have to be avoided. If it runs on an eigenfrequency or nearby it breaks explosively. I. e. the speed has to be increased fast in the neighbourhood of an eigenfrequency and slowly in areas which definitely do not contain any eigenvalues. The eigenfrequencies are found as zeros of an expression involving hyperbolic and trigonometric standard functions and a good number of matrix vector products. A conventional computation exhibits an accumulation of eigenfrequencies at high rotation speeds (light curve). An accuracy controlled computation (dark curve) reveals them as numerical artifacts. The engine can be driven at higher rotation speeds.

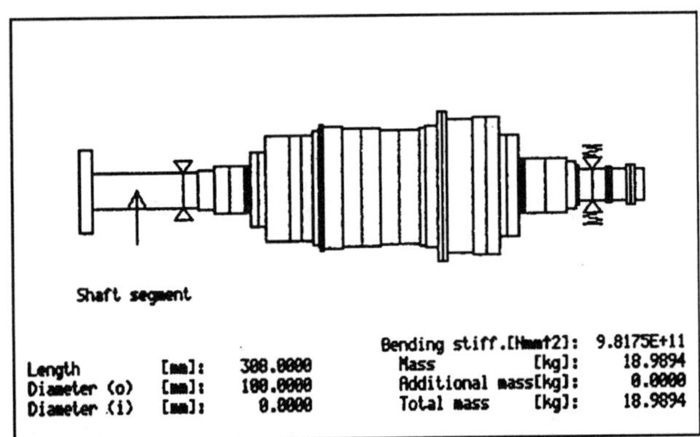

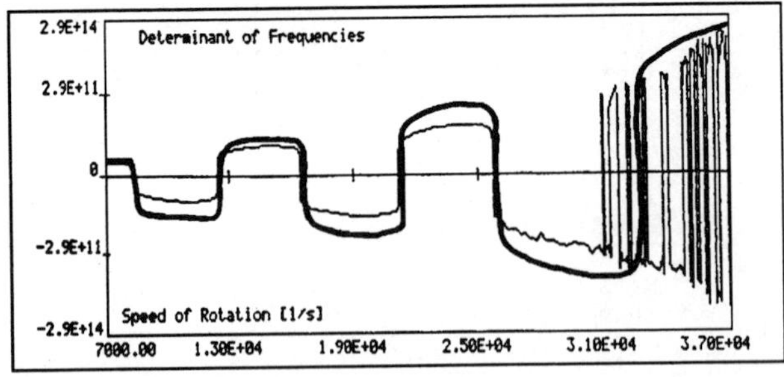

Fig. 2, Eigenfrequencies of a shaft

The problem is closely related to the former example. The main question is: Which ranges of speed do definitely not contain any eigenvalue in a strict mathematical sense?

1.3 Verified solution of systems of equations and other algebraic problems

We briefly sketch a method by which the correctness of a computed solution of a linear system of equations $A \cdot x = b$ can be verified. At first, an approximation of the solution is computed by some favorite method, for example by Gaussian elimination. This approximation $\tilde{x}$ is then expanded by augmenting it with a small number $\pm\varepsilon$ in each coordinate direction.

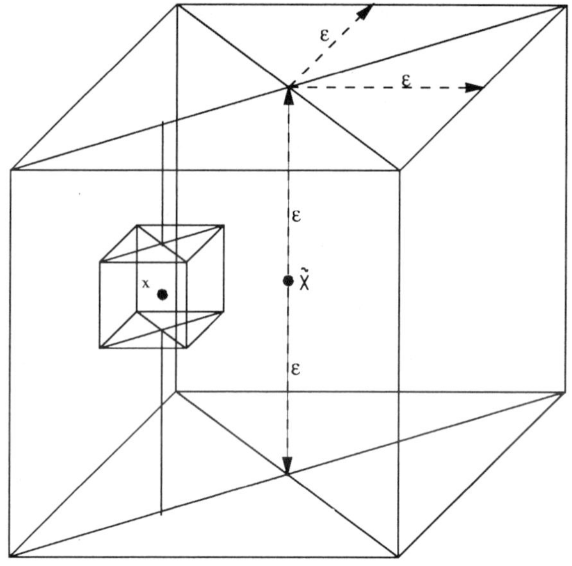

Fig. 3, Verified solution of systems of equations

This results in an n−dimensional cube X which has the computed approximation $\tilde{x}$ as its midpoint, Figure 3. This cube X is just an interval vector consisting of an interval in each coordinate direction. Now, we transform the system of equations in an appropriate way into a fixed point form $x := D \cdot x + z$. Here, the fact that a presumed approximate solution has already been computed can be used in many natural ways. Now we compute the image $Y := D \cdot X + z$ which is an interval vector again. If Y is contained in X (see Figure 3), which can be verified simply by comparison of endpoints, then Y contains a fixed point of the equation $x := D \cdot x + z$ by Brouwer's fixed point theorem[2]. If Y is strictly contained in X, i. e. the end points do not touch,

---

[2]Brouwer's fixed point theorem asserts the following: Let $\varphi : \mathbb{R}^n \to \mathbb{R}^n$ be a continuous mapping,

then the fixed point is unique. Thus, it has been verified computationally that the original matrix of the system of equations is nonsingular. Using this information, the fixed point can then be determined to high accuracy in practice by an iteration using an optimal dot product. In this way, the user obtains a mathematically certain and reliable result. In rather rare cases the reliable answer may be that the verification step failed. In this case another solution method or refined technique can automatically be called by the computer. It may consist in a repetition of the computation using higher precision.

Similar techniques of automatic result verification can be applied to many other algebraic problem areas, such as the solution of nonlinear systems of equations, the computation of zeros, the calculation of eigenvalues and eigenvectors of matrices, optimization problems, etc. In particular, valid and highly accurate evaluation of arithmetic expressions and functions on the computer is included. These routines also work for problems with (small) interval data.

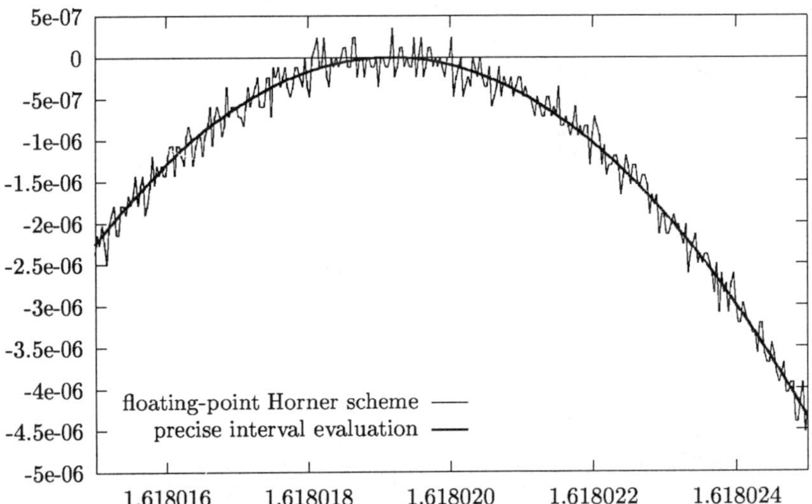

Polynomial $p(x) = 223200658x^3 - 1083557822x^2 + 1753426039x - 945804881$

Fig. 4, Function evaluation in the neighbourhood of a zero

The evaluation of a function is reduced to the solution of a simple system of equations, an approximate solution of which is corrected if necessary by defect correction and an optimal dot product to a guaranteed maximal accuracy. This technique of enclosing function values with high accuracy is an essential aid for numerics. Newton's method for the computation of roots of a system of equations is frequently unstable in the neighbourhood of a root, Figure 4. That is to say, a root results from the mutual

---

$X \subseteq \mathbb{R}^n$ be nonempty, closed, convex, and bounded, and $Y := \varphi(X)$. If $Y \subseteq X$, then φ has at least one fixed point in Y.

cancellation of positive and negative terms. In the immediate vicinity of a root, these terms are only approximately equal and cancellation occurs in ordinary floating-point arithmetic. However, in the case of insufficiently accurate function evaluation, Newton's method can easily overshoot its goal, possibly into the domain of attraction of another root. An iteration process which converges to a solution x in the real numbers may thus not converge to the value x in ordinary floating-point arithmetic. It can swing over to an entirely different root. A conclusive verification step is thus also indicated for iteration processes.

A useful and often applied method to strive for high accuracy consists of enclosing the error of an approximation instead of the solution itself.

The techniques sketched here for numerics with automatic result verification use an optimal dot product and interval arithmetic as essential tools in addition to floating-point arithmetic. Interval arithmetic permits computation of certain bounds for the solution of a problem. One obtains high accuracy by means of the optimal dot product. The combination of both instruments is indeed a breakthrough in numerical analysis. Naive application of interval arithmetic certainly always leads to reliable bounds. However, these will often be too wide to be useful. This effect was observed already 20 years ago by many numerical analysts, leading frequently to the rejection once for all of this useful, necessary and blameless tool. Recently, all processors which provide the IEEE arithmetic standard do provide also the four basic arithmetic operations with the directed roundings. They are the essential ingredients for the four interval operations. However, they are hardly used in numerical analysis. An additional feature for controlling the accuracy is missing.

1.4 Global optimization

As the next example we consider the task to determine the global minimum of a function in a given domain. Again, we restrict our discussion to the most simple, the one dimensional case, Figure 5. The given domain is subdivided into a number of subintervals. In each subinterval we now evaluate the function in interval arithmetic. This delivers a superset of the range of values of the function in each subinterval. Now we select the subinterval with the lowest lower bound for the range of function values. For an inner point of this subinterval we evaluate the function in interval arithmetic. This delivers a guaranteed upper bound of a function value. Those subintervals, the lower bound of which is greater than this function value can now be eliminated from further treatment, because they cannot contain the global minimum. The remaining subintervals (two in the example in Figure 5) now are further subdivided and treated by the same technique.

The method also works and succeeds very well for higher dimensions up to 30 or 40, since whole continua can quickly be excluded from the search. In [4], [5] and [9] more refined methods are used. Frequently, safe bounds for the global minimum can be computed faster than an approximation delivered by conventional techniques the quality of which is still uncertain.

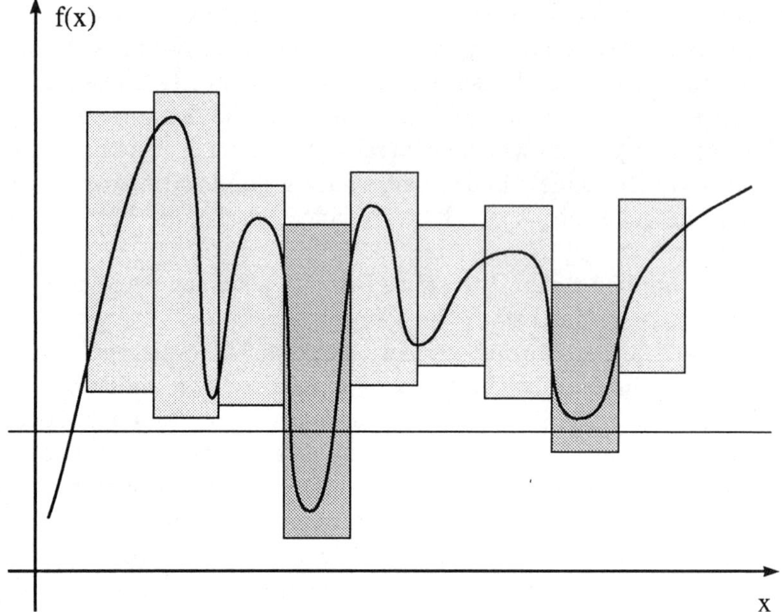

Fig. 5, Global optimization

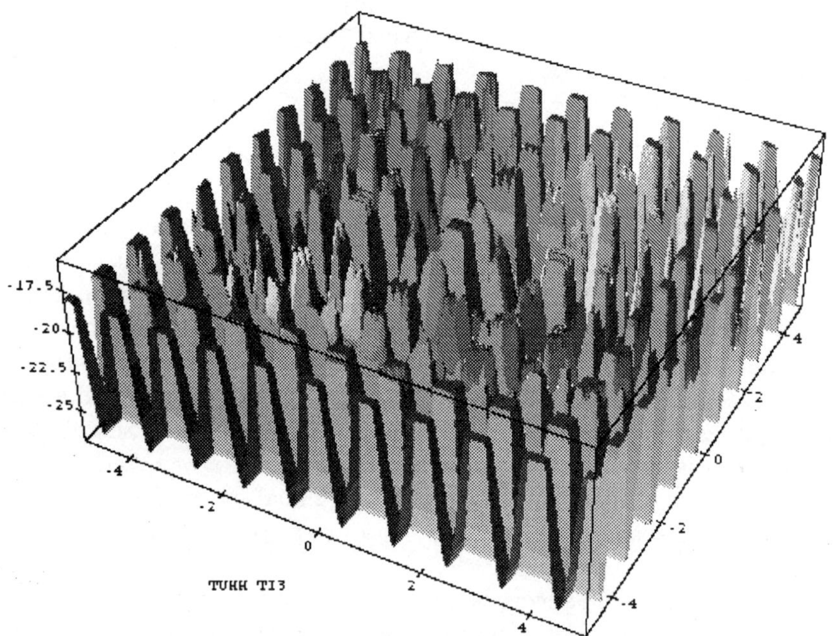

Fig. 6, Global optimization of a non differentiable function

Figure 6 shows a two dimensional example. Here, we do the computation of the global maximum of a function which is not differentiable. The picture is taken from a paper by Siegfried M. Rump.

These methods also can be used for a fast and highly accurate computation of the range of values of a function in a given domain.

1.5 The double pendulum

The behaviour of the double pendulum can be described by a system of ordinary differential equation of second order which are heavily nonlinear. We integrate this system with two different numerical solvers starting with different but identical initial conditions. Figure 7 shows the computed solutions. The solutions computed in case of the left pendulum look reasonable. Verification by visualization tells us, that we probably have got a good answer. The same conclusion can be drawn when looking at the solutions computed for the right pendulum. But if we stop both integrations after a fixed time and compare the computed values, we see that in the second and third case the final values differ. If we go on with the integration the differences are getting bigger and bigger. So at least one of the computed solutions must be incorrect or even wrong. Both can be wrong also. Which one is the better solution? In such a situation, which is not atypical in floating-point computations, the user tries to find an answer by trial and error starting again and again with different initial conditions. He tries to develop a feeling which one might be the better solution perhaps also by mathematical estimations. None of these is a safe method.

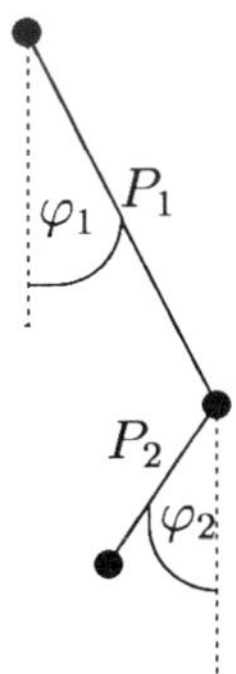

In our case the decision is simple, since the solution on the right hand side has been computed with a verifying solver. It delivers continuous upper and lower bounds for the solution which coincide to at least 10 decimal digits. Of course, the computation of a particular solution with a verifying solver is more costly and it requires a more powerful arithmetic. But it definitely solves the problem by a single run what the trial and error method does not.

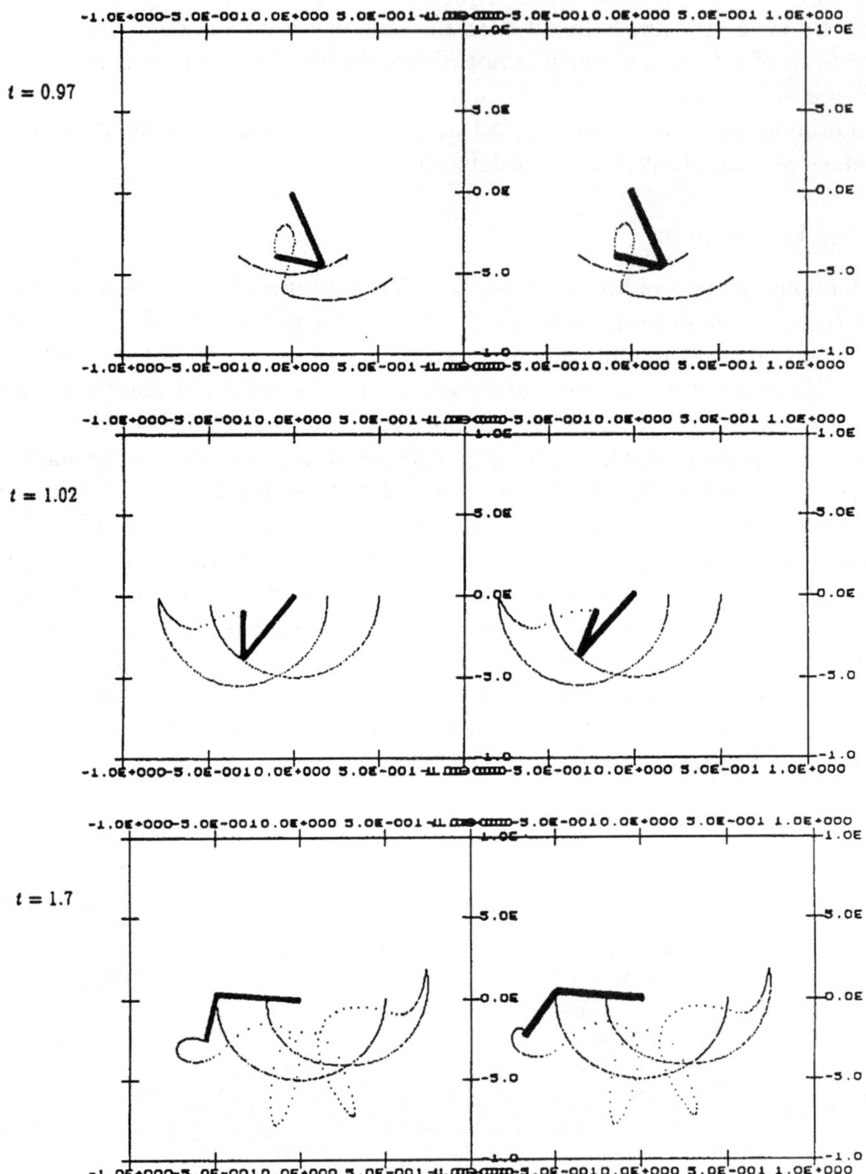

Fig. 7, The double pendulum

1.6 Dynamical systems

We consider a simple, nonlinear iterative scheme, the so called logistic equation:

$$x_{n+1} := 3.75 * x_n * (1 - x_n), \quad n = 0, 1, 2, \dots \tag{1}$$

Similar schemes occur frequently if chaotic systems are considered. They have the property to be very sensitive with respect to changes in the data. (1) is an iteration scheme for real numbers. We compute the sequence, defined by (1), in floating-point arithmetic and start the iteration with the initial value $x_0 = 0.5$. Each step of the iteration enlarges the number of digits. Soon, all 16 digits of the floating-point mantissa are filled.

n	floating point	floating point interval	long interval
0	0.5000000000000000	0.5000000000000000	0.5000000000000000
1	0.9375000000000000	0.9375000000000000	0.9375000000000000
10	0.6453672908309288	$0.6453672908309^{382}_{231}$	$0.645367290830930^{4}_{2}$
20	0.8259709787108499	$0.82597097871^{11930}_{03328}$	$0.825970978710775^{6}_{4}$
30	0.7180965684239893	$0.7180965568^{4481460}_{3876004}$	$0.718096568418754^{8}_{6}$
40	0.4163493160568014	$0.4163493^{223190691}_{118996043}$	$0.416349316957635^{7}_{6}$
50	0.3604395431283658	$0.3604^{401742922880}_{391241304220}$	$0.360439633921698^{7}_{5}$
60	0.7990957294909673	$0.799^{1391103575510}_{0303938934906}$	$0.799086334308309^{6}_{4}$
70	0.4520542523856749	$0.45^{31654999449893}_{12732497659017}$	$0.452195299859730^{2}_{1}$
80	0.8492167613301459	$0.{}^{9469520868573489}_{7677390706389728}$	$0.856177996629336^{6}_{4}$
90	0.8481094253763446	$[0,1]$	$0.73991374860737^{11}_{09}$
100	0.8358933994881687	$[0,1]$	$0.88829399228403^{50}_{48}$
150	0.8504150611290033	$[0,1]$	$0.702820413487813^{7}_{5}$
200	0.9131976055921124	$[0,1]$	$0.82355732113045^{71}_{69}$
250	0.9310785715888734	$[0,1]$	$0.219734861549818^{2}_{1}$
300	0.3198944553532391	$[0,1]$	$0.496432049973525^{8}_{6}$
350	0.9287366136360928	$[0,1]$	$0.859325800193209^{2}_{0}$
400	0.2202363784969592	$[0,1]$	$0.69393944643472^{21}_{19}$
450	0.9345460787353592	$[0,1]$	$0.61598678450720^{31}_{11}$

$$x_{n+1} \; := \; 3.75 \cdot x_n \cdot (1 - x_n), \quad n = 0, 1, 2, \ldots$$
$$x_0 \; := \; 0.5$$

Fig. 8, The logistic equation

If we continue to compute in floating-point arithmetic and round the result of each operation into a floating-point number, we are leaving the sequence defined by x_0 and (1). The computed approximation $\tilde{x}_v$ could be considered as an initial value for a new iteration. But also the sequence defined by this iteration is left during the next step if the calculation is performed in floating-point. The approximations which are actually calculated jump from one sequence to another one in each step of the iteration. If the different sequences of (1) differ essentially, the floating-point iteration computes a sequence which has not much to do with the sequence defined by x_0 and (1).

In Figure 8 the second column shows the *approximations* computed in double precision floating-point arithmetic, while the fourth column shows highly accurate enclosures of the correct values of the sequence defined by x_0 and (1). The numbers in

the third column are computed in double precision interval arithmetic. They show
drastically how fast the number of correct digits decreases if the calculation is done in
floating-point arithmetic (second column).

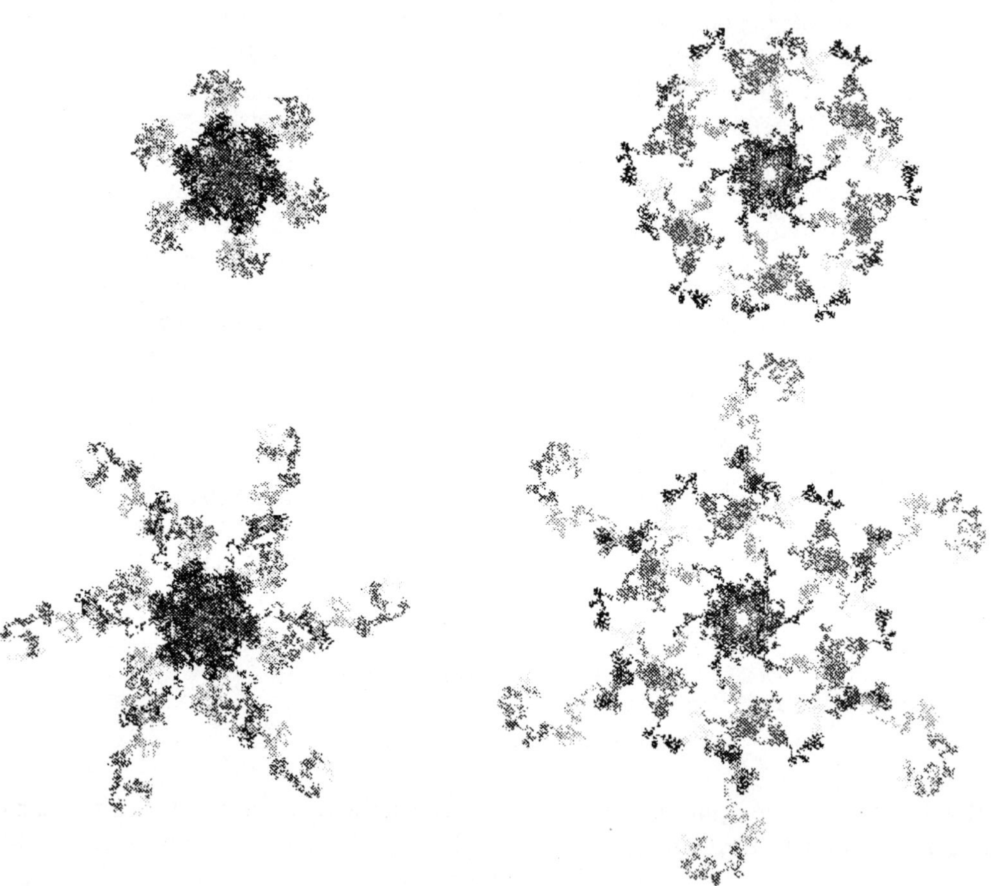

Fig. 9, Fractal sequence

Figure 9 shows a two dimensional similar example (G. Bohlender). A fractal sequence (x_n, y_n) is computed in floating-point by the following iteration process:

$$
\begin{aligned}
x_{n+1} &:= C * x_n + S * y_n - \sin\left(-3 * x_n - 36\right) \\
y_{n+1} &:= 12 - S * x_n + C * y_n, \quad x_0 := 0, \quad y_0 := 0, \\
&\text{with } C = \sin\tfrac{\Pi}{3}, \quad S = \sin\tfrac{\Pi}{3}
\end{aligned}
$$

The computed values (x_n, y_n) are printed. After 2000 iterations the colour is changed. The pictures are showing the result of two identical calculations in single precision (left figure) and double precision (right figure) floating-point arithmetic. The upper figures show the first 28000 the lower figures show 45000 points. Obviously the figures differ essentially. As in the former example the computed values have not much in common with the correct iterates of the sequence.

In case of dynamical systems of differential equations the situation is very similar in principle. For given initial conditions there exists a unique solution. A computation of the solution in real arithmetic is not possible. In a floating-point computation of the solution by any discretization method in general the approximation jumps from one solution to another one in each step of the integration. In areas where the correct solutions heavily deviate from one another, the computed solution may show a behaviour which has nothing to do with the correct solution of the problem.

It is very popular nowadays to deal with chaotic solutions of dynamical systems. Can such solutions be computed at all? In general such solutions do not have a representation in closed form by a mathematical formula. So, a computer algebra system can not help. The only remaining possibility is an integration by a numeric method. But chaotic solutions of dynamical systems are very sensitive with respect to changes of the data. A numerical integration necessarily changes the data in each step of the integration. Thus, a seemingly chaotic behaviour of a numerically computed solution might be caused by the numerics. See the lower picture in Figure 10. It is showing the computed solution of a satellite in the gravitational field of two suns. Does the correct solution perhaps look like in the upper picture of Figure 10? Only a verifying computation of the solution which contains its own error analysis can help to answer this question. Actually, the lower picture in Figure 10 has been computed with such a verifying solver. It computes upper and lower bounds of the solution which coincide within ten digits of the mantissa. The computation also establishes existence and uniqueness of the solution within the computed bounds. An orbit computed by a verifying solver and a picture drawn for it are correct like a mathematically proved theorem. Indeed, only verifying methods can numerically prove the properties of chaotic solutions of dynamical systems. Naturally, the range of these methods is limited by the available computing power.

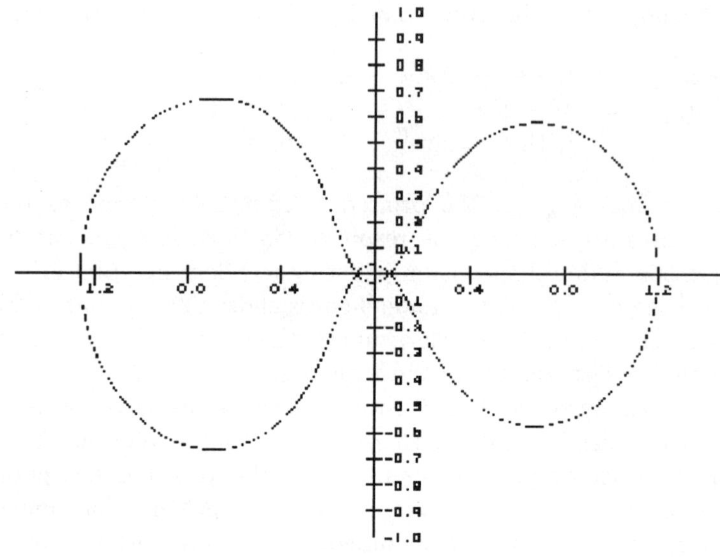

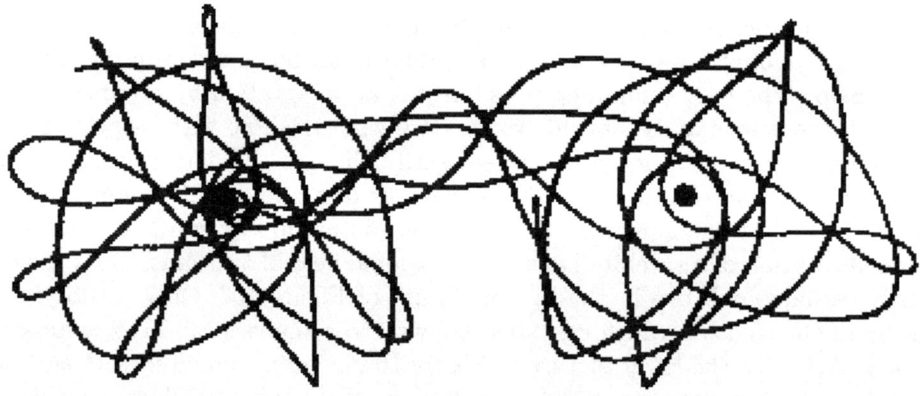

Fig. 10, Dynamical systems

Fortunately not all solutions of differential equations are extremely sensitive with respect to little changes in the data. Otherwise a reliable computation of the orbit of satellites or planets by approximating numerical methods would not be possible.

2. UNIVERSAL FLOATING-POINT ARITHMETIC

A disappointing feature is the failure of the numerical analysts to influence computer hardware and software in the way they should.

Now there is a regrettable tendency for numerical analysts to opt out of any responsibility for the design of the arithmetic facilities and a failure to influence the more basic features of software.

It is often said that the use of computers for scientific work represents a small part of the market and numerical analysts have resigned themselves to accepting facilities designed *for other purposes and making the best of them.*

J. H. Wilkinson: Turing Lecture 1970, J.ACM 18 (1971), 146

This remark by J. H. Wilkinson is true today as it was 25 years ago. We have seen by the samples discussed in the previous section that with a more powerful computer arithmetic many problems can be solved more easily and others can be solved at all. Additionally there is still the credibility-gap problem involved in a conventional floating-point computation. In a general application we do not know how much of the computers answers to believe. The aim must be to develop the computer any further so that not just approximate calculations but real mathematics can be done with the computer in the sense that safe and highly accurate bounds of solutions can be computed or even existence and uniqueness properties of problems can be proved by use of the computer. The tremendous advances in computer technology should be used to develop numerical analysis and floating-point calculations, as part of mathematics, finally to the same standard as the rest of mathematics.

2.1 Definition and properties of Universal Floating-Point Arithmetic

Arithmetic is the basis of mathematics. To approach that goal the computer has to be made arithmetically more powerful than is ordinarily the case. Its arithmetic basis has to be extended. Instead of reducing all calculations to the four elementary operations for floating-point numbers we require twelve fundamental data types (or mathematical spaces) in a computing environment.

Besides of the real numbers, the complex numbers form the basis of analysis. If computed results are to be guaranteed one needs the intervals over the real and complex numbers as well. The intervals bring the continuum on the computer. An interval with two floating-point bounds represents the continuous set of real numbers between these two floating-point bounds. So the twelve fundamental data types consist of the four basic data types real, complex, interval and complex interval and the vectors and matrices over these four types. The theory of computer arithmetic requires that all arithmetic operations within and between elements of these data types are to be defined by a general mapping principle which is called *semimorphism*:

If M is any one of these twelve data types or mathematical spaces and N is its computer-representable subset, for every arithmetic operation $\circ$ in M a corresponding computer approximation $\boxdot$ in N is defined by

$$(RG) \quad a \boxdot b := \square(a \circ b) \quad \text{for all } a, b \in N \text{ and all operations } \circ \text{ in } M,$$

where $\square : M \to N$ is a mapping from M onto N which is called a rounding if it has the following two properties:

$$(R1) \quad \square a = a \qquad \text{for all } a \in N \quad \text{(projection)}$$
$$(R2) \quad a \leq b \Rightarrow \square a \leq \square b \quad \text{for all } a, b \in M \quad \text{(monotonicity)}$$

The concept of semimorphism requires additionally that the rounding should be antisymmetric, i. e. that it has the property

$$(R3) \quad \square(-a) = -\square a \quad \text{for all } a \in M \quad \text{(antisymmetry)}$$

In case of interval spaces — the intervals over the real and complex numbers as well as the intervals over the real and complex vectors and matrices — the order relation is the subset relation $\subseteq$.

Additional important roundings are the downwardly and upwardly directed roundings with the property

$$(R4) \quad \nabla a \leq a \quad \text{respectively} \quad a \leq \triangle a \quad \text{for all } a \in M \quad \text{(directed)}$$

The directed roundings are uniquely defined by $(R1), (R2)$ and $(R4)$, see [10], [11].

With these five rules (RG) and $(R1, 2, 3, 4)$, a large number of arithmetic operations are defined in the computer-representable subsets of the twelve data types or spaces mentioned above. (RG) means that every computer operation should be performed in such a way that it produces the same result as if the mathematically correct operation were first performed and the exact result then rounded into the computer-representable subset. In contrast to the traditional approximation of the arithmetic operations in the product spaces by floating-point arithmetic, the new operations defined by the five rules (RG) and $(R1, 2, 3, 4)$ are optimal in the sense that there is no better computer-representable approximation to the true result (with respect to the prescribed rounding), i. e. between the correct and the computed result of an operation there is no other element of the corresponding computer-representable subset. This can easily be seen: *If $a, b \in N$ and $\alpha \in N$ is the greatest lower and $\beta \in N$ the least upper bound of the correct result $a \circ b$ in M, i. e.*

$$\alpha \leq a \circ b \leq \beta,$$

then

$$\underset{\square\alpha}{\overset{(R1)}{=}} \ \underset{\alpha}{\overset{(R2)}{\leq}} \ \underset{\square(a \circ b)}{\overset{(RG)}{=}} \ \underset{a \boxdot b}{\overset{(R2)}{\leq}} \ \underset{\square\beta}{\overset{(R1)}{=}} \ \beta.$$

left operand \ right op.	integer real complex	interval cinterval	rvector cvector	ivector civector	rmatrix cmatrix	imatrix cimatrix
monadic	$+, -$	$+, -$	$+, -$	$+, -$	$+, -$	$+, -$
integer real complex	$+, +<, +>$ $-, -<, ->$ $*, *<, *>$ $/, /<, />$ $+*$	$+, -, *, /$ $+*$	$*, *<, *>$	$*$	$*, *<, *>$	$*$
interval cinterval	$+, -, *, /$ $+*$	$+, -, *, /$ $+*, **$	$*$	$*$	$*$	$*$
rvector cvector	$*, *<, *>$ $/, /<, />$	$*, /$	$+, +<, +>$ $-, -<, ->$ $*, *<, *>$ $+*$	$+, -, *$ $+*$		
ivector civector	$*, /$	$*, /$	$+, -, *$ $+*$	$+, -, *$ $+*, **$		
rmatrix cmatrix	$*, *<, *>$ $/, /<, />$	$*, /$	$*, *<, *>$	$*$	$+, +<, +>$ $-, -<, ->$ $*, *<, *>$ $+*$	$+, -, *$ $+*$
imatrix cimatrix	$*, /$	$*, /$	$*$	$*$	$+, -, *$ $+*$	$+, -, *$ $+*, **$

Fig. 11, Predefined Operators of PASCAL–XSC

Figure 11 shows a table of the twelve basic arithmetic data types and corresponding operators of the programming language PASCAL-XSC. The data types are predefined. The operators can be called by the symbol contained in the table. An operator followed by a less or greater symbol denotes an operation with rounding downwards or upwards, respectively. The operator $+*$ takes the interval hull of two elements. $**$ means intersection. A count of all inner and outer operations in the table leads to a number of about 600.

Figure 12 lists the same data types in their usual mathematical notation. $\mathbb{R}$ denotes the real, $\mathbb{C}$ the complex numbers. A heading letter V, M and I denotes vectors, matrices and intervals respectively. R stands for the set of floating-point numbers and D for any set of higher precision floating-point numbers. $\mathbb{P}$ means passage to the corresponding power set. The set of operations defined by $(RG), (R1, 2, 3, 4)$ for the computer-representable sets of Figure 12, defines Universal Computer (Floating-Point) Arithmetic.

$$
\begin{array}{ccccccc}
 & & \mathbb{R} & \supset & D & \supset & R \\
 & & V\mathbb{R} & \supset & VD & \supset & VR \\
 & & M\mathbb{R} & \supset & MD & \supset & MR \\
\mathbb{P}R & \supset & \mathbb{IR} & \supset & ID & \supset & IR \\
\mathbb{P}V\mathbb{R} & \supset & IV\mathbb{R} & \supset & IVD & \supset & IVR \\
\mathbb{P}M\mathbb{R} & \supset & IM\mathbb{R} & \supset & IMD & \supset & IMR \\
 & & \mathbb{C} & \supset & CD & \supset & CR \\
 & & V\mathbb{C} & \supset & VCD & \supset & VCR \\
 & & M\mathbb{C} & \supset & MCD & \supset & MCR \\
\mathbb{P}C & \supset & I\mathbb{C} & \supset & ICD & \supset & ICR \\
\mathbb{P}V\mathbb{C} & \supset & IV\mathbb{C} & \supset & IVCD & \supset & IVCR \\
\mathbb{P}M\mathbb{C} & \supset & IM\mathbb{C} & \supset & IMCD & \supset & IMCR \\
\end{array}
$$

Fig. 12, The spaces of numerical computations

In the introduction of the books [10], [11] on computer arithmetic it is already stated that the concept of semimorphism should and *"can be used as an axiomatic definition of computer arithmetic in the context of programming languages"*. A careful analysis of the requirements is given in these books. The resulting algebraic and order structure in the computer representable subsets are studied under the mapping properties $(RG), (R1, 2, 3, 4)$. Because of $(R2)$ with respect to $\leq$ the order structure is not changed if we move from a set into a subset in any row of Figure 12, while the algebraic structure is considerably weakened. All computer operations defined by $(RG), (R1)$ and $(R2)$ are of 1 ulp accuracy (or even $1/2$ ulp accuracy in the case of rounding to nearest).

In the theory of computer arithmetic [10], [11] it is ultimately shown that all operations in the twelve spaces (real and complex numbers, real and complex intervals as well as vectors and matrices over these four basic data types) which are shown in Figure 11 can be provided by a modular technique, if on a low level, possibly in hardware, 15 fundamental operations are available: the five operations $+, -, *, /, \cdot$, each one with the three roundings $\square, \nabla, \triangle$. Here $\cdot$ means the dot product of two vectors, $\square$ is a monotone, antisymmetric rounding (e. g. to nearest) and ∇ and $\triangle$ are the monotone downwardly and upwardly directed roundings. All 15 operations $\boxdot, \underline{\nabla}, \underline{\triangle}$ with $\circ \in \{+, -, *, /, \cdot\}$ are defined by (RG). In case of the dot product a and b are vectors of any finite dimension, $a = (a_i), b = (b_i)$.

$$a \,\square\, b \;=\; \square \sum_{i=1}^{n} a_i * b_i$$

$$a \,\triangledown\, b \;=\; \triangledown \sum_{i=1}^{n} a_i * b_i$$

$$a \,\triangle\, b \;=\; \triangle \sum_{i=1}^{n} a_i * b_i$$

The IEEE 754 standard prescribes 12 of these 15 operations $\square$, $\triangledown$ and $\triangle$ for $\circ \in \{+, -, *, /\}$, but fails to define the dot product operations, which are the most crucial operations in the common vector spaces. By adding just the three dot product operations $\square$, $\triangledown$ and $\triangle$, all arithmetic operations in all the basic product spaces of numerical mathematics can be performed with 1 (or 1/2) ulp accuracy in each component.

In Numerical Analysis, the dot product is ubiquitous. It appears in complex arithmetic, vector and matrix arithmetic, multiple precision arithmetic, iterative refinement techniques, range reduction of elementary functions and many other places.

Two serious disadvantages of most existing IEEE arithmetic processors are that full double-length products cannot be obtained from the outside even though they are computed internally and that the operations with directed roundings $\triangledown$ and $\triangle$ where $\circ \in \{+, -, *, /\}$ are not provided as single instructions with the rounding mode incorporated. The former restriction complicates the implementation of accurate dot products whereas the latter slows down interval arithmetic considerably. Many processors need as many machine cycles to set the rounding mode as to perform the rounded arithmetic operation itself. Since in interval calculations, the rounding mode has to be switched frequently, this introduces a severe and unnecessary penalty. Another hindrance to further progress in numerical analysis is the fact that the directed roundings as well as the corresponding arithmetic operations $\triangledown$ and $\triangle$, $\circ \in \{+, -, *, /\}$, are not supported by the traditional programming languages.

A *Proposal for Accurate Floating-Point Vector Arithmetic* was drafted by an *GAMM / IMACS*[3] *Working Group on Enhanced Computer Arithmetic* and extensively discussed between 1989 and 1992. In its final form, the proposal was unanimously approved and adopted by GAMM and by IMACS in 1993 [14]. It is now an official document of the two mathematical societies. In essence, it requires universal computer arithmetic and gives some hints on its realization.

2.2 Implementation and realization of Universal Floating-Point Arithmetic

Universal computer arithmetic has been implemented in software, in firmware and in hardware on many different platforms and supported by powerful programming languages (PASCAL-XSC, ACRITH-XSC, C-XSC, FORTRAN-XSC). It has been avail-

---

[3]GAMM = Gesellschaft für Angewandte Mathematik und Mechanik
  IMACS = International Association for Mathematics and Computers in Simulation

able since 1980. We briefly comment on its implementation. As already mentioned, the fifteen operations $\boxdot, \triangledown, \triangle, \circ \in \{+, -, *, /, \cdot\}$ are the basic building blocks. Among these the implementation of the twelve operations $\boxdot, \triangledown, \triangle, \circ \in \{+, -, *, /\}$ is state of the art technology and common place nowadays. So we only comment on the realization of the three optimal dot products

$$s := \bigcirc(x \cdot y) = \bigcirc \sum_{i=1}^{n} x_i * y_i, \quad \bigcirc \in \{\Box, \triangledown, \triangle\}$$

for vectors $x = (x_i)$ and $y = (y_i)$. I. e. s has to be computed as if an intermediate result $x \cdot y$ correct to infinite precision and with unbounded exponent range were first produced and then rounded to the desired floating-point destination format according to the selected rounding $\Box, \triangledown$ or $\triangle$.

This definition of the dot product operation guarantees highest possible accuracy (for the given rounding and the given floating-point destination format). It is important to note that the order in which the elementary operations are performed when determining the dot product is not specified, and that this accommodates pipelined (and potentially parallel) processing. Reordering of the summands has no influence on the computed result since all intermediate results are exact.

By way of contrast, a traditional computation of the dot product of two vectors with n components each (in ordinary floating-point arithmetic with rounded multiplications and additions) involves $2n - 1$ roundings. If catastrophic cancellation occurs a large number of significant digits may be lost. If executed in a pipeline, what most vector processors do, the sequence of the summands is exchanged which can cause additional, unexpected errors. All this may happen even if an extended precision data format is used for the accumulation. Loss of accuracy aside, a considerable amount of processing time may be required to perform gratuitous intermediate steps – such as composition, decomposition, normalization, and rounding of intermediate floating-point values. Furthermore, unnecessary load and store operations connected with pack into and unpack out of a floating-point format may have to be performed in the traditional mode.

A natural way of adding n numbers or products is via fixed-point accumulation. A most natural hardware implementation has been in use for more than ten years. This device employs a long fixed-point register (long data word or accumulator), which is kept as local memory on the arithmetic unit, covering twice the exponent range of the floating-point format in which the vector components are given (see for instance [17], [18]). If a relatively small number of extra digits is provided at the front (high significance) end of such a register to collect extra carries, the intermediate result in such a register is always exact and cannot overflow or underflow for any feasible number of summands. Only one rounding occurs, and this at the end of the accumulation process. This method has the advantage of being rather simple and straight forward, and it always provides the desired accurate answer. A fixed-point accumulation of n floating-point numbers or products is the only computer operation which never looses a bit and never fails. An error analysis is never necessary. For Numerical Analysis this absolutely reliable compound operation is extremely helpful.

Existing implementations and detailed studies show that a full hardware implementation of the long register (long data word) is routine and inexpensive in units of additional silicon and can be made to perform at least as fast as conventional floating-point accumulations. In case of the /370 data format double precision the long data word consists of 672 bits. For the IEEE arithmetic standard about four times as many bits are needed. In modern computer technology this is rather inexpensive.

The fundamental operations that are needed are: *initialization* of the long register, *addition of a product* of two floating-point numbers to the register, and *rounding* the contents of the register to a floating-point number (using the prescribed rounding). Other desirable operations include: addition of a floating-point number to the register, addition, and comparison of two long registers. Existing programming language extensions provide and exploit these operations [6], [7], [16].

From the user's point of view, the well-defined, predictable behaviour of the long register or long data word on the arithmetic unit is highly desirable. Any partial hardware support requires complementary software and may cause substantial performance penalty in essential applications.

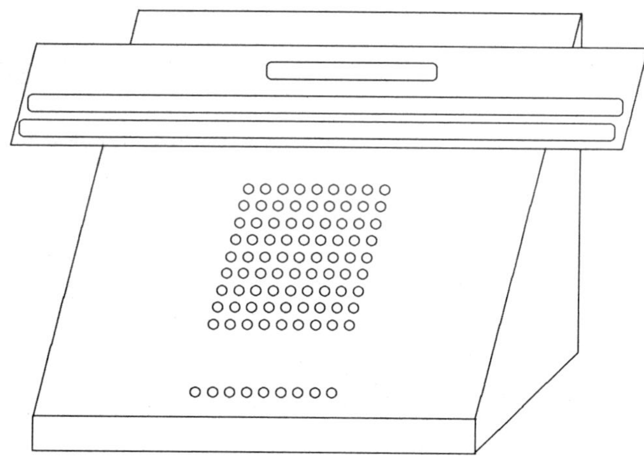

Fig. 13, Sketch of an old calculator

It is an interesting observation that the better mechanical calculators already had all the arithmetic capabilities that are needed today. They were able to calculate dot products exactly. The techniques used then can be considered as direct predecessors of the electronic circuits available today. Figure 13 shows one such calculator. It had a keyboard of say 9 decimal digit. The result register was essentially larger, and had perhaps 25 decimal digits. It was sitting on a movable carriage and was a fixed-point register. The decimal point always stayed in the same place during a calculation. This assigned a unique exponent to each digit. The summands were first positioned correctly and then added. Products were accumulated by adding multiples of one factor corresponding to the digits of the other one. As long as no underflow or overflow of the result register occurred, which was obvious and visible, all digits of the result were correct. This operation was called simply a *running total*. It was used as often as possible, because it was the fastest way to carry out calculations. Intermediate results did not need to be read out and typed in again for subsequent operations.

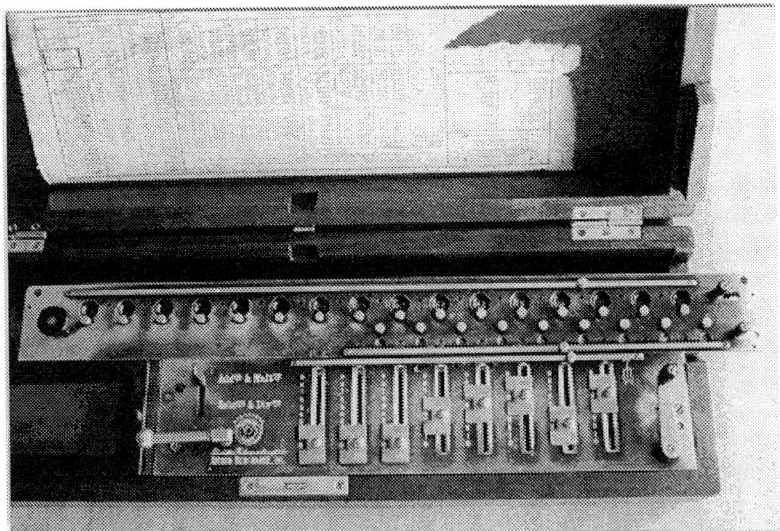

Fig. 14, Two old mechanic calculators

Rounding was done by hand as needed downwards, upwards or to the nearest number.This extremely useful, fast and accurate *running total* was lost in the transition to the first electronic computers. The early technology was too expensive. It did not allow registers on the arithmetic unit, which were essentially wider than the data format as kept in the memory. Figure 14 shows two other calculators designed and built in the last century. Again the result register was much wider than the input register which allowed the error free operation of running total.

A completely wrong direction was taken with the development of modern vector processors. In order to gain speed, accumulation is not done simply in a fixed-point register, but instead in a pipeline in floating-point arithmetic. Thus the sequence of summands is altered, which in floating-point gives rise to additional computational errors. One really has to wonder why this has not led to an outcry from all mathematicians. Customers should not overlook such defects when purchasing such installations which generally cost several million dollars.

The entirely correct accumulation of floating-point numbers as well as products of such numbers into a fixed-point register is considerably simpler and even faster than an accumulation in floating-point arithmetic. For all kinds of computers such as personal computers, workstations, mainframes and super computers, electronic circuits have been developed so that the arithmetic does not contribute to the computing time at all any more. In the calculation of the dot product in a pipeline, the multiplication, a possible shift and the corresponding addition can be carried out without further ado in the time required to read the two factors of a product from the memory whatever the memory may be, main memory, cache or register memory. This means, that there is no other way to perform dot products which can be faster.

2.3 Multiple-precision arithmetic

Besides of the arithmetic operations defined for the twelve numerical data types of Figure 11 or mathematical spaces of Figure 12, multiple-precision (*long*) arithmetic for real and complex numbers and intervals of varying length can easily be provided if the long fixed-point register for accumulation of dot products is available on the arithmetic unit. With double-, triple-, quadruple-... precision arithmetic for intervals often critical or unstable parts of an algorithm easily can be detected, analyzed and controlled. A quadruple precision variable, for instance, simply is an array of four floating-point numbers the mantissas of which succeed one another, Figure 15. Variables of such type can easily be added, subtracted or accumulated into the long data word on the arithmetic unit. Multiplication of two such variables $(a_1+a_2+a_3+a_4) \cdot (b_1+b_2+b_3+b_4) = a_1 b_1 + a_1 b_2 + a_1 b_3 + a_1 b_4 + a_2 b_1 + \cdots + a_4 b_3 + a_4 b_4$ is nothing else than a dot product. Division can be performed by a simple iterative process.

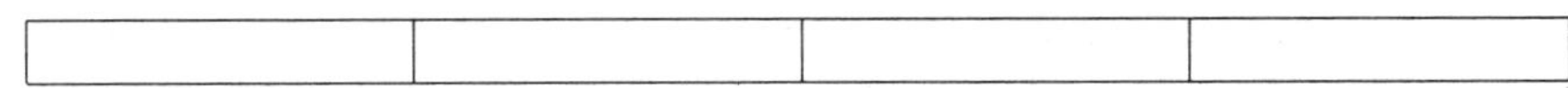

Fig. 15, A quadruple precision floating-point number

In practice many programs contain elementary functions. If a long real arithmetic is used in a program and the elementary functions are only available in a single precision on the computer, the computation is no longer in balance. Programming environments like PASCAL-XSC and C-XSC, therefore, also provide the elementary functions for arguments of the data types long real and long real interval. They deliver a highly accurate result of the same data type. All functions can be called by their generic name. The compiler first checks the type of the argument and then selects the required routine. If, for instance, the data type long real consists of an array of four reals, the mantissa of a variable of this type has $4 \cdot 53$ bits $= 212$ bits. In such a case the function call $\cos(x)$ in PASCAL-XSC, for instance, delivers a result which is correct to about 209 bits. The accuracy of the result can easily be checked by calling the function for a point interval.

In C-XSC, for instance, the data type *long real* is controlled by a global variable called *staggered*. If staggered is set to be 1, then long real is just real. If staggered is set to be 4, a variable of type long real consists of an array of four reals. The precision of the type long real, i. e. the value of staggered, can be increased or decreased at any place in a program. So if an unstable part is detected in a program by a verification step, a loop around it for enlarging the precision is often a simple cure.

3. INTERVAL ARITHMETIC, A SIMPLE BUT POWERFUL CALCULUS TO DEAL WITH INEQUALITIES

Analysis deals with three kinds of structures, the algebraic structure, the order structure and the topological structure. On the computer the first and the latter are heavily changed while for computation with inequalities (the order structure), with respect to $\leq$ because of (R2), the same rules hold as in the basic spaces [10], [11]. In practice this means that properties of a method or an algorithm which are solely derived by using the order structure are also valid on the computer. In this sense the order structure or computation with inequalities is of particular importance. With interval arithmetic a very powerful calculus has been developed during the last decades which systematizes the computation with inequalities.

Consider the two inequalities $a_1 \leq a \leq a_2$ and $b_1 \leq b \leq b_2$. For the sum and the difference we obtain $a_1 + b_1 \leq a + b \leq a_2 + b_2$ and $a_1 - b_2 \leq a - b \leq a_2 - b_1$. In case of multiplication the rule is much more complicated. Nine cases are to be distinguished depending on whether a_1, a_2, b_1, b_2 are less or greater than zero. For division the situation is similar [10], [11], [2]. Because of these many case distinctions an explicit computation with inequalities is very difficult. For complicated expressions it is practically not executable.

Interval arithmetic summarizes these complicated rules for the computation with inequalities. With $A = [a_1, a_2]$ and $B = [b_1, b_2]$ we obtain for all operations $\circ \in \{+, -, *, /\}$, $(0 \notin B$ in case of division):

$$A \circ B = \{a \circ b \mid a \in A \wedge b \in B\} = \left[\min_{i,j=1,2} (a_i \circ b_j), \max_{i,j=1,2} (a_i \circ b_j) \right].$$

If executed on a computer the lower bound of the result has to be rounded downwards and the upper bound upwards. The many case distinctions are programmed once and for all. In an appropriate programming language henceforth they disappear in the runtime system of the compiler. The user does not have to take care of them anymore.

We illustrate the efficiency of this calculus by a simple example. Consider a system of linear equations in fixed point form $x = A \cdot x + b$ with a contracting matrix A. Let the interval vector X be a rough initial enclosure of the solution $x^* \in X$. Now we could formally write down the Jacobi method, the Gauss-Seidel method and the relaxation method and interpret all data as being intervals. Additional methods can be obtained by taking the intersection of two successive approximations. If we now decompose all these methods in formulas for the bounds of the intervals we obtain a major number of methods for the computation of bounds for the solution of linear systems which have been derived by well known mathematicians painstakingly about 50 years ago [2]. The calculus of interval arithmetic reproduces these and other methods in the simplest way. The user does not have to take care of the many case distinctions occuring in the matrix vector multiplications. The computer executes them fully automatically by the preprogrammed calculus. Additionally also the rounding errors are captured. The calculus develops its own dynamics.

What are other advantages of the calculus of interval arithmetic? We have derived it as calculus to compute with inequalities. It is clear, therefore, that it can be used to compute the range of arithmetic expressions and functions in dependence of intervals occurring in the data. If this is done naively just executing the rules of interval arithmetic, the bounds for the range may become very large and imprecise. The range of the function may be heavily overestimated. This has brought interval arithmetic a negative image in its early days. It was argued that safe bounds which always can be computed are the interval from minus infinity to plus infinity.

In interval arithmetic it is shown that for continuous functions the overestimation of the range of a function converges to zero if the diameters of the argument interval do so. I. e. interval evaluation of an expression converges to its range, in the limit to the value of the function. If a special representation of the expression or the function is used, the so called centered form, the convergence is even quadratic. In case of differentiable functions the centered form can fully automatically be produced by the computer by use of automatic differentiation. In other words: in case of small interval arguments, interval evaluation of a function or an arithmetic expression is a quadratically convergent method for the computation of the range of the function and, if we include automatic differentiation into our consideration also for the computation of bounds of the range of derivatives, Taylor coefficients, gradients, Jacobian- and Hessian matrices and so on. These bounds are evaluated automatically by the computer without any additional work to be done by the user. The true intelligence is transferred into the run time system of the compiler. In case of product spaces (vectors, matrices, complex numbers etc.) the dot product with only one rounding serves as a single operation. In particular the connection with automatic differentiation opens a huge field of applications to interval arithmetic (enclosures of the solution of linear and non linear systems of equations, verified adaptive numerical quadrature by computing enclosures of the remainder term of quadrature formulas, verified integration of ordinary differential equations and of integral equations and so on).

By what has been said already, an obvious thing to apply interval computations are problems with small intervals or to use methods of subdivision. This is not a particular restriction, since in numerics many methods proceed in small steps anyhow. We illustrate this by a few examples:

3.1   Numerical quadrature

The definite integral of a function between two bounds a and b has to be computed. As usual the interval $[a, b]$ is subdivided into n subintervals. Then, the integral can be represented as a sum consisting of a quadrature rule plus a remainder term. The quadrature rule is a dot product consisting of certain weights and function values computed at the points x_i of the partition of the interval $[a, b]$. If the integrand is sufficiently smooth the remainder term in general can be represented in a form which contains a higher derivative of the function at one or several unknown intermediate points in the interval $[a, b]$. Instead of such intermediate points the whole small interval which contains it is now considered. The higher derivative in the remainder term is now evaluated over this whole interval by automatic differentiation. Thus the unknown intermediate points as well as the values of the higher derivatives at these points are definitely enclosed. Thus safe bounds for the remainder term are obtained. Since the intervals in question are small and the remainder term contains factors like $h^n/n!$ in general these bounds are very good enclosures of the remainder term. In summary one obtains close bounds for the error of the used quadrature rule. In case of the trapezoidal rule the following formulas describe the process:

$$\int_{x_i}^{x_{i+1}} f(x)dx = \frac{h}{2}(f(x_i) + f(x_{i+1})) - \frac{h^3}{12}f''(\xi_i), \quad \xi_i \in [x_i, x_{i+1}]$$

$$\int_{a}^{b} f(x)dx = h \sum_{i=0}^{n}{}' f(x_i) - \frac{h^3}{12} \sum_{i=0}^{n-1} f''(\xi_i), \quad \xi_i \in [x_i, x_{i+1}]$$

$$\int_{a}^{b} f(x)dx \in h \sum_{i=0}^{n}{}' f(x_i) - \frac{h^3}{12} \sum_{i=0}^{n-1} f''([x_i, x_{i+1}])$$

Therein, the dash after the sigma symbol indicates that the first and the last summand have to be divided by two.

The good and true enclosure of the remainder term of the quadrature rule can be used to adapt the step size to the properties of the integrand. In case the function is smooth and flat one would like to proceed with a relatively large step size while a smaller step size has to be chosen in areas where the function is steep. In each step of the integration the step size has to be chosen in such a way that the width of the remainder term is under a given accuracy threshold. Since by these methods always the error term itself is evaluated and used to control the step size, the method can also be used in case of Gauss quadrature. The conventional technique to use a second quadrature formula in order to estimate the error is no longer necessary. A verifying adaptive numerical quadrature, therefore, can be faster than a traditional approximate computation of the integral. By choosing appropriate weight functions safe enclosures of singular integrals can be computed by Gauss quadrature.

For the computation of the following two integrals Romberg integration with adaptive step size control has been used.

$$f_1(x) \;=\; \frac{1}{0.1^2+(3x-1)^2} - \frac{1}{0.1^2+(3x-4)^2} + \frac{1}{0.1^2+(3x-7)^2} - \frac{1}{0.1^2+(3x-10)^2}$$

$$I_1 \;=\; \int_0^4 f_1(x)dx$$

$$f_2(x) \;=\; 2xe^{x^2}\sin(e^{x^2})$$

$$I_2 \;=\; \int_0^2 f_2(x)dx$$

The error term is obtained by use of the Euler-MacLaurin sum formula. Using double precision floating-point interval arithmetic the following bounds are obtained:

$$I_1 \;\in\; -0.1519639422329030_5^6$$
$$I_2 \;\in\; 0.9109640392659322_8^9$$

Both integrals are not easy to compute numerically using traditional approximate methods, Figure 16, [1], [8].

The computation of a highly accurate enclosure of a difficult integral by Romberg integration may require to proceed to the 7th or 8th row in the extrapolation table. Evaluation of the remainder term then requires the computation of an enclosure of the range of perhaps the 18th derivative of the function over an interval. How can this be obtained? In an appropriate and advanced programming environment this is very easy. Figure 17 shows a PASCAL-XSC program for the computation of enclosures of the range of the 36th and the 40th Taylor coefficient of the function $e^{\frac{5000}{\sin(11+(x/100)^2+30)}}$ over the interval $[1.001, 1.005]$. The program uses a predefined module for the computation of enclosures of derivatives and Taylor coefficients. With it the arithmetic operations in the expression, i. e. the operators and elementary functions are interpreted according to the rules of the Taylor arithmetic by the run time system of the compiler. In the program the function can be written as if it would be a function with variables of type real. However, the argument x as well as the function value are specified as being of type interval Taylor. A variable of this type is an array of 41 intervals.

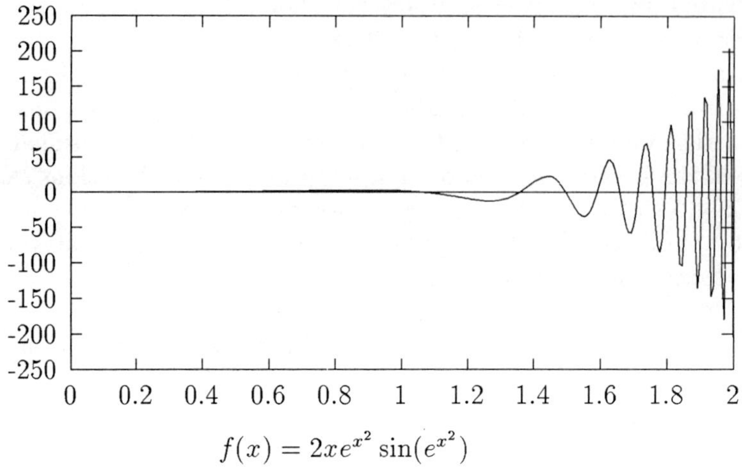

$$f(x) = 2xe^{x^2}\sin(e^{x^2})$$

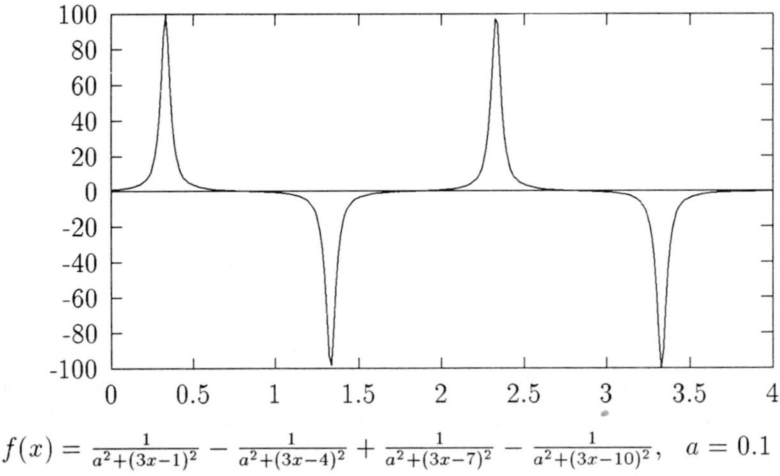

$$f(x) = \frac{1}{a^2+(3x-1)^2} - \frac{1}{a^2+(3x-4)^2} + \frac{1}{a^2+(3x-7)^2} - \frac{1}{a^2+(3x-10)^2}, \quad a = 0.1$$

Fig. 16, Verified Romberg integration

```
program sample;

use itaylor;

function f ( x : itaylor ) : itaylor[lb(x)..ub(x)];
begin
   f := exp( 5000 / (sin(11 + sqr(x/100)) + 30));
end;

var
  a     : interval;
  b, fb : itaylor[0..40];

begin
   read(a);
   expand(a,b);
   fb := f(b);
   writeln('36. Taylor coeff.: ',fb[36]);
   writeln('40. Taylor coeff.: ',fb[40]);
end.


Test results:

a = [ 1.001 , 1.005 ]
36. Taylor coeff.: [ -2.4139E+002, -2.4137E+002 ]
40. Taylor coeff.: [  1.0759E-006,  1.0760E-006 ]
```

Fig. 17, Computation of Taylor coefficients in PASCAL–XSC

The routine expand(a, b) expands the interval variable "a" into the array of its interval Taylor coefficients. With it, then the expression for f is evaluated over the interval $[1.001, 1.005]$ according to the rules of the Taylor arithmetic. Enclosures of the ranges of the 36th and 40th Taylor coefficient over this interval are printed.

3.2 Numerical integration of ODEs

More sophisticated methods have been developed for a verified numerical integration of initial value problems of ordinary differential equations. There, in case of a system of first order, the right hand side of the differential equation is developed into a Taylor polynomial with remainder term by automatic differentiation. Now the remainder term is evaluated using an appropriate step size. Since an enclosure of the solution has to be computed, the computation has to be done in interval arithmetic. In contrast to numerical quadrature, in case of a differential equation, the remainder term besides of the independent variable also contains the unknown function. In order to compute an enclosure of the remainder term in interval arithmetic, therefore, one needs already a safe enclosure of the unknown function for the particular step of integration. So it seems as if one is caught in a circle. The circle yet can be opened. At first one only needs a rough enclosure of the unknown function in the integration interval. It can be obtained in the following way. We transform the differential equation into an equivalent integral equation. The integrand then besides of the independent variable, again contains the unknown function. In order to enclose the integral into safe bounds by interval arithmetic one again needs an enclosure of the unknown function in the whole integration interval. In the simplest case such an enclosure is obtained by estimation. Since the differential equation fulfills a Lipschitz condition, there is always such an enclosure. To keep this estimated enclosure small it may be that the step size has to be reduced. This first, still rudimentarily estimated initial enclosure is now made safe by an application of Banach's fixed point principle. With the initial enclosure the right hand side of the integral equation is evaluated in interval arithmetic and it is checked whether an expected enclosure is obtained. If this is not the case the step size has to be reduced and/or the estimated initial enclosure has to be changed. As soon as an enclosure is obtained one has a safe initial enclosure of the unknown solution in the integration interval. With it now the remainder term of the Taylor polynomial is evaluated. The remainder term is now small of higher order. So a much better enclosure of the unknown solution in the integration interval is obtained by a polynomial with interval coefficients. Continuation of this method over several such steps causes the well known wrapping effect. It can be controlled by use of local coordinates [1].

Boundary and eigenvalue problems of ordinary differential equations can be reduced to initial value problems by shooting methods. In case of boundary and eigenvalue problems existence and uniqueness of a solution are often not a priori known. These properties are automatically verified by the enclosure algorithm, i. e. by the computer. These methods have successfully been applied to detect and compute periodic solutions of ordinary differential equations. In such a case it is proved by use of the computer that a given differential equation has periodic solutions. Such solutions are computed and enclosed in safe and narrow bounds.

3.3   Computation of the cut of two surfaces

Many conventional methods to compute the cut of two surfaces are available and known as continuation methods, marching or path following methods. If two different parts of the cut come close to each other all these approximating methods may easily fail. The computed cut can jump from one part of the cut to the other one pretending a completely wrong topology. A verified computation of the cut avoids such difficulties. First a system of differential equations for the cut is derived. Now a verified initial enclosure of the cut is computed by the Banach step as in case of differential equations. This step controls the step size already. The value at the right end of the step is then contracted to a very small interval by an interval version of Newton's method using the parameter representations of the surfaces. This procedure establishes existence and uniqueness of the cut within the computed bounds, Figure 18.

Fig. 18, Computation of the cut of two surfaces

3.4   Integral equations

Automatic differentiation, fixed point theorems and interval arithmetic are also the main tools for verified numerical integration of integral equations. In case of Fredholm integral equations of the second kind, for instance, the kernel is developed into a two dimensional Taylor polynomial with remainder term by automatic differentiation. Then it consists of a sum of a degenerate and a contracting kernel. Both parts can then be treated separately which leads to a verified enclosure of the solution [1].

These methods for a verified computation of enclosures of solutions of differential and integral equations in general are more costly than a traditional computation of an approximate solution by simple floating-point arithmetic. The new methods, however, deliver much more. They deliver a strict mathematical answer. Existence and uniqueness of a solution within the computed close bounds are established by the computer. In many applications mathematical safety is essential, and a safe method is often the fastest way to the goal. Unnecessary experiments can also be very costly.

For many years interval arithmetic was condemned to mark time. Essential progress was only possible at places where adequate programming tools could be provided which required a major manpower. In order to be fast, compilers directly produced machine code with the result that with the next computer the whole work had to be done anew. Now with C as a wide spread intermediate language, powerful programming tools like PASCAL-XSC, C-XSC or FORTRAN-XSC [6], [7], [16] are available for a wide variety of platforms. Many of the most sophisticated methods of interval computation which are available today, are not programmable and not thinkable and can not be developed in programming languages where each operation requires a procedure call.

4. AVAILABILITY OF THE NECESSARY ARITHMETIC IN HARDWARE AND SUITABLE PROGRAMMING SUPPORT

Already in the sixties, an ALGOL extension was created and implemented at the author's Institute which had a type for real intervals including provision of the corresponding arithmetic and relational operators. This language influenced the provision of an operator concept in ALGOL-68. Except for the necessary arithmetic, ALGOL-68 had all the necessary language concepts for implementation of numerical algorithms with automatic result verification. Subsequent languages which provide similar language capabilities are Ada, C++ and FORTRAN-90. Ada has not yet caught on so well. Fast compilers for FORTRAN-90 are still under development. Three languages and corresponding compilers which contain all programming **and** arithmetic concepts for programming numerical routines with automatic result verification have been developed at the author's Institute. These are PASCAL-XSC, ACRITH-XSC and C-XSC. ACRITH-XSC is an extension of FORTRAN-77 [16]. It was developed with the sponsorship and collaboration of IBM in the eighties. Unfortunately, its use is limited to machines with IBM/370-architecture and here only to those which operate under the VM CMS operating system.

PASCAL-XSC and C-XSC are extensions of the widely-distributed programming languages PASCAL and C++. PASCAL-XSC was already developed in the seventies. Software for PASCAL-XSC and C-XSC is provided for a large number of machines of various manufacturers. Language descriptions have been published by Springer-Verlag [6], [7]. A Russian edition of PASCAL-XSC is also available.

The language description also contains advice pertaining to application of the compiler[4], including some problem solving routines with automatic result verification. In addition to a large number of predefined arithmetic data types and the corresponding operations with maximal accuracy, the XSC-languages include among others the

---

[4]Compilers for PASCAL-XSC and C-XSC are obtainable from Numerik Software GmbH, Postfach 2232, D-76492 Baden-Baden, Germany, fax (+49) 7221 - 949218, phone (+49) 7221 - 949217, or FBSoftware, P.O. Box 44666, Madison, WI 53744-4666, United States, phone (608) 273-3702. Compilers and documentation can also be downloaded from the web sites `http://www.xsc.de/` and `http://www.uni-karlsruhe.de/~iam`

following language concepts, which were not present in the relatively basic versions of the languages. A module concept, an operator concept, functions and operators with general result types, overloading of functions, procedures, and operators, overloading of the assignment operator as well as read and write, dynamic arrays, access to subarrays, control of rounding by the user, and exact evaluation of expressions. For example, with these language concepts, modules and operators for complex arithmetic, matrix and vector arithmetic, rational arithmetic, double- or multiple-precision arithmetic or interval arithmetic can be developed in the language itself. Arithmetic expressions and algorithms with these data types can be written down in ordinary mathematical notation. This simplifies programming very greatly. Many programs can be read like a technical report.

The definition of computer arithmetic on the basis of mathematical criteria should be an inherent feature of every programming language. Users must know exactly what happens when they call an arithmetic operation in their program. Only in this way will the computer be a mathematical instrument. Such a definition is very simple. One only needs to require that for a given data format and data type, the arithmetic operations given by the hardware or callable from the programming language satisfy the conditions of a semimorphism. This can be written down mathematically in four lines. All operations defined in this way, including those in product spaces, are then maximally accurate. That is, the computed result differs from the correct result by at most one rounding. The techniques of implementing these mathematically strict conditions differ only marginally from what is done now, that is, they are not essentially more complicated. If the vector and matrix operations are built into the design of the hardware, then the computer will consequently be a good bit faster.

The IEEE, ANSI and ISO microprocessor arithmetic standards 754 and 854 promulgated around 1985 are basically a step in the right direction. The operations given in them realize a semimorphic floating-point and interval arithmetic. Unfortunately, except for the XSC languages, there are still no programming languages and compilers available which support simple use of interval arithmetic. This is all the more lamentable since prototypes for them were already available 25 years ago. By 1985, technology had finally advanced so far and the necessary mathematical insight was available, so that one should not have stopped with a standardization of ordinary floating-point arithmetic. The always correct accumulation in a fixed-point register should have been included unconditionally. Now, many manufacturers believe with respect to arithmetic that realization of the microprocessor standards is all that is necessary to do. In this way, the existing standards also prove to be a great hindrance to further progress.

The programming languages PASCAL-XSC, ACRITH-XSC and C-XSC bring in an essential advance here. They provide a universal computer arithmetic and permit simple handling of the semimorphic operations by means of ordinary mathematical operator symbols also in product spaces such as of complex numbers, as well as for vectors and matrices of types real, complex, interval and complex interval. If the latter operations are not supported by hardware they have to be simulated in software. Thus, at a place where an increase in speed is really to be expected, a loss of speed results into the bargain.

Finally, we mention how far the required arithmetic properties have been realized in hardware. In 1982-1983, IBM furnished all required operations on machines with /370 architecture (with the exception of the Vector Facility), on the larger machines by assembly software, and by microcode and later in VLSI on the smaller machines. These operations are addressable in the programming language ACRITH-XSC. Following the development by IBM, Siemens apparently commenced offering the optimal dot product in microcode in about 1985 on all the computers with /370 architecture developed and built by Siemens themselves. Hitachi and NAS followed IBM and supported all the required operations on some of their machines in hardware (probably by microcode). However, support of these operations by means of a programming language is lacking.

Unfortunately, these advances achieved in the eighties have nearly dissolved into nothingness, since for some time scientific computation has already shifted to workstations and supercomputers. Here, there is no hardware support available at present, so that again one has to resort to software simulation. We would like to strengthen the insight that an essential step forward has to be made in the field of computer arithmetic, as well as in computer hardware and also in standards. Basically, it is not comprehensible why guaranteed maximal accuracy is required of the four fundamental operations for real variables, but not for complex variables, vectors and matrices. The additional hardware cost is small in the case of a unified design of the arithmetic unit. However, it makes a big difference if one of these operations always gives a correct result, or only mostly. In the latter case, the user has to make an account of the desired accuracy in the case of each application, and certainly under conditions of many million times per second. In the first case, this is never necessary.

In 1993 and 1994 a vector arithmetic coprocessor for the PC, the XPA 3233, Figures 19 and 20, has been developed and built at the author's Institute in collaboration with two other Institutes in CMOS VLSI technology. On a small board of about half the size of a post card, Figure 20, the chip can be used on any PC with industry-standard PCI bus. The chip performs dot products of vectors with components of the IEEE double data format to full accuracy or with only one final rounding. The products are accumulated into a long fixed-point register which is kept on the chip. This is the fastest way to perform dot products. In high level programming languages like PASCAL-XSC or C-XSC the chip's functionality is directly coupled to the operator symbols in vector and matrix expressions or, for instance, to the operators *dotproduct* or *matmul* in FORTRAN-90. Access to the coprocessor from other programming languages is achieved by subroutine calls to a special C subroutine library. In the computer the chip plays the role of a catalyst in a double sense. It speeds up the computation and eliminates many unnecessary rounding errors.

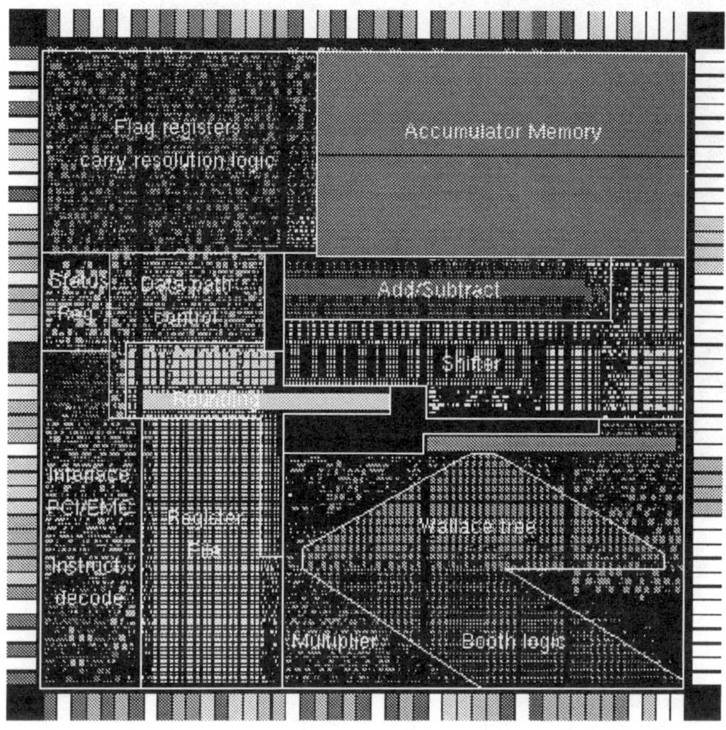

Fig. 19, Function units and cells of the vector arithmetic coprocessor XPA 3233

In conclusion, to recall the theme of this paper once more, it does not follow in any way that if an algorithm which has been verified to be correct in the sense of computer science, then a numerical result computed with it will be correct. So numerics with automatic result verification is an important additional yet simple tool. Naturally, with it alone, one is not absolutely certain, because the verification step could still lead to a positive result because of a programming error. Verification of the algorithm, the compiler, the operating system and also the computer hardware can never be made superfluous by it. In case of success, numerical result verification at least implies with high probability that all these components have furnished a correct result. Additionally, this result is certainly generally independent of whether a program verification has been carried out or not. On the other hand, a negative outcome of result verification, that is, when the computed result cannot be verified to be correct, also has positive value. This result is established quickly and automatically by the computer, without anything having to be done externally. The user can provide program alternatives in this case, for example, choice of different algorithms or methods, or a change to higher precision. Thus, something can often succeed even for extremely unstable cases.

Fig. 20, PCI-bus compatible board with vector arithmetic coprocessor XPA3233

REFERENCES

1. Adams, E., Kulisch, U. (eds.): *Scientific Computing with Automatic Result Verification*, Academic Press, New York, 1993.

2. Alefeld, G., Herzberger, J.: *An Introduction to Interval Computations*, Academic Press, New York, 1983.

3. de Beauclair, W.: *Rechnen mit Maschinen*, Vieweg, Braunschweig, 1968.

4. Hammer, R., Hocks, M., Kulisch, U., Ratz, D.: *Numerical Toolbox for Verified Computing I, Theory, Algorithms and PASCAL-XSC Programs*, Springer, Berlin, 1993.

5. Hammer, R., Hocks, M., Kulisch, U., Ratz, D.: *C++ Toolbox for Verified Computing I*, Springer, Berlin, 1995.

6. Klatte, R., Kulisch, U., Neaga, M., Ratz, D., Ullrich, Ch.: *PASCAL-XSC-Sprachbeschreibung mit Beispielen*, Springer, Berlin, 1991, engl. Übersetzung, Springer, Berlin, 1992.

7. Klatte, R., Kulisch, U., Wiethoff, A., Lawo, Ch., Rauch, M.: *C-XSC, A C++ Class Library for Extended Scientific Computing*, Springer, Berlin, 1993.

8. Lohner, R., Krämer, W., Kulisch, U.: *Numerical Toolbox for Verified Computing II, Theory, Algorithms and PASCAL-XSC Programs*, Springer, Berlin, 1999.

9. Ratz, D.: *Automatische Ergebnisverifikation bei globalen Optimierungsproblemen*, Dissertation, Universität Karlsruhe, 1992.

10. Kulisch, U.: *Grundlagen des Numerischen Rechnens*, BI Wissenschaftsverlag, Mannheim, 1976.

11. Kulisch, U., Miranker, W.L.: *Computer Arithmetic in Theory and Practice*, Academic Press, New York, 1981.

12. Ungerer, T.: *Mikroprozessoren - heute, morgen und übermorgen*, Jahrbuch Überblicke Mathematik 1996, Vieweg.

13. Zuse, K.: *The Plankalkül*, Oldenbourg, München, 1989.

14. GAMM-IMACS, *Proposal for Accurate Floating-Point Vector Arithmetic*, in Rundbrief der GAMM, 1993, Brief 2, und in *Mathematics and Computers in Simulation*, Vol. 35, No. 4, IMACS, 1993.

15. IBM, *High Accuracy Arithmetic Subroutine Library (ACRITH)*, General Information Manual, 3rd ed., GC33-6163-02, IBM, 1986.

16. IBM, *High Accuracy Arithmetic–Extended Scientific Computation (ACRITH-XSC)*, General Information, GC33-6461-01, IBM, 1990.

17. IBM: *System /370 RPQ, High Accuracy Arithmetic.* SA22-7093-0, IBM Corp., 1984.

18. Zeitschrift *Elektronik* 26, 1994, *Genauer und trotzdem schneller, Ein neuer Coprozessor für hochgenaue Matrix- und Vektoroperationen*, Titelgeschichte.